Fundamentals of Photoinduced Electron Transfer

Fundamentals of Photoinduced Electron Transfer

by
George J. Kavarnos

Dedicated to the memory
of my late brother

SPIRO KAVARNOS

George J. Kavarnos, Ph.D.
121 Riverview Avenue
New London, CT 06320

This book is printed on acid-free paper. ∞

Library of Congress Cataloging-in-Publication Data

Kavarnos, George J.
Fundamentals of photoinduced electron transfer / by George J. Kavarnos.
p. cm.
Includes index.
ISBN 0-89573-751-5
1. Photochemistry. I. Title.
QD708.2.K38 1993
541.3′5—dc20 93-1291
CIP

Printed in the United States of America
ISBN 0-89573-751-5 VCH Publishers, Inc.

Printing History:
10 9 8 7 6 5 4 3 2

Published jointly by

VCH Publishers, Inc.
220 East 23rd Street
New York, New York 10010

VCH Verlagsgesellschaft mbH
P.O. Box 10 11 61
69451 Weinheim
Federal Republic of Germany

VCH Publishers (UK) Ltd.
8 Wellington Court
Cambridge CB1 1HZ
United Kingdom

Preface

Although numerous review articles on photoinduced electron transfer have appeared in recent years, there is yet no pedagogical text expressly written for undergraduate and graduate students. *Fundamentals of Photoinduced Electron Transfer* has been written for this readership, and is intended as an introductory text for students of chemistry, biology, physics, and material sciences, as well as readers with a general background in chemistry with an interest in the field. The book is meant as a primary or secondary text in introductory and advanced photochemistry courses at the undergraduate and graduate level. As a self-contained text that includes a brief introduction to the principles of photochemistry, this book can also be used in a course exclusively devoted to photoinduced electron transfer, or it can be used as a supplementary source in chemistry and biology courses for advanced undergraduates or beginning graduate students. *Fundamentals of Photoinduced Electron Transfer* requires, at the minimum, 2 or 3 years of college chemistry (general, organic, and physical). Accordingly, the material is within the grasp of junior or senior undergraduates with no prior knowledge of photochemistry or electrochemistry.

The author has included material that should appeal to chemists, physicists, and biologists who may want to familiarize themselves with this vital subject. The book is not intended as a comprehensive review. For those requiring additional and specialized information, the review articles listed at the end of each chapter should suffice. In writing on such a rapidly expanding field, it was virtually impossible to include all of the excellent research that has been done. Consequently, it was found necessary to exclude several lines of research. An effort, however, was made to limit these omissions.

The material in the first few chapters is presented at a level to allow the student to become comfortable with the subsequent subject matter. The first chapter introduces the reader to the basic principles and terminology of photochemistry. Chapter 1 also introduces the student to the Weller equation and the role of energetics. In Chapter 2, the properties of ion pairs and exciplexes are discussed as well as experimental procedures to study them. Chapter 3 covers photoinduced electron-transfer reactions of organic substrates; Chapter 4 deals with intramolecular and supramolecular photoinduced electron transfer. A number of topics are covered in Chapter 5, including photoinduced electron transfer in organized environments, photocatalysis with semiconductors, solar energy capture and utilization, and photoimaging. The last chapter is an introduction to classical and nonclassical theories of electron transfer and is written at a level to help the student understand the underpinnings of photoinduced electron transfer. The author has included a set of problems at the end of each chapter. Answers to most of the problems can be found in the references provided.

The author has received considerable support from several colleagues who over the years have stimulated his interest in the subject matter of this book. Above all, he owes a special note of gratitude to his mentor, Professor Nicholas Turro of Columbia University, for his encouragement and inspiration. The author recalls with appreciation that unique gift of Professor Turro to challenge his students by asking the "right" questions. It was this "prodding" and "questioning" that helped motivate the author to think about certain issues in this field and eventually to write this book. Warm thanks also go to Sister Claire Markham and Professor Harold McKone of Saint Joseph College and to Professor Bruno Vittimberga of the University of Rhode Island, who generously gave of their time to read initial drafts of the text. It is a distinct pleasure to recognize Dr. Günter Grampp of the Institut für Physikalische und Theoretische Chemie of the University of Erlangen in Nuremberg for his perceptive comments.

The author also wishes to acknowledge the assistance of several individuals who offered assistance during the long and tortuous ordeal of bringing this project to fruition. The staff of VCH Publishers offered much guidance for which the author is most grateful. James Powers, Thomas Pantelis, Dr. Roger Richards, and the Gilardi's, both Chris and Steve, are thanked for help with the figures at a critical time. The author extends his heartfelt thanks to his colleagues at the Naval Undersea Warfare Center in New London for their good-natured understanding and encouragement. And, finally, the author acknowledges with deep gratitude the support of his mother and late father, who offered much to teach and educate him.

George J. Kavarnos

New London
June 1993

Contents

CHAPTER

1

Introductory Concepts

1.1. Scope of Photoinduced Electron Transfer

Photoinduced electron transfer is the branch of photochemistry dealing with the property of certain photoexcited molecules to act as strong oxidizing or reducing species. A photoexcited molecule is often a better electron donor or electron acceptor than its ground state. The photoexcited species is an electron-transfer *photosensitizer* in the sense that it induces or "sensitizes" permanent chemical changes in a neighboring ground-state molecule by an electron-transfer mechanism.

Photoinduced electron transfer plays a central role in many areas of science. The reader is undoubtedly aware of biological photosynthesis where sunlight is harnessed for the growth and nourishment of plants. The early events in photosynthesis involve light absorption by an antenna system followed by a series of electron transfers. These ultrafast electron-transfer processes are known as *charge separations* since they involve the development of a large separation of positive and negative charges within photosynthetic *reaction centers*. The maintenance of this charge separation is critical for ensuing biochemical reactions. By "mimicking" the light-harvesting ability of green plants, chemists have attempted to duplicate the events in photosynthesis with model compounds. These models have been used as artificial photosynthetic systems for harnessing solar energy. Since photoinduced electron transfer is fundamental to these systems, much effort has been expended to understand the principles of photoinduced electron transfer, with the ultimate goal of achieving the efficiency and economy of natural light-harvesting systems.

A closely related area is the study of electron transfer between excited and ground-state species embedded in proteins. This knowledge has contributed to our understanding of the key features of biological electron-transfer processes such as oxidative phosphorylation. In proteins and other macromolecules such as DNA, the transfer of electrons is often over large distances. In an effort to rationalize the factors influencing the rates of electron transfer in these systems, chemists have attempted to modify biological macromolecules such as blue copper proteins, cytochrome, and DNA by covalently attaching light-active electron acceptors and donors to these molecules. By exposing these light- and redox-active complexes to short-lived bursts of light and analyzing the subsequent transformations, they have learned much about the structural and environmental factors controlling the passage of electrons over long distances.

Photoinduced electron transfer is of great interest to organic chemists concerned with the synthesis of novel organic compounds that may be difficult to synthesize by other routes. After being oxidized or reduced by a photosensitizer, an organic substrate can be transformed to a reactive intermediate which may be capable of undergoing further reaction. These photosensitized electron-transfer reactions often culminate in the formation of stable products.

Photoinduced electron transfer also plays a key role in several emerging technologies. In semiconductor photocatalysis, for example, a light-exposed semiconductor surface provides the environment for complex but often useful chemical transformations. Another area of interest concerns imaging systems such as silver halide photography, spectral sensitization, and xerography. There have also been proposals to design molecule-size electronic devices capable of writing and processing information. Electron transport in molecular electronic devices might be triggered by exposure to short-lived light pulses from a laser source. Strategies to design and synthesize such devices on the molecular scale are beginning to emerge and hopefully will someday be successful.

The ability to exploit the full potential of photoinduced electron transfer requires an understanding of certain principles of photochemistry and electron transfer. This book is an introduction to these principles and is intended to complement courses in photochemistry, biochemistry, and related courses. The reader of this book, it is assumed, has a background in organic and physical chemistry and has at least some familiarity with basic concepts of molecular orbital theory, spectroscopy, and chemical kinetics.

1.2. Review of Photochemical Principles and Definitions

All photochemical reactions start with the absorption of light (Fig. 1.1). Photoexcitation of ground-state species leads to formation of excited states. Excited states differ in many respects from their respective ground states. To distinguish an excited state from its ground state, chemists usually use a superscript asterisk placed to the left or right of the symbol for the ground state (i.e., M^* is the excited state of molecule M). Generally, excited states are more chemically reactive. In fact, the unique property of some excited molecules to induce transitory or permanent changes in neighboring molecules by an electron-transfer pathway is one example of their

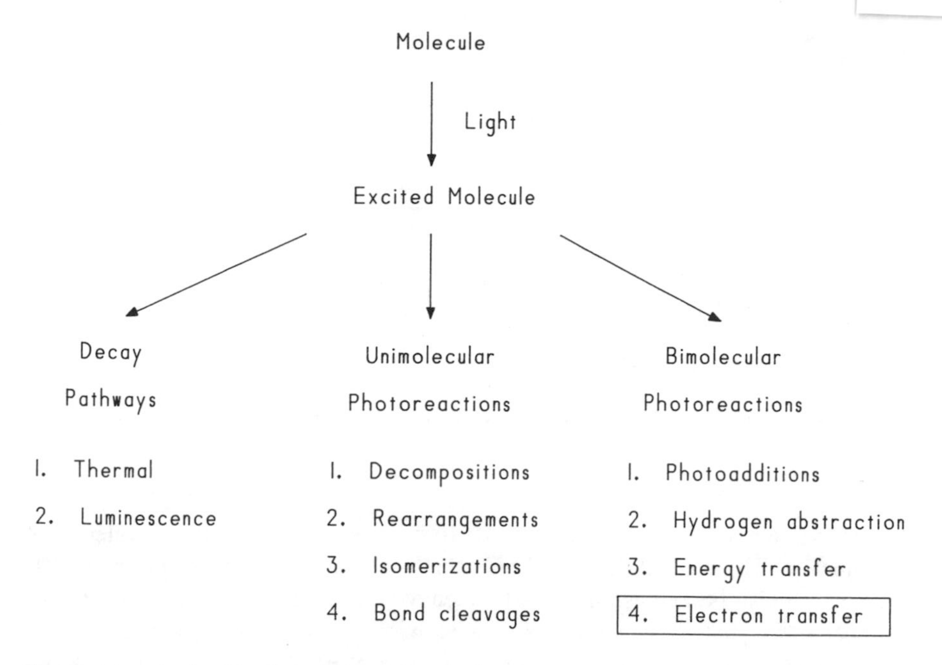

Figure 1.1 A classification of photochemical pathways.

chemical reactivity. To appreciate the role of excited states in photoinduced electron transfer, it is desirable to review several basic photochemical concepts. This section briefly surveys some features and properties of electronically excited states.

1.2.1. Electronic Excitation

When ground-state molecules absorb visible or ultraviolet light, electrons in the highest occupied orbitals undergo transitions to unoccupied orbitals lying at higher energies (Fig. 1.2). By absorbing a photon of light, the ground state is converted into a higher energy state, or electronically excited state. $h\nu$ is used to represent a photon, which is the designation for a discrete quantum of light. The energy (E) of the photon is given by

$$E(\text{erg}) = h\nu \tag{1.1}$$

where h is Planck's constant, which is equal to 6.626×10^{-27} erg-s quantum^{-1}. ν, the frequency of the photon, is expressed as reciprocal seconds, or s^{-1}. Occasionally, the chemist needs to know the relationship between the excitation wavelength and the energy of the exciting photon. Equation 1.2 is a useful aid for this purpose:

$$\nu = \frac{c}{\lambda} \tag{1.2}$$

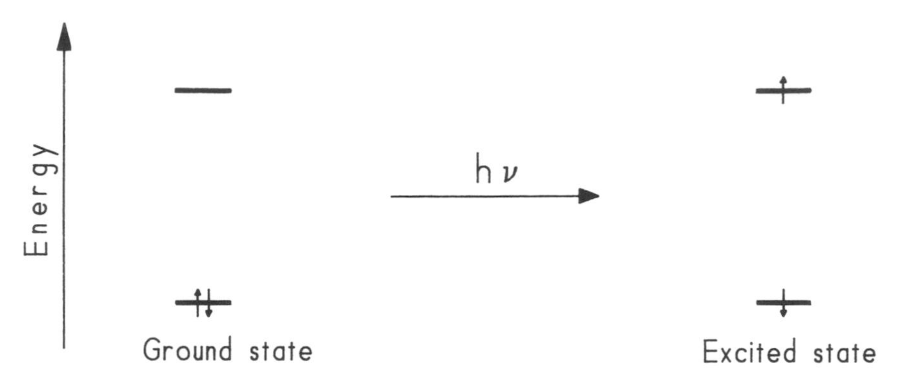

Figure 1.2 Photoexcitation results in an electronic transition.

where c is the velocity of light or 3×10^{10} cm s^{-1}, and λ is the wavelength in centimeters. The wavelength of light is usually given as nanometers (1 nm = 10^{-7} cm); it can also be expressed by its frequency (Eq. 1.2) or by its reciprocal wave number, $\bar{\nu} = \lambda^{-1}$. The energy associated with $h\nu$ is usually given in units of kilocalories or kilojoules per mole. To convert between the various units of energy, the reader may want to refer to Appendix I.

The tightly held electrons in the lower energy orbitals, such as the "core" 1s orbitals, are normally not perturbed by the absorption of light. However, electrons in the higher lying orbitals are susceptible to light excitation in the spectral regions in the far ultraviolet, near ultraviolet, and visible regions of the electromagnetic spectrum. These regions range, respectively, from about 100 to 250, 250 to 350, and 350 to 700 nm (Table 1.1).

Table 1.1 ELECTRONIC ABSORPTION MAXIMA OF SELECTED COMPOUNDS[a]

Compound	Transition	λ_{mx} (nm)
Ethylene	π, π^*	~170
1^0, 2^0, and 3^0 amines	n, σ^*	190–200
Benzene	π,π^*	180–260
Naphthalene	π,π^*	170–320
Carbonyl compounds	π,π^*	180–190
	n,π^*	230–340
Metalloporphyrins	π,π^*	400–600
UO_2^{2+}	π, f	420
$Ru(bpy)_3^{2+}$	d,π^*	452
$Rh_2(dicp)_4^{2+}$	d,π^*	553

[a] bpy, 2,2′-bipyridine; dicp, 1,3-diisocyanopropane.

1.2.2. Electronic States of Organic Molecules

It is useful to consider electronic transitions within the framework of simple molecular orbital theory. Molecular orbitals can be regarded as linear combinations of atomic orbitals of the atoms which comprise the molecule. For purposes of this discussion, several atomic orbitals are depicted in Fig. 1.3. When two atomic orbitals merge to form a molecular orbital, "bonding" and "antibonding" molecular orbitals are formed. The bonding molecular orbital is formed by the combination of orbitals of the same wave-like character, whereas the antibonding molecular orbital is formed by two atomic orbitals of opposite wave-like character (Fig. 1.4).

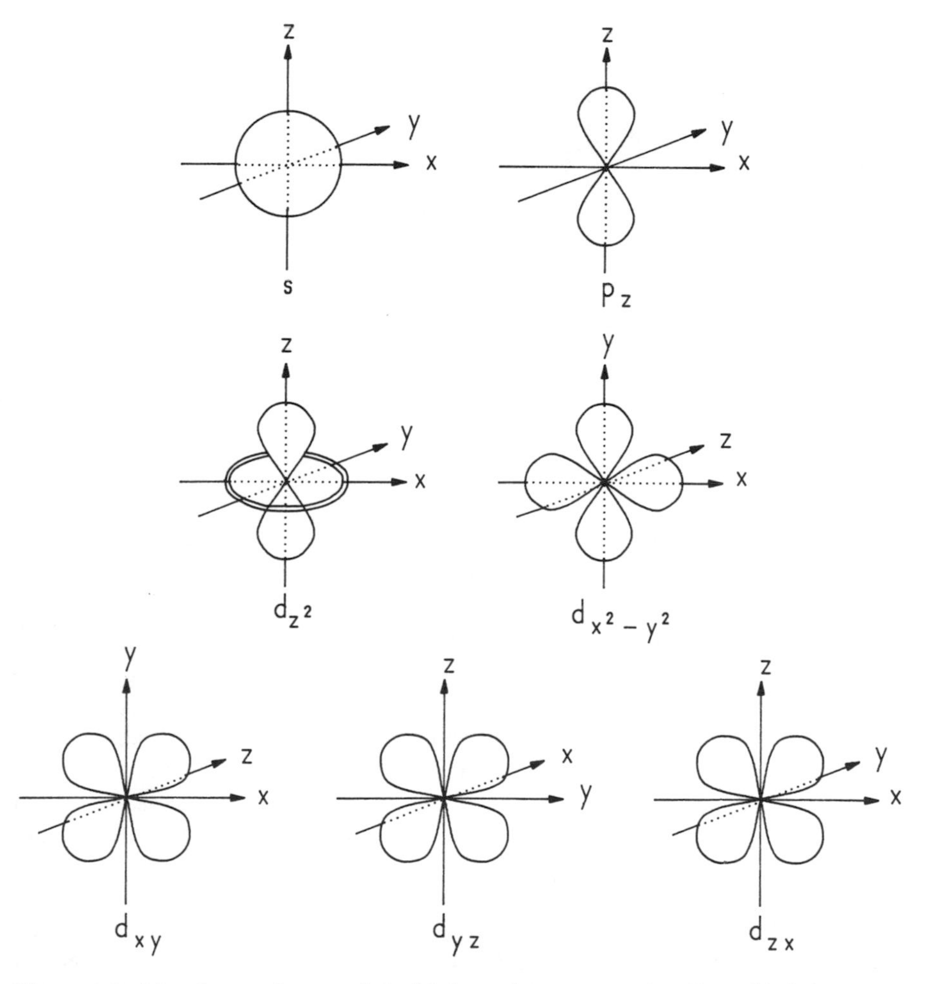

Figure 1.3 The shapes of s, p and d orbitals used to construct bonding orbitals important in photochemistry.

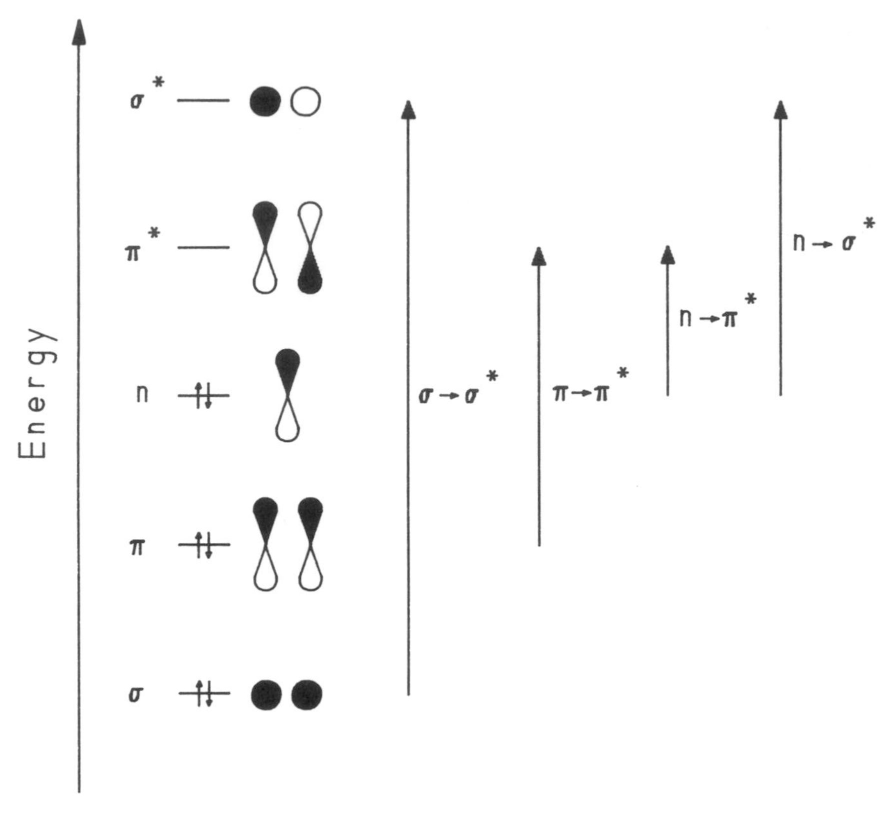

Figure 1.4 An orbital diagram showing the energy ordering of the most common types of electronic transitions in organic molecules.

On a vertical energy scale, the antibonding orbital lies at a higher energy than the bonding orbital. Electrons residing within the atomic orbitals will populate the newly formed molecular orbitals, the maximum number of electrons in one molecular orbital being two. To complete the molecular orbital representation, we must take into account the spins of the two electrons. We use directional arrows to designate spin. ↑ represents $+1/2$ spin; ↓ is used for $-1/2$ spin. The two electrons populating the same orbital are depicted with two antiparallel direction arrows, i.e., ⇅. This is done because two electrons occupying the same orbital must have their spins oriented in opposite directions (a consequence of the Pauli principle).

Let us now consider electronic transitions in simple organic molecules. In the formation of a carbon–carbon single bond, 2s orbitals of two carbon atoms combine to form a bonding σ-orbital and antibonding σ*-orbital (Fig. 1.5). The two electrons enter the σ-orbital leading to a stabilized electronic configuration. The σ-orbital is symmetric around the molecular axis. The electron density is greatest between the two nuclei. On interaction with a photon, an electron in the bonding σ-orbital is

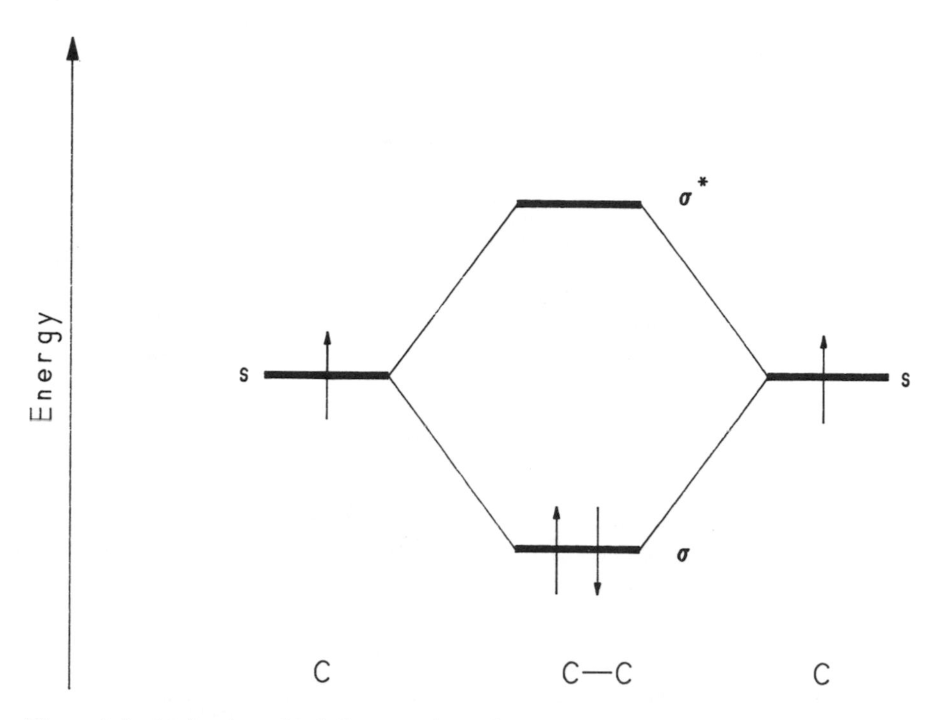

Figure 1.5 Molecular orbital diagram of a carbon-carbon σ-bond.

excited to the antibonding σ*-orbital (Fig. 1.4). The promotion of the "bonding" electron into the higher energy orbital is referred to as a σ → σ* transition. The newly formed state is designated a σ,σ* excited state. Since an electron now occupies an antibonding orbital, there is less "bonding" character in the σ,σ* excited state. σ → σ* transitions require a large amount of energy, usually about 200 kcal mol^{-1}.

In another type of electron transition, the π → π* transition, an electron from a bonding π-orbital is promoted to an antibonding π*-orbital (Fig. 1.4). π-orbitals, which are formed by the interaction of two 2*p*-orbitals (Fig. 1.6), are found in unsaturated organic molecules. In contrast to the electrons occupying σ-orbitals, the electrons populating π-orbitals are more delocalized. Also, the overlap between the 2*p*-orbitals is less than the overlap between the *s* atomic orbitals comprising the σ-bond. Therefore, the strength of a π-bond is less than that of a σ-bond. As compared to the σ-orbitals, which are separated by a large energy gap, the gap between π-orbitals is smaller since the energy of the bonding π-orbital is raised relative to the bonding σ-orbital, and the energy of the matching π*-orbital is lowered with respect to the σ*-orbital. As a result, photon excitation of an electron in the bonding π-orbital to the antibonding π*-orbital requires less energy than a σ → σ* transition.

A third type of transition is found in organic molecules containing the carbonyl group such as ketones, aldehydes, esters, and carboxylic acids, and is known as

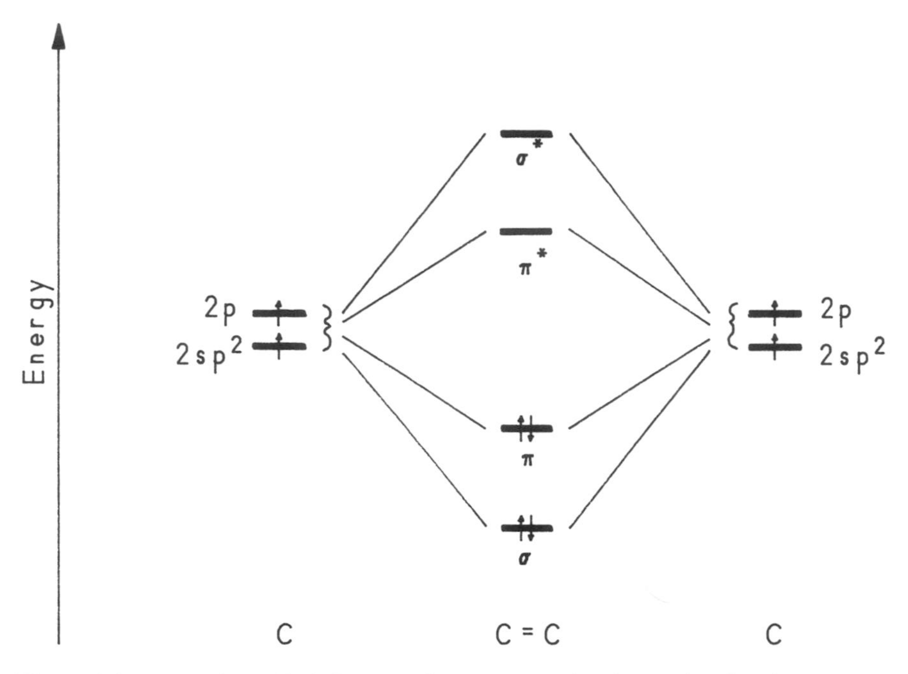

Figure 1.6 Molecular orbital diagram of an unsaturated carbon-carbon bond.

the $n \rightarrow \pi^*$ transition (Fig. 1.4). The molecular orbitals of the carbonyl group are constructed by combinations of carbon and oxygen atomic orbitals (Fig. 1.7). In oxygen, the 2*p*- and 2*s*-orbitals are hybridized into two linear *sp*-orbitals. One orbital combines with the carbon sp^2-orbital to form a σ-bond; the other orbital is directed away from oxygen colinear with the axis of the molecule. An oxygen 2*p*-orbital combines with a carbon 2*p*-orbital to generate a π-bond. The remaining oxygen 2*p*-orbital is designated an *n*-orbital, containing two "lone pair," or "nonbonding" electrons. The *n*-orbital does not participate in bonding. Since the *n*-orbital occupies a position higher in energy than the σ- and π-bonds, the $n \rightarrow \pi^*$ transition to create an n,π^* excited state usually involves less energy than the $\pi \rightarrow \pi^*$ transition (Fig. 1.4), although in some carbonyl compounds, the π-orbital may lie at a higher energy than the *n*-orbital. These differences in energy ordering may be due to the structure of the molecules or to the effects of solvent on stabilizing the electronic structure (the reader is urged to consult textbooks on photochemistry for more discussion on this topic).

A fourth type of transition involves the promotion of an *n*-electron to an antibonding σ*-orbital. These transitions can take place in compounds containing heteroatoms such as aliphatic amines, alcohols, and halogens.

In summary, the most common types of electronic transitions in organic molecules are represented by the formation of n,π^*, π,π^*, n,σ^*, and σ,σ^* excited states. Although these transitions have been visualized here in terms of simple molecular orbital theory, this approach conveys the spirit of orbital occupancy. We

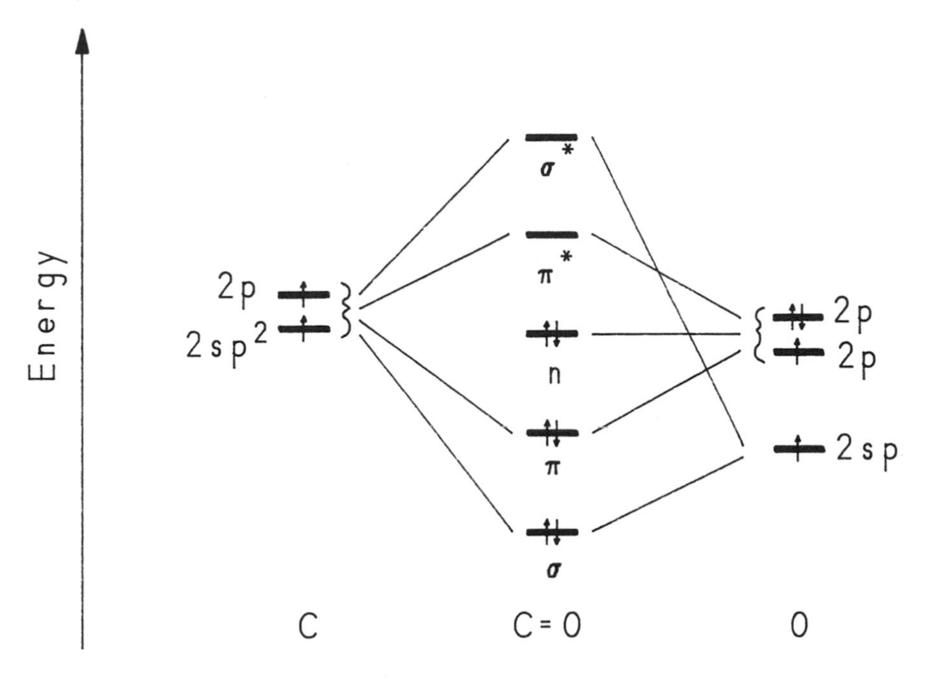

Figure 1.7 Molecular orbital diagram of a carbonyl group.

shall note in a later section that we can also describe the ability of the excited state to accept or donate electrons by invoking simple orbital representations.

1.2.3. Electronic States of Transition Metal Complexes

In constructing the electronic configurations of inorganic complexes, it is useful to consider an octahedral complex, a prototype of many transition metal complexes. An octahedral complex consists of a transition metal bonded to six groups or ligands positioned at the corners of an octahedron (Fig. 1.8). In the ground states of octahedral complexes, the "ligand" molecular orbitals are completely occupied by electrons. In contrast, the metal orbitals may or may not be completely filled. The molecular orbitals in octahedral complexes are made up of *s*, *p*, and *d*-orbitals of the metals and ligands. Figure 1.9 depicts the combination of these orbitals to form form σ- and π-bonds [the subscripts in Fig. 1.9 signify whether the major contribution to an orbital is from a metal (M) orbital or ligand (L) orbital]. These combinations are influenced by the symmetries of the orbitals. Metal *d*-orbitals (Fig. 1.3) belong to two symmetry sets known by their spectroscopic designations as t_{2g} and e_g. From a comparison of the geometries of the *d*-orbitals in Fig. 1.3 with the directional drawing of the octahedral complex in Fig. 1.8, we can draw certain conclusions concerning the orbital combinations which result in bonding. d_{xy}-, d_{yz}-, and d_{zx}-orbitals should be particularly suited for bonding with *p*-orbitals of ligands to form π-bonds; these *d*-orbitals belong to the t_{2g} symmetry set. On the

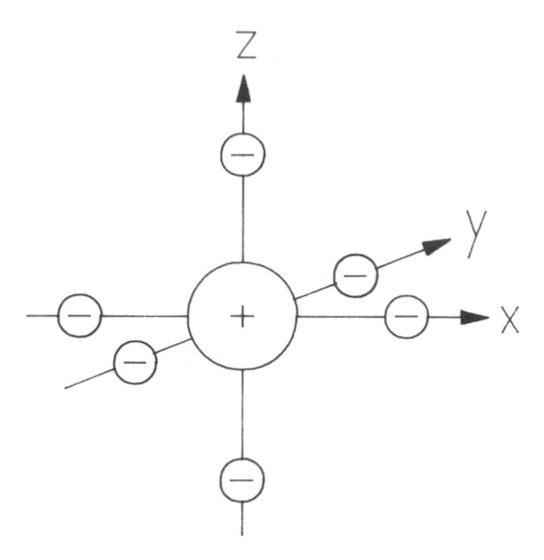

Figure 1.8 In an octahedral complex, a central atom is surrounded by six ligands situated at the corners of an octahedron.

other hand, the d_{z^2}- and $d_{x^2-y^2}$-orbitals are directed along the *x, y,* and *z* axes in octahedral complexes and combine with *s*-orbitals of ligands to generate σ-bonds. The interaction of negative ligands with metal *d*-orbitals causes the energies of the latter to split. Since the doubly degenerate orbitals in the e_g set lie along the in-bond axis of the metal–ligand bond, these orbitals experience more destabilization than the triply degenerate t_{2g} orbitals.

The electronic transitions for an octahedral complex are summarized in Fig. 1.10. These transitions may simply involve the electrons in metal-centered molecular orbitals. These are called *metal-centered* transitions and usually involved $d \rightarrow d$ transitions from t_{2g}- to e_g-orbitals. Transitions taking place in molecular orbitals with predominantly ligand character are *ligand-centered* transitions. Typical examples are $\pi \rightarrow \pi^*$ transitions of electrons localized in the ligands. Electron promotions from metal-centered to ligand-centered orbitals, or vice versa, are *ligand-to-metal charge transfer* (LMCT) or *metal-to-ligand charge-transfer* (MLCT) transitions, respectively. A $d \rightarrow \pi^*$ charge-transfer transition takes place when a *d* electron in a t_{2g} orbital is promoted into an antibonding ligand orbital.

Metal-centered transitions in metal complexes may also involve spin flips. If the excitation energy corresponds to the energy required for a spin flip, a spin reversal may take place with subsequent spin pairing of two electrons.

1.2.4. State Descriptions

A *state description* is a shorthand description containing essential information of the electronic configuration of a molecular species. A state may be described using spectroscopic nomenclature or, in the case of complicated molecules, by valence orbitals. These descriptions provide information on the energy ranking of the molecule and the net electronic spin.

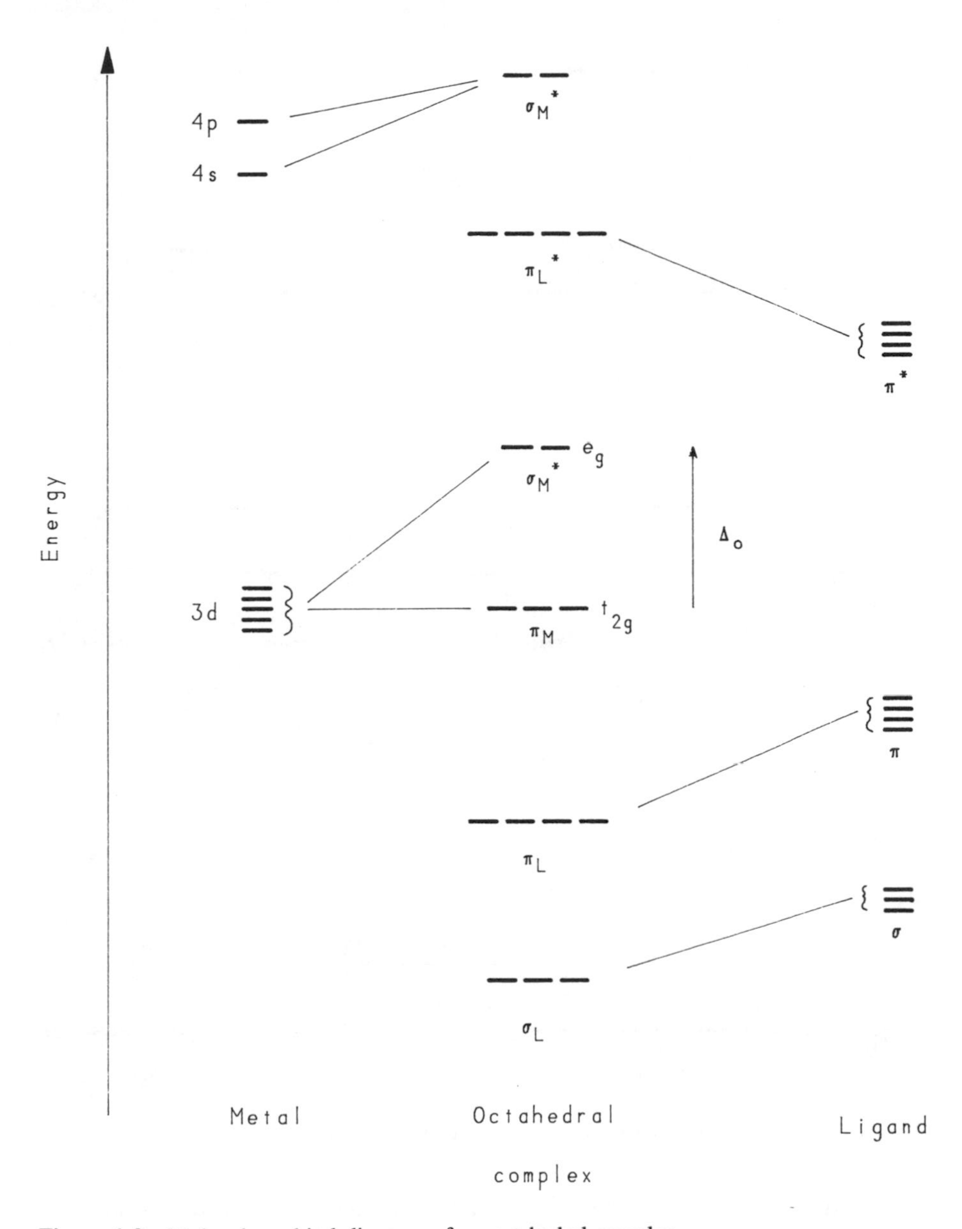

Figure 1.9 Molecular orbital diagram of an octahedral complex.

A state description of any molecule can be constructed from a knowledge of its electronic configuration. In ground-state molecules, all electrons occupy the lowest energy molecular orbitals. The spins of the electrons in each orbital are paired (i.e., the directional component of the spin of one electron is oriented in an antiparallel direction with respect to the spin of the other). This spin configuration places the molecule at the lowest point or "ground state" of the vertical energy scale.

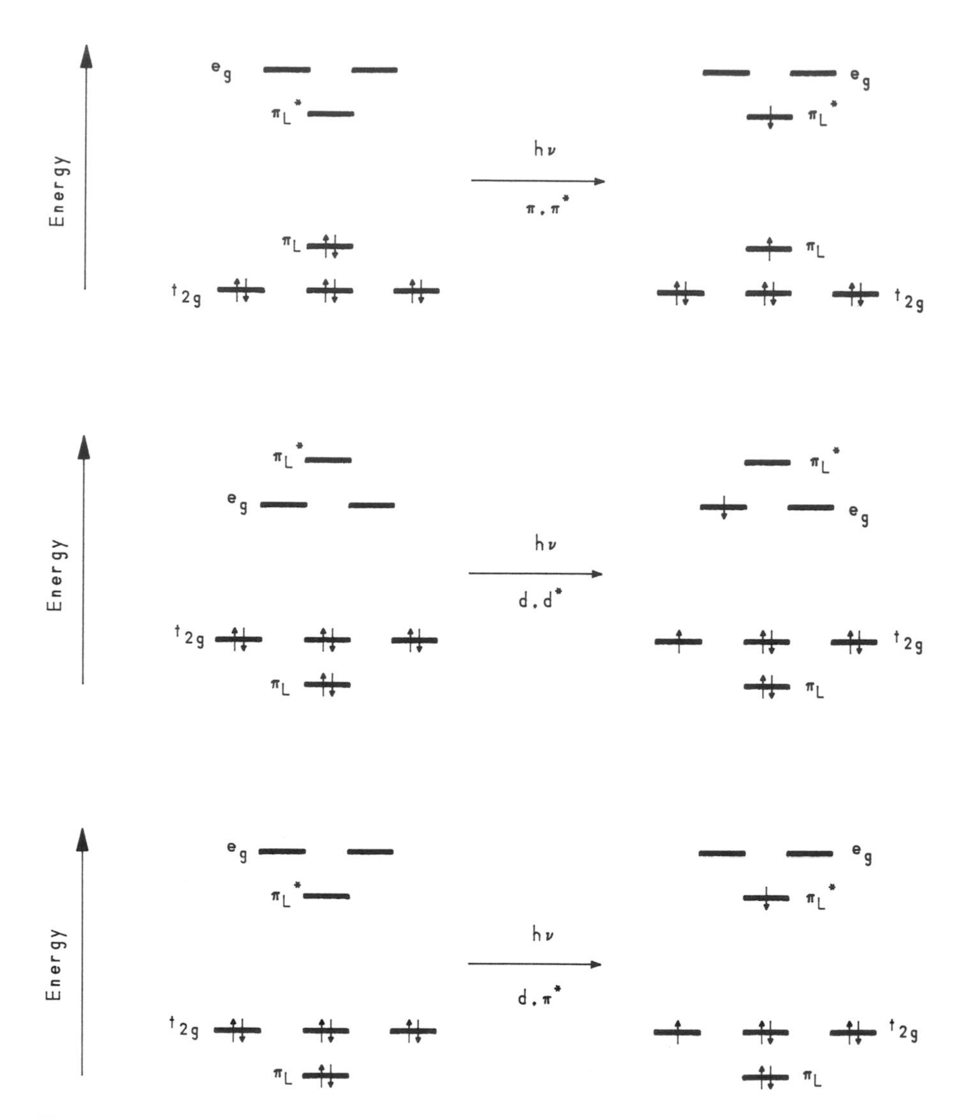

Figure 1.10 Electronic transitions in an octahedral complex.

Accordingly, we designate this ground-state molecule as a *singlet ground state*. "Singlet" defines the spin multiplicity of a state, which can be deduced from Eq. 1.3:

$$m = 2S + 1 \tag{1.3}$$

where m is the spin multiplicity and S is the total spin of all electrons. m represents the possible ways by which electrons can "line-up" with a magnetic field (Fig. 1.11). For a ground-state molecule with all electron spins paired, $S = 0$ and $m = 1$. Most molecules have in fact an even number of electrons, two electrons occupying each orbital with opposed spins.

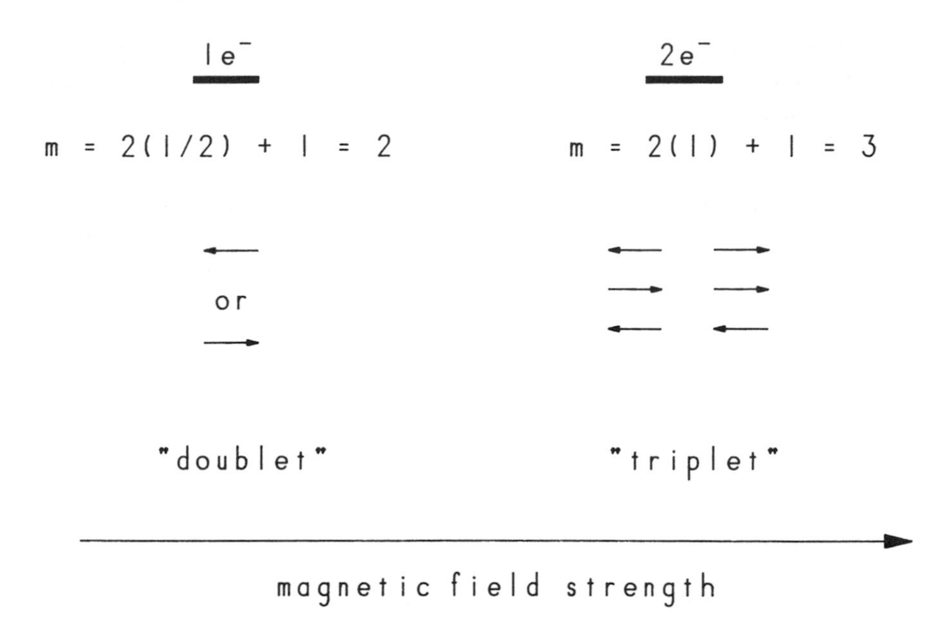

Figure 1.11 This figure shows the ways one or two unpaired electrons can line up with an external or neighboring magnetic field. This is the basis for m = $2S + 1$.

The reader should note that "singlet" is a spectroscopic state description given to the ground state. It means that all of the electron spins are paired. The spectroscopic notation, S_0, is employed for a ground state singlet, the subscript referring to "zero" energy or ground state. A species with one unpaired electron such as a free radical has a spin of $S = 1/2$, and a spin multiplicity of $m = 2$. Hence, this species exists in two *degenerate* states, which we call a *doublet*. The doublet state will lose its degeneracy in the presence of a magnetic field where the electron spins can line up in two opposing directions. In a species with two unpaired electrons having parallel spin, $S = 1$ and $m = 3$. In a magnetic field, the spin vector assumes three levels, $+1$, 0, and -1. This species splits into three states, hence, the designation *triplet state*.

Spectroscopic state descriptions are constructed for excited states, depending on the way the directional components of the electron spins are oriented. If its electron spins are parallel, or spin-paired, this state is designated an *excited* singlet state, or S_n, where the subscript n denotes the level of the singlet state. In some excited states, the electron spins of the excited state may be unpaired. Additional spectroscopic states, each with its own spin multiplicity, can be constructed. Thus, an excited state with two unpaired electrons is an *excited triplet state,* or T_n.

It is common practice to describe an excited state on the basis of its electronic configuration. In the case of organic molecules, singlet and triplet states may be designated as $^1n,\pi^*$, $^3n,\pi^*$, $^1\pi,\pi^*$, and $^3\pi,\pi^*$ states, the left superscript referring to singlet or triplet. This notation does not necessarily imply anything about the pathway by which the state is formed. For example, a $^3n,\pi^*$ state can be formed directly by a singlet → triplet $n \rightarrow \pi^*$ transition, or indirectly by a singlet → singlet $n \rightarrow \pi^*$ absorption followed by a spin flip.

The designations used for transition metal complexes are formulated in a similar manner. A ground-state metal complex with three unpaired electron spins [e.g., $(t_{2g})^3$] is classified as a quartet. $(t_{2g})^3$ is given the spectroscopic notation, $^4A_{2g}$, a symmetry symbol representing an orbital having the identical symmetry of the whole complex. An example of this electronic configuration is the d^3 Cr^{3+} octahedral complex. When excited by light, this complex is transformed into an excited state with only one unpaired electron. The spin multiplicity of this excited state (^2E_g) is a doublet (Eq. 1.3). ^2E_g is a symmetry symbol used to specify two equivalent orbitals with different orientations.

1.3. An Overview of Photochemical and Photophysical Processes

A *state energy diagram* is a convenient tool used to trace the pathways for the formation and disappearance of excited states. In this section, we analyze the construction of a state energy diagram for singlet and triplet states.

A state consists of closely spaced energy *vibrational levels* (Fig. 1.12). All ground- and excited-state molecules exist in a variety of nuclear geometries. These geometries may consist of numerous bond stretches or bond deformations, each of which corresponds to a unique vibrational frequency. We may consider each nuclear vibration as possessing a distinct energy and occupying a unique position on a vertical energy scale. Energy levels corresponding to closely spaced vibrational levels are superimposed onto the energy diagram depicting the energies of singlet and triplet electronic states. The population of the various vibrational levels follows a Boltzmann distribution. The lowest vibrational level, the zero-point level, is normally populated at room temperature to a much greater extent than higher energy orbitals.

In the formation of excited singlet states, photoexcitation normally takes place from the lowest vibrational level of the ground electronic state into any one of the higher vibrational levels of the excited singlet state. After population of the upper excited singlet states, conversion to the first excited state, S_1, is quite rapid, usually within $\sim 10^{-11}$–10^{-13} s. During absorption, the nuclear geometry of the molecule normally does not change, although the electrons may undergo rapid motions. Hence, the transition is "vertical" in the sense that this transition can be drawn in a vertical fashion as shown in Fig. 1.13. Chemists often refer to this transition as a *Franck–Condon* transition—after the Franck–Condon principle. Because the mass of an electron is so much less than the mass of a nucleus (by $\sim$1836), the electronic transition proceeds so much faster (within $\sim 10^{-16}$ s) than the typical nuclear vibration ($\sim 10^{-12}$–10^{-14} s). During the electronic transition, the nuclei act as "spectators": they do not move during the rapid motion of the electron. The state formed by this "vertical" transition is called the *Franck–Condon state*. It possesses the

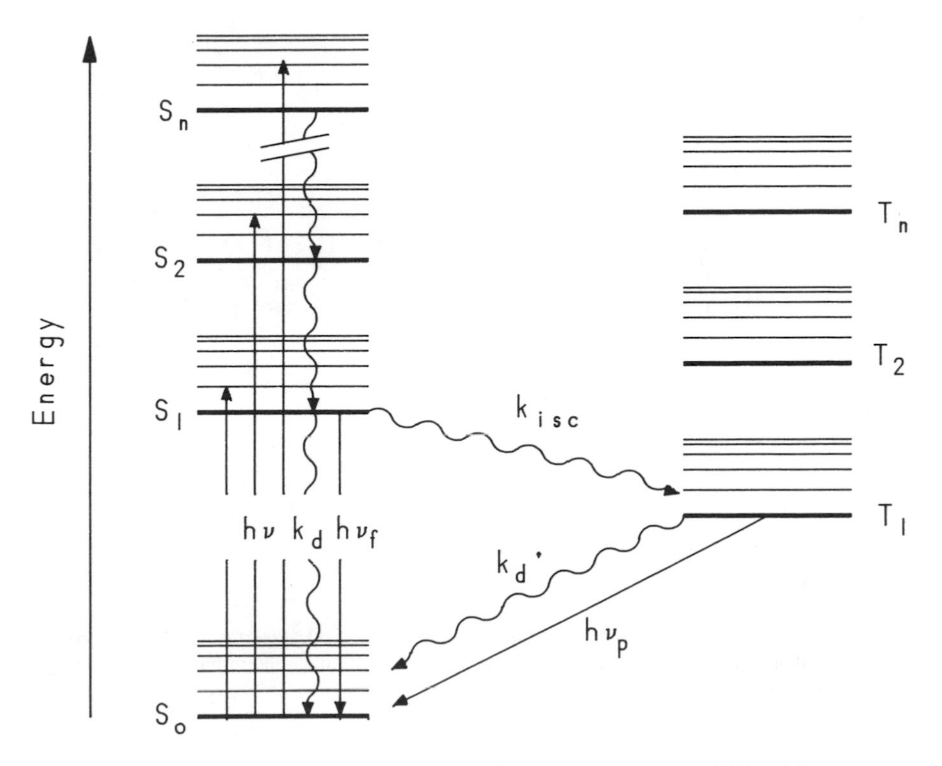

Figure 1.12 A state energy diagram. $h\nu_f$ and $h\nu_p$ are the quantum yields of fluorescence and phosphorescence emission, respectively. k_d, k_{isc}, are k_d' the rates of radiationless conversions (see text).

geometry of the ground state but a "new" electronic structure. After its formation, the Franck–Condon state undergoes *vibrational relaxation* to an equilibrium geometry. Vibrational relaxation involves changes in the bond lengths and interbond angles as well as changes in the solvation sphere of the excited state. During vibrational relaxation, excess vibrational energy can be quickly absorbed by collisions with neighboring solvent molecules to form the lowest vibrational level of the singlet state. Because it involves nuclear motions, vibrational relaxation usually proceeds within $\sim 10^{-12}$–10^{-14} s. Relaxation results in a thermally equilibrated, long-lived excited state. The lifetime of the equilibrated excited singlet state, S_1, is relatively long (as compared to the short-lived Franck–Condon state). The energy corresponding to the equilibrated singlet state is the *zero spectroscopic energy* and is written as E_{00}, as it represents the energy difference between the lowest vibrational level in the ground electronic state and the lowest vibrational level in the excited state (the subscript refers to this $0 \rightarrow 0$ vibrational transition). The energy of an absorbed photon may be larger than E_{00} if the transition leads to a state lying above the equilibrated state. The high energy state nonetheless rapidly relaxes to the equilibrated state. Unless otherwise specified, it is the equilibrated excited state that participates in a photochemical reaction.

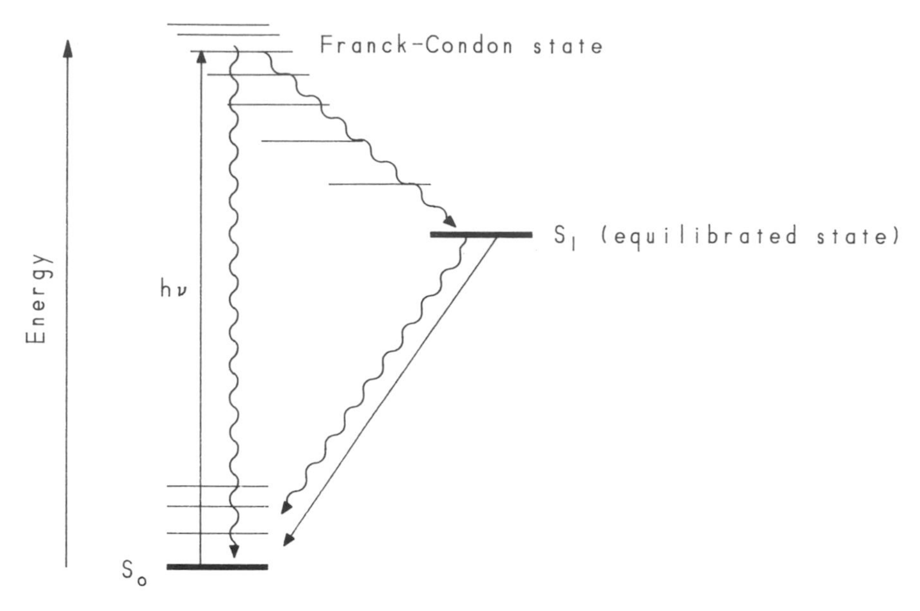

Figure 1.13 Excitation of a ground state involves a vertical transition to a high-energy Franck-Condon state which subsequently undergoes vibrational relaxation to the thermally equilibrated state.

The excited singlet state is an energy-rich and reactive species. It may release its energy or undergo chemical transformation. The energy-releasing pathways can be classified as *radiative* (i.e., transitions to lower states involving light emission) and *nonradiative* (i.e., transitions involving the release of heat). *Fluorescence* is light emission accompanying the transition from the lowest vibrational state of the excited singlet state to any of the vibrational levels of the ground state. Fluorescence emission is identified by its rate constant, k_f, by the energy of the released photon, $h\nu_f$, or by its *quantum yield,* Φ_f, which is defined as the number of photons emitted divided by the number of photons absorbed. From Fig. 1.12, note that $h\nu_f < h\nu$. Accordingly, the wavelength of fluorescence is usually longer than the absorption wavelength. The extent to which the emission and absorption wavelengths are separated is called a *Stokes shift* and is a measure of the structural distortion between the ground and excited state.

In a nonradiative transition to the ground state (represented by k_d), the energy of the excited singlet state is dissipated as heat to the surrounding medium. This transition is referred as an *internal conversion*. One nonradiative transition of special importance in photochemistry is the isoenergetic formation of vibrationally excited triplet states from excited singlet states ($S_n \rightarrow T_n$). Like its singlet counterpart, the vibrationally excited triplet can decay to a "relaxed" state. The overall pathway by which the triplet state is formed from a singlet state is called *intersystem crossing* and involves a spin flip of an orbital electron (Fig. 1.14). In triplet states, the electrons must occupy different orbitals, since pairs of electrons with parallel spins are forbidden from occupying the same region of space (the Pauli exclusion principle). Another characteristic of triplet states is that their state energies tend to be lower than those of the singlet states from which they originate. This follows from Hund's rule, which postulates that spectroscopic states with the higher spin multiplicity (i.e., triplet states) are lower in energy than states of lower spin multiplicity (i.e., singlet states). Light emission from the triplet state is called *phosphorescence* and is identified with $h\nu_p$, Φ_p, or k_p.

In attempting to characterize the decay pathways of excited states, one should observe that $S_n \rightarrow S_0$, $S_n \rightarrow T_n$, and $T_1 \rightarrow S_0$ radiative and nonradiative transitions do not take place with equal probability. It is preferred practice to regard radiative and nonradiative transitions as *allowed* or *forbidden* pathways. "Allowed" and "forbidden" are quantum mechanical descriptions. A "forbidden" transition is improbable and proceeds *more slowly* than an "allowed" transition. A forbidden transition may take place because of perturbations between states. For example, a "forbidden" transition involving a spin flip is observable because of the coupling between the spin and orbital motion of an electron. This perturbation is known as *spin-orbital* coupling and may be the only mechanism available for intersystem crossing. Since, in any isolated system, the total momentum must be conserved, a change in an electron's spin angular momentum (which takes place during a spin flip) must be accompanied by a corresponding change in orbital momentum. This spin-orbital coupling may be accelerated by the presence of "heavy-atom" substituents (atoms of high atomic number) in the excited state or in the solvent. This phenomenon is known as the "heavy-atom" effect.

To summarize, absorption from the ground state to the excited singlet state ($S_0 \rightarrow S_n$), and fluorescence and internal conversion from the excited singlet to

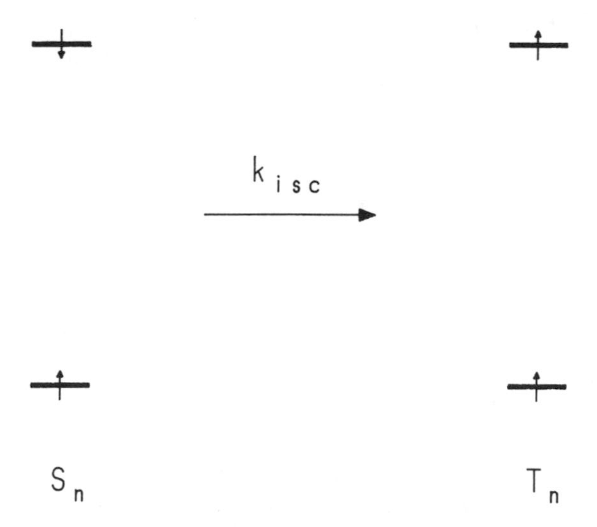

Figure 1.14 Intersystem crossing from a singlet or triplet state (or vice versa) involves the spin flip of an electron.

ground singlet state ($S_n \rightarrow S_0$) are allowed transitions. Intersystem crossing from an excited singlet to a triplet state ($S_n \rightarrow T_n$) is a forbidden process, as are phosphorescence and internal conversion from the first excited triplet to ground singlet state ($T_1 \rightarrow S_0$). Direct excitation of ground-state singlets to excited triplets is also a forbidden process. As described earlier, triplets are usually formed by intersystem crossing from the excited singlet state, even though intersystem crossing is formally a forbidden pathway.

The allowed and forbidden pathways available to excited singlet and triplet states may have a pronounced influence on their lifetimes. Because transitions involving the relaxation processes of excited singlet states are spin-allowed, their lifetimes are normally shorter than the lifetimes of triplet states. Triplet states generally have longer lifetimes since transitions involving these states are usually spin forbidden. Experimentally, it is found that in fluid solution, the lifetimes of excited singlet states range from a few picoseconds to a few microseconds, in contrast to the lifetimes of triplet states which range from several microseconds to seconds.

1.4. General Features of Quenching by Electron Transfer

In addition to radiative and nonradiative transitions described in the previous section, the excited state can participate in numerous inter- and intramolecular reactions (Fig. 1.1). The study of these processes forms the basis of photochemistry. Examples of intramolecular processes may include ejection of an electron (photoionization), decomposition into smaller fragments (photodecomposition), and spontaneous isomerization (photoisomerization). Intermolecular pathways involve reactions with ground-state molecules. These processes may include (1) an addition reaction where the excited state combines with a ground-state molecule to form a stable product, (2) hydrogen abstraction, (3) energy transfer where the excited state donates its electronic energy to a ground-state species, and (4) electron transfer where the excited state operates as an electron donor or acceptor when interacting with a ground-state species.

1.4.1. Quenching Pathways

Energy and electron transfer are classified as *quenching pathways*. *Quenching* is defined as the deactivation of an excited sensitizer by an external component. By convention, the external component is called the *quencher,* this species usually being a ground-state molecule. During quenching, the sensitizer induces permanent or temporary chemical changes in surrounding ground-state molecules. Quenching processes may take place between a sensitizer and quencher completely separated or attached to one another via a flexible or rigid spacer molecule. In fluid media, the viscosity of the molecular environment controls the motions of the sensitizer and quencher and how close these molecules can approach for quenching to take place. In the solid state or in organized assemblies, the sensitizer and quencher are fixed or held rigidly in a molecular environment. In such environments the freedom of the reactants to collide or attain the ideal distance for quenching is determined by the molecular motions of the surrounding molecules. To the extent that the environment or linking molecule allows a close approach of the sensitizer and quencher during the lifetime of the sensitizer, quenching may be a major pathway for the deactivation of an excited sensitizer.

Quenching generates reactive intermediates that rapidly undergo transformation to stable products. In the case of energy transfer, when the quencher receives the excess energy of the excited state, the quencher itself becomes excited and subsequently undergoes the same physical and chemical processes as if excited directly:

$$D^* + A \xrightarrow{k_{en}} D + A^* \tag{1.4}$$

In reaction 1.4, k_{en} refers to the rate constant of energy transfer. Here we imply that D is a donor of *electronic energy* and A, the quencher, the acceptor of this excess energy. In photoinduced electron transfer, D is the *electron donor* and A the *electron acceptor*. Since the excited state can function as an electron donor or acceptor, we can write

$$D^m + A^n \xrightarrow{h\nu} [D^n]^* + A^m \xrightarrow{k_{el}} D^{m+1} + A^{n-1} \tag{1.5}$$

or

$$D^m + A^n \xrightarrow{h\nu} D^m + [A^n]^* \xrightarrow{k_{el}} D^{m+1} + A^{n-1} \tag{1.6}$$

Here k_{el} is the rate constant of electron transfer. *m* and *n* refer to the original charges on D and A, respectively. For convenience, we have not specified the spin multiplicity of the excited state.

1.4.2. Charge-Transfer Intermediates: Nomenclature and Definitions

In the case of electron transfer between neutral organic reactants ($n,m = 0$), the initial ion pair consists of the oxidized donor and reduced acceptor. Very often we use the generic term *charge-transfer intermediate* to describe these ion pairs. The ion pairs formed when D and A are organic molecules are designated $D^{\stackrel{+}{\cdot}}$ for a radical cation and $A^{\stackrel{-}{\cdot}}$ for a radical anion, respectively. Radical ions are charged intermediates having an odd number of electrons. (We shall use the symbols, $D^{\stackrel{+}{\cdot}}$ and $A^{\stackrel{-}{\cdot}}$, only when we explicitly intend to indicate radical ion species. Otherwise, D^+ and A^- refer to oxidized and reduced species that may or may not be radical ions.) Some examples are shown in Fig. 1.15. Note that we can describe an ion either by one of its resonance stabilized structures or by the dotted circle and line notation commonly used by organic chemists to symbolize electron delocalization. We will employ both representations in this book. It will also be our practice to enclose a radical ion in brackets and to designate its charge outside of the upper portion of the right bracket.

The distance separating $D^{\stackrel{+}{\cdot}}$ and $A^{\stackrel{-}{\cdot}}$ is an important structural feature of radical

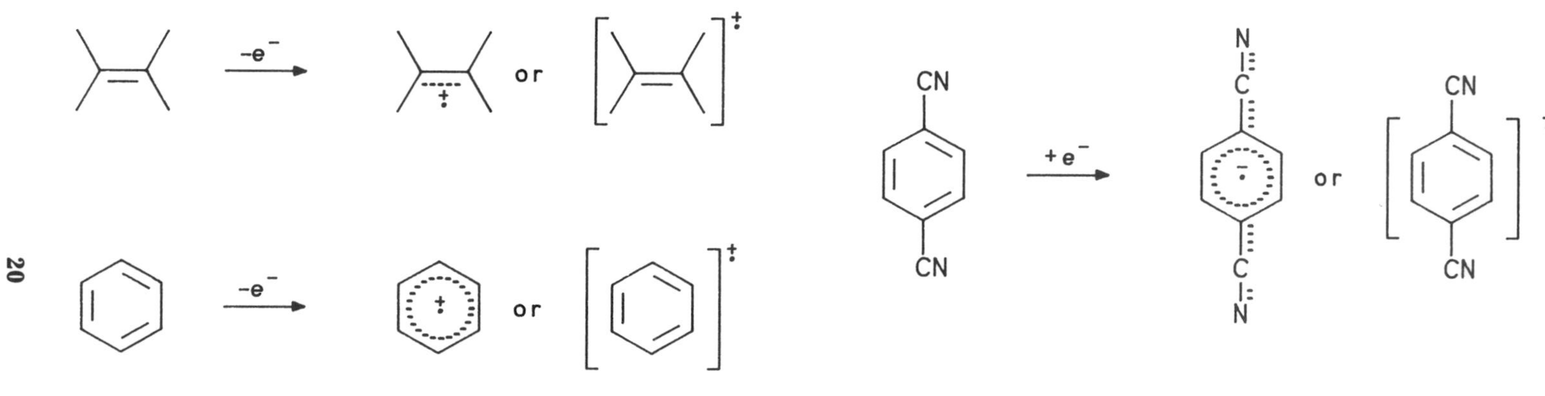

Figure 1.15 The formation of radical cations and anions. Radical ions can be written as resonance structures or in bracket notation.

ion pairs. When $D^{\dot{+}}$ and $A^{\dot{-}}$ are in contact, the ion pair is described as a *contact ion pair*. If several solvent molecules separate $D^{\dot{+}}$ and $A^{\dot{-}}$, the species is called a *solvent-separated ion pair*. When the ions separate to large distances, they are referred to as *free ions*.

Very often, quenching by electron transfer in solution leads initially to the formation of a complex between the excited-state and the ground-state quencher. This complex we call an *exciplex*. This species is an emitting charge-transfer complex that is held together by favorable orbital interactions as well as by Coulombic binding forces. In addition, solvent molecules may stabilize the exciplex by aligning themselves favorably around the exciplex. The exciplex can be regarded as a charge-transfer complex in which charge and electronic excitation energy are shared by the donor and acceptor molecules. Exciplexes are distinct intermediates in their own right and possess unique properties. The most conspicuous property of exciplexes is their fluorescence, which almost always is at longer wavelengths (lower energy) than the fluorescence of the excited state. The nature of an exciplex depends significantly on the structure of the reacting partners. Planar organic molecules are ideal choices for exciplex formation since these reactants can adapt face-to-face geometries at distances of 3–4 Å. In nonpolar solvents, exciplexes rapidly decay by light emission; in polar solvents, they are most likely to dissociate into solvent-separated ions or free ions.

For operational purposes, we shall use the following notation to designate the various kinds of ion-pair intermediates: EX for an exciplex, CIP for a contact ion pair, SSIP for a solvent-separated ion pair, and FIP for a free ion pair. In the reaction schemes in this book, these charge-transfer species will be designated as follows: an exciplex with brackets together with an asterisk to note that it is an excited-state species (i.e., $[D^+A^-]^*$), a contact ion pair with brackets but without an asterisk (i.e., $[D^+A^-]$), solvent-separated ions as $(D^+ \cdots A^-]$, and free ions as D^+ and A^-. In cases where the identity of the radical ion species is not known, the designation $[D^+ + A^-]$ shall be used.

1.5. Quantum Yields, Efficiencies, and Lifetimes

It is desirable for a photochemist to have a way of comparing the yield of ion pairs generated during the course of a photoinduced electron transfer. To accomplish this, there are several approaches. First, the rates of electron transfer, k_{el}, can simply be compared. This is useful for examining a group of related photoinduced electron-transfer processes. For example, if one wishes to evaluate the effect of several related electron donors or acceptors on the quenching of an excited state, k_{el} can be measured and compared. Alternatively, the effect of quenching of a series of related photosensitizers by one electron donor or acceptor can be investigated.

The rates of electron transfer tell us something about the velocity of the process. However, it must be kept in mind that frequently there may be other processes in competition with the electron-transfer step. As we learned in previous sections, these may include radiationless deactivation, emission, energy transfer, and other competing photochemical reactions. For this reason, it is desirable that electron transfer be at least competitive with these processes. To evaluate the "competitiveness" of an electron-transfer pathway, we need an expression that allows us to quantitate the ratio of the rate of electron transfer against the rates of the competitive pathways. For this purpose, the *quantum efficiency* for generation of the ion pair, ϕ_{IP}, is defined as

$$\phi_{IP} = \frac{k_{el}}{k_{el} + k_{en} + k_d + k_{other}} \tag{1.7}$$

where the rate constants in the denominator represent the competitive pathways (i.e., k_{en} for energy transfer, k_d for radiationless deactivation and emission pathways, and k_{other} for photochemical pathways). A lower case Greek ϕ is used for quantum efficiency, in contrast to Φ, which is reserved for describing quantum yield. As used in this book, quantum efficiency, ϕ, refers to the rate of electron transfer vs. the rates of the competitive processes. Quantum yield, Φ, is the amount of species formed divided by the number of moles of photons (quanta) absorbed by the reaction system.

Figure 1.16 illustrates the relationship between k_{el} and the competitive pathways.

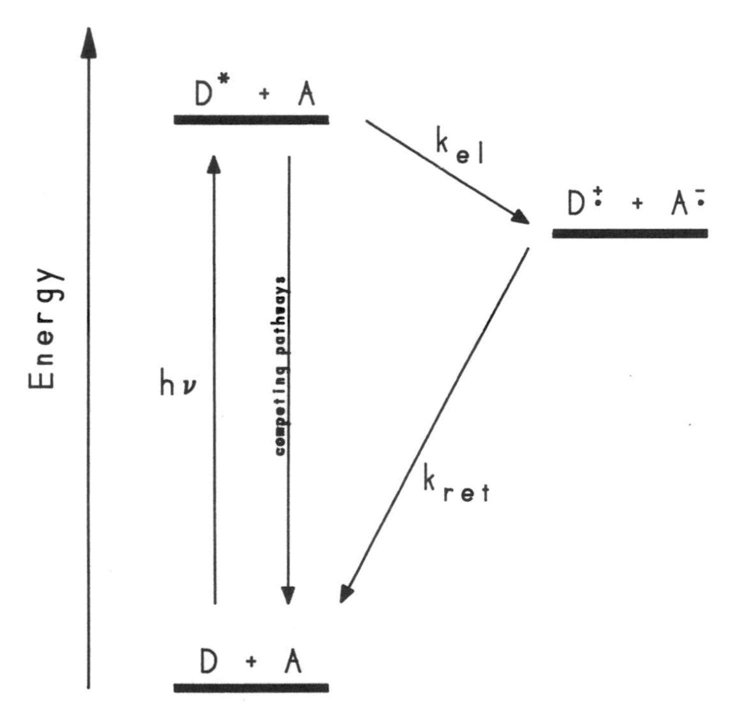

Figure 1.16 An energy diagram showing the rates of electron transfer and electron return. Electron transfer competes with other photochemical and physical pathways.

Another useful expression is τ_{IP}, the lifetime of the ion pair formed by electron transfer:

$$\tau_{IP} = \frac{1}{k_{ret}} \tag{1.8}$$

where k_{ret} is the rate of *electron return*. Electron return is the pathway where the electron is "returned" to its origin generating the original ground-state reactants (Fig. 1.16). Since it destroys the ion pair, electron return can be regarded as a nonproductive pathway. Much of the research effort of photochemists working in photoinduced electron transfer is in fact directed to discovering improved ways to maximize the yield and lifetimes of ion-pair intermediates by minimizing electron return. Throughout this text, we shall be describing approaches designed to enhance both ϕ_{IP} and τ_{IP}.

1.6. Energetics of Photoinduced Electron Transfer

Photoinduced electron transfer can be regarded as a process where absorbed light energy is transformed into chemical energy. After formation of the equilibrated excited data, the subsequent event is the actual transfer of the electron. This step is crucial for the successful generation of intermediate charge-transfer species in such processes as photosynthesis. The property of some excited states to act as strong electron donors and/or electron acceptors depends on *thermodynamic* factors. Understanding these factors is critical if we are going to be able to exploit photoinduced electron transfer for certain applications. In this section, we consider the energetics that determine the feasibility of photoinduced electron transfer.

1.6.1. Excited-State Ionization Potentials and Electron Affinities

The energy contained in the excited state has an important effect on the ability of the excited state to donate or accept electrons. In this section, we shall show that excited states can be more effective at accepting or donating electrons than their ground states.

First, consider a simple model of an electron moving within the Coulombic field of the positive nucleus. Because it possesses a negative charge, the electron can be visualized as "tightly" bound to the nucleus by an attractive force. This interaction can be described by Coulomb's law:

$$\text{Coulombic energy} = \frac{e^2}{d} \tag{1.9}$$

Here, e is the atomic unit of charge (4.8×10^{10} electrostatic units) and d is the distance separating them. We can deduce that the closer the electron is to the nucleus, the stronger will be the attractive forces holding these particles together.

Since the electron moves within an orbital, the simple model of an electron bound to a nucleus can be depicted by an *orbital* energy diagram such as that shown in Fig. 1.17. In this figure, orbital energy, plotted vertically, is shown to *increase* for orbitals displaced farther away from the nucleus. That is, as the electron moves away from the nucleus, the energy of the electron increases. Eventually, at an infinite distance ($n = \infty$), the energy reaches a maximum. To state this another way, the electron occupies discrete energy states. These states increase in energy with increasing separation between electron and proton. Eventually, these states merge into a continuum of states at large separation distances. For our purposes, we will assume that these states represent an infinite distance, and that electrons in orbitals below the continuum are bound electrons (the energy associated with the bound electrons has a negative sign). In the ground-state hydrogen atom, the electron resides in the lowest orbital corresponding to an energy of -13.6 eV (1 eV is defined as the energy that one electron gains as it accelerates through a potential of 1 V). If the bound electron is ejected from the orbital to an orbital

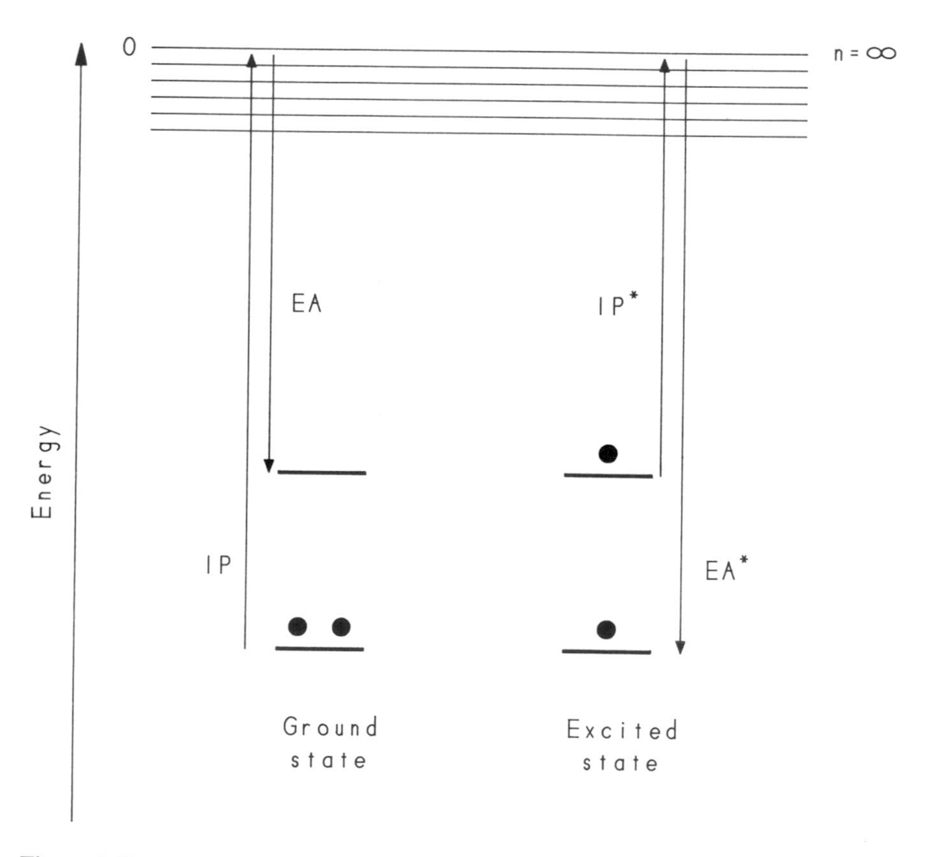

Figure 1.17 The ionization potential (IP) and electron affinity (EA) of an excited state are decreased and increased, respectively, as compared with the ground state.

within the continuum (i.e., at infinite distance), the energy change associated with this ejection is known as the *ionization potential,* or IP, which is the difference between orbital energy at infinite distances (E_∞) and the energy of the lowest orbital, E_1:

$$\text{IP} = E_\infty - E_1 = 0 - (-13.6) = 13.6 \text{ eV} \quad \text{(hydrogen)} \tag{1.10}$$

More strictly, the ionization potential is defined as the energy required to remove an electron from an atom or molecule in the gas phase. According to our simple model, the process of ionization from orbitals located at larger separation distances requires *less* energy since these electrons are not as strongly bound as the core electrons residing near the nucleus.

The magnitude of IP reflects the energy required to ionize an atom or molecule. The donating species is called an *electron donor* or D. The half-reaction is written as

$$\text{D} \rightarrow \text{D}^+ + e^- \qquad +\text{IP}_\text{D} \tag{1.11}$$

In the excited state, the excited electron populates an antibonding orbital, which is positioned at a greater separation distance from the nucleus. Hence, its orbital energy is greater than the orbital energies of the bonding electrons. From this simple picture, we immediately note that IP of the electron expelled from the antibonding orbital of an excited state is less than the IP of that same electron in the ground state (Fig. 1.17). This is simply a manifestation of a unique property of excited states: their ionization potentials are smaller than the ionization potentials of their respective ground states, the difference in the ionization potential of excited state (IP*) and the ionization potential of the ground state (IP) being equal to E_{00}:

$$\text{IP}_{\text{D}^*} = \text{IP}_\text{D} - E_{00} \tag{1.12}$$

Thus, for the half-reaction

$$\text{D}^* \rightarrow \text{D}^+ + e^- \qquad +\text{IP}_{\text{D}^*} \tag{1.13}$$

IP_{D^*} can be estimated from IP_D and E_{00}, which are measurable quantities.

We can also use this model to describe the reverse process where the electron moves from an orbital at infinite separation to a vacant orbital near the nucleus. In this case, energy is *released* because of the favorable attraction between the electron and positive nucleus. The energy released as the electron moves from infinity to the vacant E_n-orbital is the energy difference between these orbital energies, which from Fig. 1.17 is simply the reverse of ionization, or $-$IP. In this case, IP refers to the ionization potential of the anion, as is shown below:

$$\text{A} + e^- \rightarrow \text{A}^- \qquad -\text{IP}_{\text{A}^-} \tag{1.14}$$

The accepting species is called an *electron acceptor*. Obviously, the process is more favorable for electrons entering vacant orbitals closer to the nucleus since more energy is released. In the case of excited states, electronic excitation creates vacant sites in low-lying bonding or nonbonding orbitals (occasionally referred to as *holes*):

$$A^* + e^- \rightarrow A^- \qquad -IP_{A^{-\prime}} \tag{1.15}$$

$IP_{A^{-\prime}}$ is calculated from Eq. 1.16

$$IP_{A^{-\prime}} = IP_{A^-} + E_{00} \tag{1.16}$$

It is customary to employ a parameter called the *electron affinity,* or EA, in place of IP_{A^-} or $IP_{A^{-\prime}}$ to describe the energy released in reactions such as 1.14 and 1.15, respectively. Since IP = EA, according to this usage, the electron affinity is assigned a *positive* sign, so that a positive sign implies an *exothermic reaction.** Thus, the EAs of excited states are greater than the EAs of ground states:

$$EA_{A^*} = EA_A + E_{00} \tag{1.17}$$

In presenting this highly idealized model, we did not take into account the effects of other orbital electrons. For example, in the transition of an electron from an outer orbital to a lower-lying half-vacant orbital, repulsive electron–electron interactions will be present. These unfavorable effects may lead to a smaller EA. Similarly, ionization of an electron from a doubly occupied orbital may be facilitated by the repulsive interactions with the other electron, resulting in a smaller IP. Nonetheless, even if we ignore these electrostatic effects, our simple model provides insight into the ability of excited states to act as better electron donors and acceptors.

Suppose now that two species interact such that one donates its electron to the other:

$$D + A \rightarrow D^+ + A^- \tag{1.18}$$

Assuming a gas phase reaction, we can deduce the corresponding energy change from the ionization potential and electron affinity of 1.18 as

$$\Delta E = IP_D - IP_{A^-} = IP_D - EA_A \tag{1.19}$$

Suppose that we excite D with a photon:

$$D \xrightarrow{h\nu} D^* \tag{1.20}$$

Now D* is the electron donor and the reaction is written as

$$D^* + A \rightarrow D^+ + A^- \tag{1.21}$$

The energy change for 1.21 is

$$\Delta E = IP_D - EA_A - E_{00} \tag{1.22}$$

*Actually, according to thermodynamic convention, the electron affinity should be assigned a negative sign. However, this practice has not been adopted. In thermodynamic calculations, the enthalpy is assumed to be −EA.

This treatment also can be applied to the case where the acceptor is the excited species. Thus, for reactions 1.23 and 1.24,

$$A \xrightarrow{h\nu} A^* \tag{1.23}$$

$$D + A^* \rightarrow D^+ + A^- \tag{1.24}$$

the energy change is

$$\Delta E = IP_D - EA_A - E_{00} \tag{1.25}$$

Ionization potentials and electron affinities can be measured by photoelectron spectroscopy. In this technique, a monochromatic beam of photons ionizes a molecule thereby generating its radical cation. If a bonding electron is removed, the Franck–Condon cation rapidly equilibrates to give a cation with a larger equilibrium bond distance. The ejected electron may have energy in excess of the photon energy. Since the excess energy is in the form of kinetic energy, the electrons can be identified and counted on the basis of their kinetic energies. The kinetic energy (K.E.) of the expelled electron, in turn, can be used to estimate IP of an electron donor, since

$$K.E. = h\nu - IP_D \tag{1.26}$$

The electron affinity of an electron acceptor can be calculated using the following relationship:

$$EA_A = -IP_D + h\nu \tag{1.27}$$

Representative values of IPs and EAs for several ground- and excited-state species are listed in Tables 1.2 and 1.3, respectively. These values are given in electron volts (eV) and apply to the *first* ionization potential and *first* electron affinity, where the *first* IP is the energy needed to abstract the most weakly held electron, and the *first* EA is the energy given off when an electron enters the lowest unoccupied orbital. In Table 1.2, the IPs of several amines and aromatic hydrocarbons are listed. Note that the species near the top of the table (e.g., hydrogen and methane) are the molecules having the largest IPs and accordingly are the poorest electron donors listed in the table. Compounds near the bottom of the bottom (e.g., amines) have the lowest IPs and are therefore the best electron donors. In the case of amines, the IPs refer to the ionization of one of the lone pair *n* electrons. Since the lone pair electrons are nonbonding, these electrons are easily abstracted. Accordingly, amines and other molecules with nonbonding electrons serve as effective quenchers of excited electron acceptors. The low IPs of naphthalene and anthracene reflect the fact that in these aromatic hydrocarbons, electrons populating π-orbitals can easily be abstracted. However, the excited states of aromatic hydrocarbons should be even better donors because of the presence of an electron in a high-lying π^*-orbital. Although the σ-orbitals of saturated alkanes lie at fairly low energies so that these compounds have large IPs and are therefore poor donors, strained cyclic

Table 1.2 IONIZATION POTENTIALS[a]

Molecule or Atom	IP[b]
Hydrogen	13.6[c]
Methane	12.99[d]
Ammonia	10.19
Benzene	9.25
Methylamine	8.97
Naphthalene	8.12
Aniline	7.70
Anthracene	7.43
Quadricyclane	7.4
N,N-dimethylaniline	7.10
Triphenylamine	6.86

[a] Unless otherwise indicated, IPs are from Murov, S. L. *Handbook of Photochemistry*. Marcel Dekker, New York, 1973, p. 197.
[b] IP represents the first ionization potential, in electron volts; ability to donate an electron increases from the top to bottom of table.
[c] Heilbronner, E., and Bock, H. *The HMO-Model and Its Application*. Wiley-Interscience, London, 1976, p. 73.
[d] Heilbronner and Bock, 1976, p. 368.

molecules are an important exception. The latter compounds are usually good donors. The fact that these molecules have σ-orbitals at higher energies makes the electrons populating these orbitals susceptible to abstraction. This is true, for example, for quadricyclane, which has a relatively low IP.

Table 1.3 lists EAs. Note, for example, that the EA of 1-cyanonaphthalene is higher than that of naphthalene. Thus, the cyano group exerts a strong electron-withdrawing effect (Fig. 1.15). This leads to a lowering of the electron density in the π-orbitals. In the π,π* excited state of a cyanoaromatic molecule, since the low-lying π-orbital (the accepting orbital) has only one electron, the orbital is even more electron deficient. In fact, the excited states of cyanoaromatic hydrocarbons are among the most powerful electron acceptors. The excited states of quinones, (e.g., *p*-benzoquinone) and carbonyl compounds (e.g., benzophenone and acetophenone) are also electron acceptors. Like the cyano compounds, these excited states have low-lying vacant orbitals.

It should now be evident that the feasibility of electron transfer depends on the relative ordering of the energies of occupied and unoccupied orbitals. This is noted with the examples in Figs. 1.18 and 1.19 where the orbitals of excited states are shown interacting with those of the quenchers. Low-lying vacant positions or "holes" in orbitals are created during photoexcitation and accordingly serve as much better electron acceptors than vacant orbitals at higher energies. This is shown in Fig. 1.18 for the reaction of singlet 1-cyanonaphthalene with an amine and with a strained cyclic molecule. Without the presence of the electron-withdrawing cyano group, the π-orbitals of the singlet aromatic ring would lie at energies too high to abstract electrons from amines (Fig. 1.19). This consideration also applies to the electron transfer from saturated alkanes to singlet aromatic molecules (Fig. 1.19).

Table 1.3 ELECTRON AFFINITIES[a]

Molecule or Atom	EA[b]
Ethylene	−1.55
Benzene	−1.15
Pyridine	−0.62
Naphthalene	0.15[c]
1-Cyanonaphthalene	0.68[d]
Hydrogen	0.754
Anthracene	0.82
O_2	1.40[e]
p-Benzoquinone	1.89

[a] Unless otherwise indicated, EAs are from Dewar, M. J. S., and Rzepa, H. S. *J. Am. Chem. Soc.* **1978,** *100,* 784 and references therein.
[b] EA represents the first electron affinity, in electron volts; ability to accept an electron increases from the top to bottom of table.
[c] Dewar, M. J. S., and Dougherty, R. C. *The PMO Theory of Organic Chemistry*. Plenum, New York, 1975, p. 505.
[d] Chowdhury, S., and Kebarle, P. *J. Am. Chem. Soc.* **1986,** *108,* 5453.
[e] Pearson, R. G. *J. Am. Chem. Soc.* **1986,** *108,* 6109.

We can apply similar reasoning to excited donors. For example, electrons promoted to high-lying orbitals during photoexcitation are more predisposed to abstraction than electrons in lower orbitals. Thus, the d,π^* excited state of $Ru(bpy)_3^{2+}$ readily donates an π-electron to methyl viologen in a thermodynamically favorable reaction.

Substituting IPs and EAs into Eqs. 1.22 or 1.25 can give the chemist an indication whether an excited-state electron transfer is *exothermic* or *endothermic*. However, IPs and EAs are one-electron properties obtained in the gas phase and do not properly describe electron-transfer processes in solution. When IPs and EAs are substituted into Eqs. 1.22 or 1.25, the estimated energies are not very applicable to the solution phase. At best, these equations are useful only for comparison purposes, so we will not direct our attention to them. We need experimentally derived parameters which realistically account for Coulombic effects and the effects of solvation. This subject is covered in the next section.

1.6.2. Excited-State Redox Potentials

Figure 1.20 is a state diagram showing the energy changes which accompany the ionization of an electron donor in its ground and excited state. Expressing each ionization as a half-reaction, we have

$$D \rightarrow D^+ + e^- \qquad (1.28)$$

and

$$D^* \rightarrow D^+ + e^- \qquad (1.29)$$

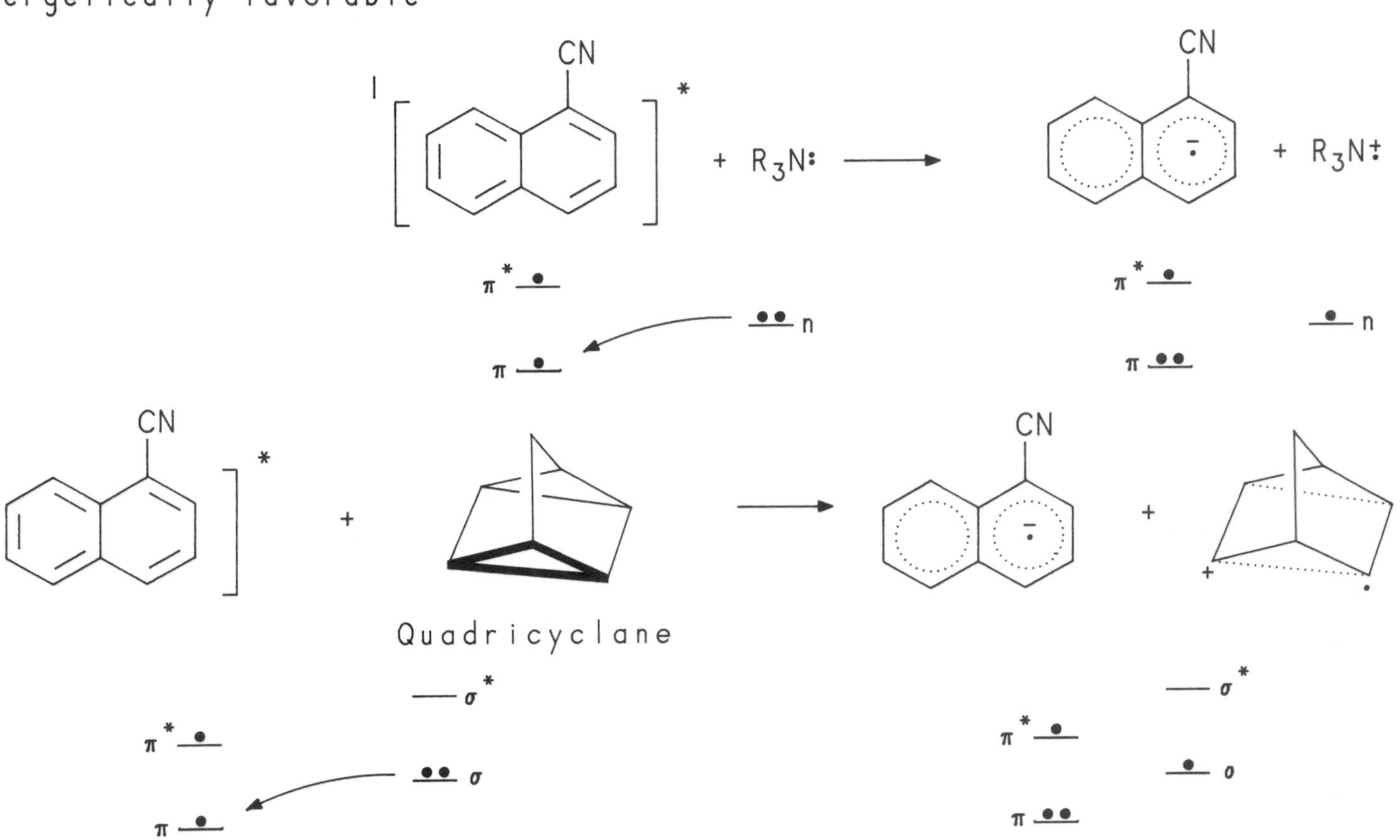

Figure 1.18 Thermodynamically favorable electron transfers.

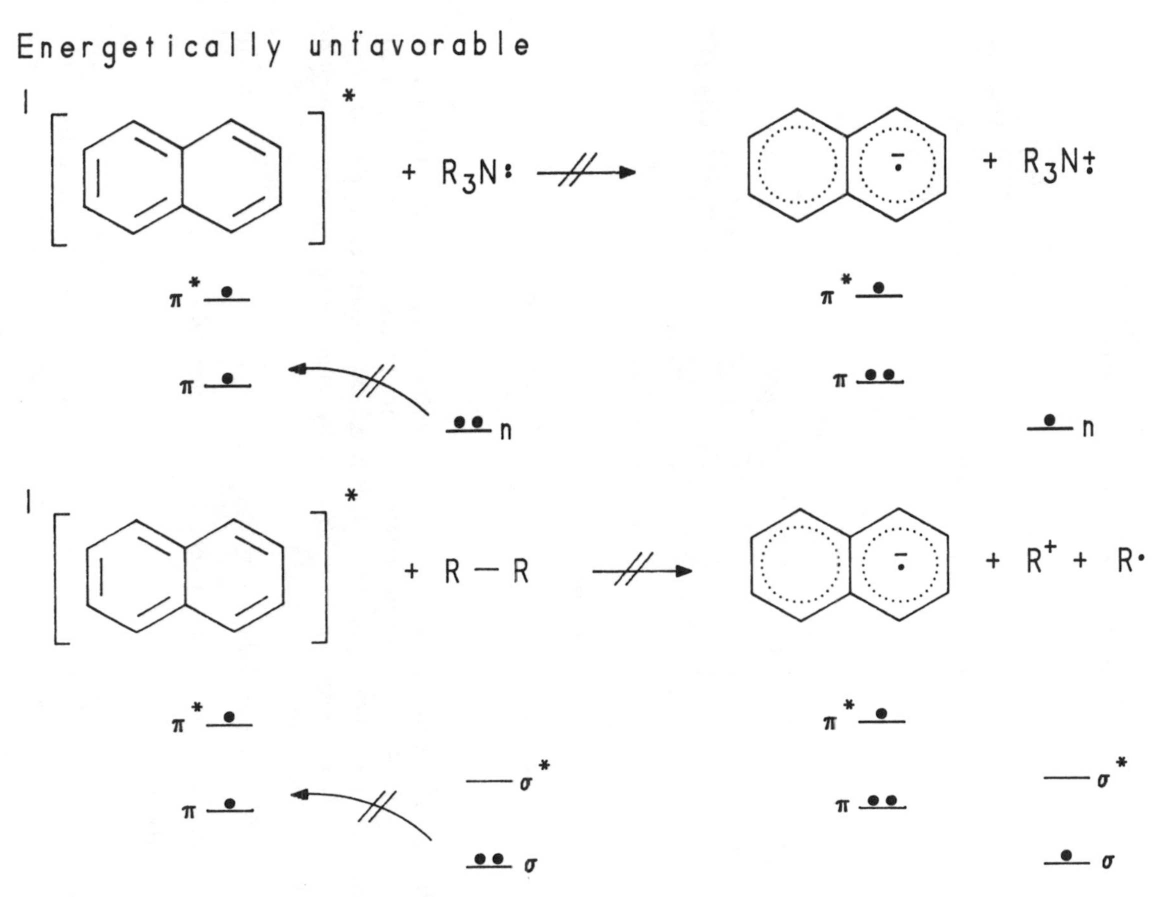

Figure 1.19 Thermodynamically unfavorable electron transfers.

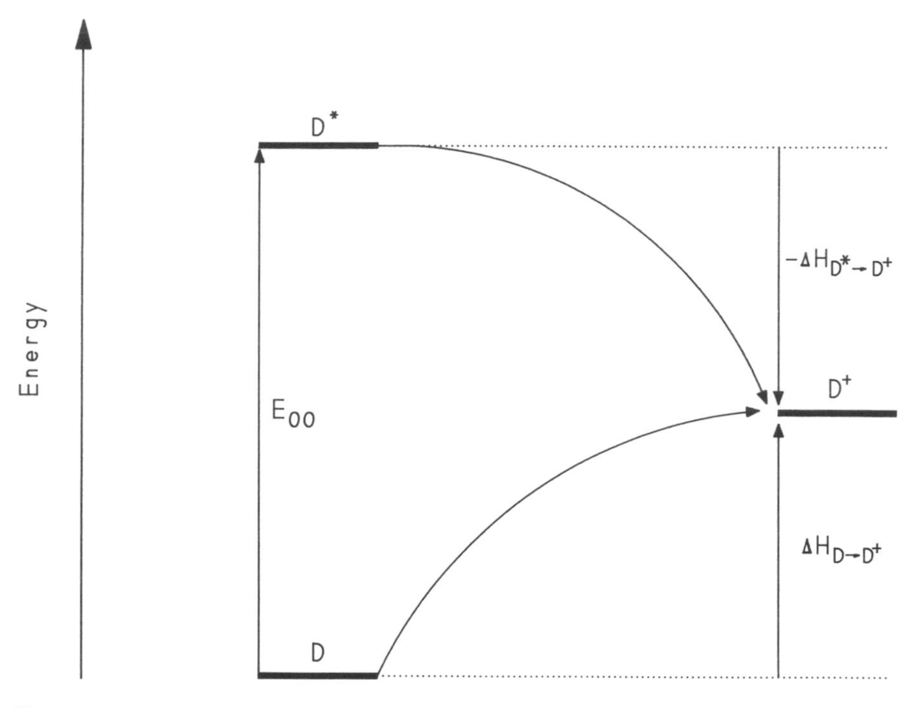

Figure 1.20 Enthalpy changes for formation of D^+ from D and D*.

$\Delta H_{D \to D^+}$ and $\Delta H_{D^* \to D^+}$ are the *heats of enthalpy* during ionization of D and D*, respectively. Note that for the electron-transfer processes shown in Fig. 1.20, $\Delta H_{D \to D^+}$ is a *positive* quantity since it refers to an endothermic process. Thus $\Delta H_{D^* \to D^+}$ is *more negative* than $\Delta H_{D \to D^+}$ by an amount equal to the energy of the excited state, or

$$-\Delta H_{D^* \to D^+} + \Delta H_{D \to D^+} = E_{00} \tag{1.30}$$

Now, the *free-energy change* accompanying a chemical process is

$$\Delta G = \Delta H - T\Delta S \tag{1.31}$$

where ΔG is the free-energy difference between the thermodynamic reactant and product states, ΔS is the *entropy of reaction,* and T is the absolute temperature. The free-energy term has been introduced because it represents the work of a chemical reaction. If the free-energy change, ΔG, is positive the system requires the input of energy; if ΔG is negative, the reaction proceeds spontaneously, giving off energy which then would be available to carry out work. Combining the energy terms in Eqs. 1.30 and Eq. 1.31 leads to:

$$-\Delta G_{D^* \to D^+} - T\Delta S_{D^* \to D^+} + \Delta G_{D \to D^+} + T\Delta S_{D \to D^+} = E_{00} \tag{1.32}$$

Variations in entropy are closely linked with the differences in structure and solvation between initial and final states in electron transfer. However, if we assume that the structures of D and D^+ and of D* and D^+ are approximately the same, the free-energy is

$$\Delta G = \Delta H \tag{1.33}$$

and

$$\Delta G_{D^* \to D^+} = \Delta G_{D \to D^+} - E_{00} \tag{1.34}$$

Thus, the free-energy change involving an electron donation from D* is more negative by an amount equal to the energy of the excited state.

Similar arguments can be used to estimate the free-energy changes for the reduction of an acceptor and its excited state. We first write the half-reactions:

$$A + e^- \rightarrow A^- \tag{1.35}$$

and

$$A^* + e^- \rightarrow A^- \tag{1.36}$$

We then derive an equation for the free-energy change for an acceptor's excited state, using the energy diagram in Fig. 1.21:

$$\Delta G_{A^* \to A^-} = \Delta G_{A \to A^-} - E_{00} \tag{1.37}$$

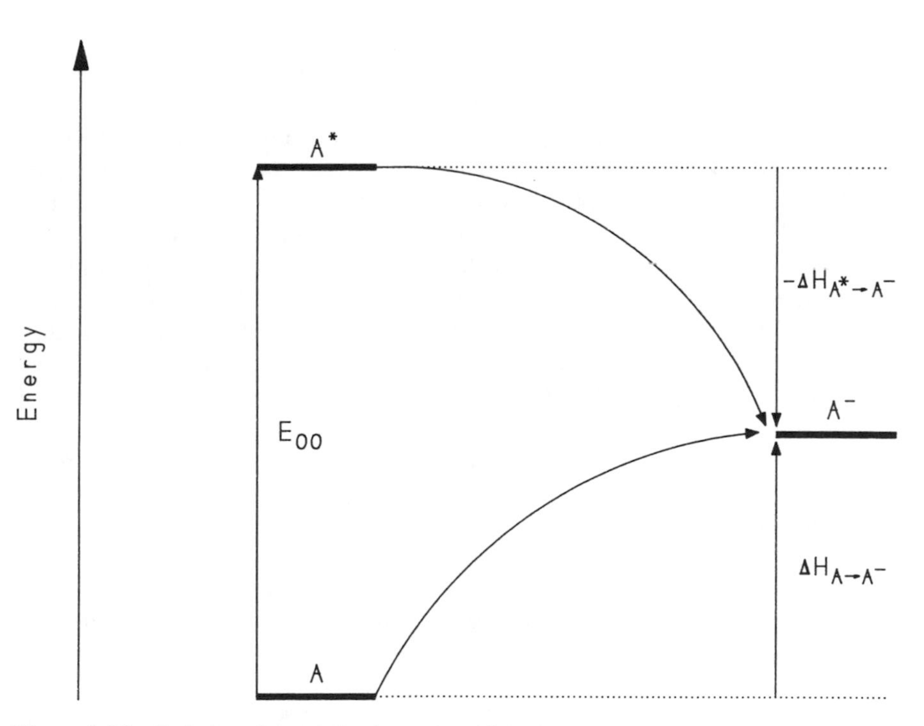

Figure 1.21 Enthalpy changes for formation of A^- from A and A*.

where $\Delta G_{A \rightarrow A^-}$ and $\Delta G_{A^* \rightarrow A^-}$ are the free-energy changes for 1.35 and 1.36. Equation 1.37 is analogous to Eq. 1.34 in that it states the thermodynamic condition for the free-energy change for the reduction of an acceptor.

Having shown the thermodynamic conditions for oxidation or reduction of an excited- and ground-state species, we now introduce the concept of *electromotive* force, or E_{redox}, of the half-reaction. E_{redox} is defined as the potential of a half-reaction and is related to the free energy, as shown below:

$$\Delta G = -nFE_{redox} \tag{1.38}$$

where n represents the number of electrons transferred and F is a unit known as the faraday, which is defined as the number of coulombs (96,493) or the charge associated with 1 mole of electrons. Since E_{redox} is sensitive to the direction of the reaction, E_{redox} is sign dependent. A *positive* E_{redox} implies an exothermic, spontaneous reaction, and a *negative* value suggests an endothermic process. In practice, when using E_{redox}, the half-reaction, whether it is an oxidation or reduction, is usually written as a reduction process. Thus, for the oxidation of an electron donor molecule and reduction of an acceptor, we write both of these processes as reductions:

$$D^+ + e^- \rightarrow D \tag{1.39}$$

$$A + e^- \rightarrow A^- \tag{1.40}$$

The emfs of reactions 1.39 and 1.40 are now written as $E_{D^+ \rightarrow D}$ and $E_{A \rightarrow A^-}$, respectively. For the oxidation of D shown in 1.28, we switch the sign of the electromotive force (i.e., $E_{D \rightarrow D^+} = -E_{D^+ \rightarrow D}$). For the reduction of A, the sign remains the same since the reaction is already written as a reduction. When writing the half-reaction as a reduction, the electromotive force is replaced by the term *redox potential*. Redox potential refers to the emf of a reduction process and is written as $E^0(D^+/D)$ for removal of an electron from a donor, and as $E^0(A/A^-)$ for addition of an electron to an acceptor [or $E^0(D^{\dot{+}}/D)$ and $E^0(A/A^{\dot{-}})$ for a radical cation and radical anion, respectively]. Since $E^0(D^+/D) = E_{D^+ \rightarrow D} = -E_{D \rightarrow D^+}$, $E^0(D^+/D)$ decreases with increasing propensity to surrender an electron, or to state this another way, for a series of donors, their $E^0(D^+/D)$s progressively decrease with increasing exothermicity of the ionization process. Conversely, for a series of electron acceptors, the ability to accept electrons increases with increasing $E^0(A/A^-)$. Redox potentials are reported in volts (V) and are always listed vs. a reference electrode, which in many cases is a standard calomel electrode (SCE) or normal hydrogen electrode (NHE). When comparing redox potentials, one should relate redox potentials to the same reference electrode. For this purpose, Table 1.4 lists conversions to report redox potentials vs. the same reference electrode.

Now the redox potentials of half-reactions involving excited states can be derived in a similar fashion. From Eqs. 1.34 and 1.38, we obtain

$$E_{D^* \rightarrow D^+} = E_{D \rightarrow D^+} + E_{00} \tag{1.41}$$

Since by convention, $E^0(D^+/D) = -E_{D \to D^+}$ and $E^0(D^+/D^*) = -E_{D^* \to D^+}$, we write

$$E^0(D^+/D^*) = E^0(D^+/D) - E_{00} \tag{1.42}$$

Now the magnitude of $E^0(D^+/D^*)$, which according to Eq. 1.42 should be smaller than $E^0(D^+/D)$, implies that the excited state is a better electron donor than the ground state.

A similar treatment can be applied to the reduction of A. Thus, from Eqs. 1.37 and 1.38, we obtain

$$E^0(A^*/A^-) = E^0(A/A^-) + E_{00} \tag{1.43}$$

In this case, since its redox potential is more positive, the excited state is a better electron acceptor than its ground state.

At this point, it will be useful to observe the relationships between ionization potentials, electron affinities, and redox potentials. As we noted in the previous section, there is limited information to be gained by considering only the IPs and EAs of any reactant pair since our main concern in this book is with excited-state electron transfers in solution. If, for example, two neutral species react in solution to form an ion pair, solvent molecules rapidly stabilize the ions within the ion pair. The solvation terms are ΔG_{D+} and ΔG_{A-} for D^+ and A^-, respectively. The following equations express the relations between the gas-phase energy terms and the redox potentials:

$$IP = E^0(D^+/D) - \Delta G_{D+} + \text{constant} \tag{1.44}$$

$$EA = E^0(A/A^-) + \Delta G_{A-} + \text{constant} \tag{1.45}$$

From these equations, we note that the variations in redox potentials parallel those in ionization potentials and electron affinities. In fact, plots of redox potentials vs. ionizations potentials or electron affinities are often linear. These relationships validate the assumptions we made in this section to derive expressions for the redox

Table 1.4 CONVERSION OF REFERENCE ELECTRODES[a,b]

A	B	C	Solvent
NHE	SCE	0.24	H_2O
SCE	NHE	−0.24	H_2O
SCE	Ferrocenium/ferrocene	0.10	CH_3 CN
Ferrocenium/ferrocene	SCE	−0.10	CH_3 CN
SCE	$Ag/AgNO_3$	0.29	CH_3 CN/$NaClO_4$
$Ag/AgNO_3$	SCE	−0.29	CH_3 CN/$NaClO_4$

[a] To convert redox potential vs. reference electrode A to an redox potential vs. reference electrode B, substract C from A (i.e., B = A − C).

[b] Fox, M. A., and Chanon, M. (eds.), *Photoinduced Electron Transfer. Part A. Conceptual Basis*. Elsevier, Amsterdam, 1988, Chapter 1.

Table 1.5 ELECTROCHEMICAL DATA OF SELECTED SENSITIZERS[a]

Sensitizer	E_{00}	E^0 (D^+/D)	$E^0(D^+/D^*)$	$E^0(A/A^-)$	$E^0(A^*/A^-)$
Acetophenone	3.41 (T_1)	—	—	−1.85	1.56
Chloranil	2.7 (T_1)	—	—	0.02	2.72
1-Cyanonaphthalene	3.88 (S_1)	—	—	−1.98	1.90
	2.49 (T_1)	—	—	−1.98	1.47
2-Cyanonaphthalene	3.68 (S_1)	—	—	−2.13	1.55
Phenanthrene	3.59 (S_1)	1.58	−2.01	−2.20	1.39
	2.69 (T_1)	1.58	−1.11	−2.20	0.49
$Ru(bpy)_3^{2+}$	2.12	1.29	−0.83	−1.35	0.77
$Cr(bpy)_3^{3+}$	1.71	>1.6	⩾0.1	−0.26	1.45
$Rh_2(dicp)_4^{2+}$	1.69	0.89	−0.80	−1.4	0.29

[a] Data based on tables in Kavarnos, G. J., and Turro, N. J. *Chem. Rev.* **1986,** *86,* 401, and references therein; all energies are in volts obtained in acetonitrile, H_2O, or other polar solvents; redox potentials are reported vs. standard calomel electrode (SCE). bpy, 2,2′-bipyridine; dicp, 1,3-diisocyanopropane.

potentials of excited states. In a later chapter, we shall examine more closely the solvation terms, ΔG_{D^+} and ΔG_{A^-}. For now it suffices to keep in mind that because redox potentials are relevant to solution-phase reactions, they, rather than IPs and EAs, are frequently preferred when evaluating the thermodynamics of excited-state electron transfers.

Table 1.5 lists the redox potentials of the ground and excited states of several organic sensitizers. The data listed in this table demonstrate the effect of photoexcitation on the oxidation and reducing properties of excited states. In this case of 1-cyanonaphthalene, for example, since the redox potentials for reduction of the excited singlet and triplet states π,π^* are both increased relative to the ground state, these states are predicted to be effective electron acceptors. As we noted in the previous section, the cyano group withdraws electrons from the ring resulting in a *decrease* in electron density. Accordingly, the ring can readily accept an electron leading to the formation of the 1-cyanonaphthalene radical anion.

The excited states of photosensitizers possessing carbonyl groups also possess large positive redox potentials. In the formation of their singlet and triplet π,π^* and n,π^* states, half-vacant π- and n-orbitals are formed that possess electroaffinic properties. The excited states of quinones, as exemplified by chloranil in Table 1.5, are also characterized by high redox potentials because of the presence of carbonyl groups.

Some photosensitizers can perform a dual role and function as both electron acceptors and donors. Phenanthrene is an example, as the redox potentials for oxidation and reduction of its excited singlet and triplet states are *decreased* and *increased,* respectively. This is also true of numerous transition metal complexes. Ruthenium(II) tris(bipyridyl), $Ru(bpy)_3^{2+}$, is probably the best known example:

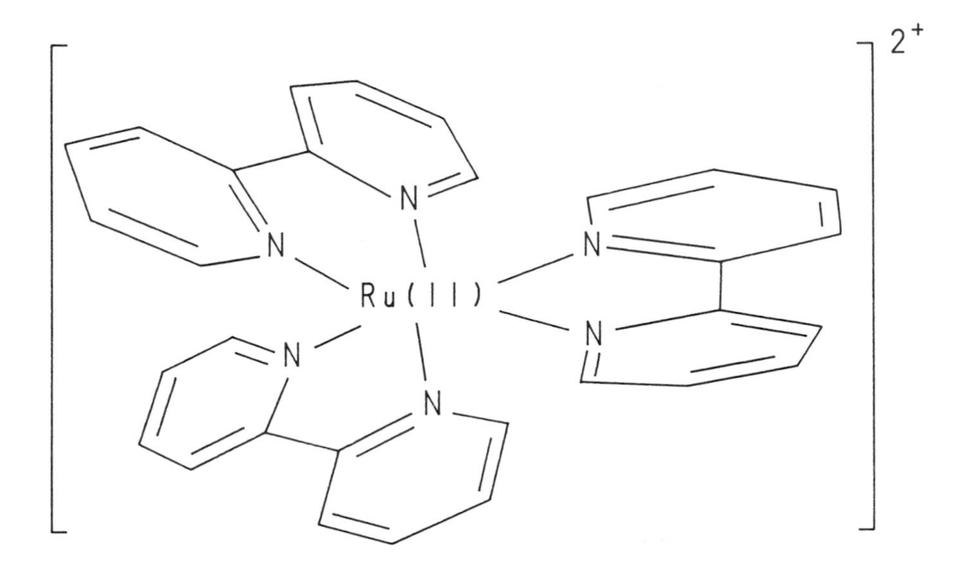

This complex was one of the first metal transition metal complexes to be the subject of extensive study, especially in regard to its ability to act as an excited-state oxidant and reductant. By themselves, solutions of Ru(II) and bipyridine are clear. When these solutions are combined, they form a deep red solution of $Ru(bpy)_3{}^{2+}$. When this solution is irradiated with visible light, $Ru(bpy)_3{}^{2+}$ forms what is called a *charge-transfer excited state*. From spectral studies, it has been determined that a metal-centered electron (t_{2g}) is promoted to a π-orbital in the formation of a d,π^* state. Detailed investigations have revealed that the d,π^* state consists of three closely space thermally equilibrated states. In the excited state, the electron is localized on the bipyridyl ligands. The hole created in the t_{2g}-orbital can act as an electron acceptor whereas the electron promoted to the π^*-orbital is available for electron donation.

However, unlike $[Ru(bpy)_3{}^{2+}]^*$, $[Cr(bpy)_3{}^{3+}]^*$ functions primarily as an oxidant. Its ^2E excited state is formed by a $^4A_2 \rightarrow {}^2E$ metal-centered transition, creating a vacant low-energy orbital resulting in a significant increase in both the electron affinity and redox potential.

1.6.3. The Rehm–Weller Equation

Having established thermodynamic relationships for deriving the electromotive forces of reductions and oxidations of excited-state donor and acceptor molecules, we are at the stage where we can derive a useful expression for calculating the free-energy change accompanying excited-state electron transfer. Figure 1.22 shows a thermodynamically uphill pathway involving D and A and a downhill pathway from D* and A. The overall free-energy change for the uphill process is equal to the

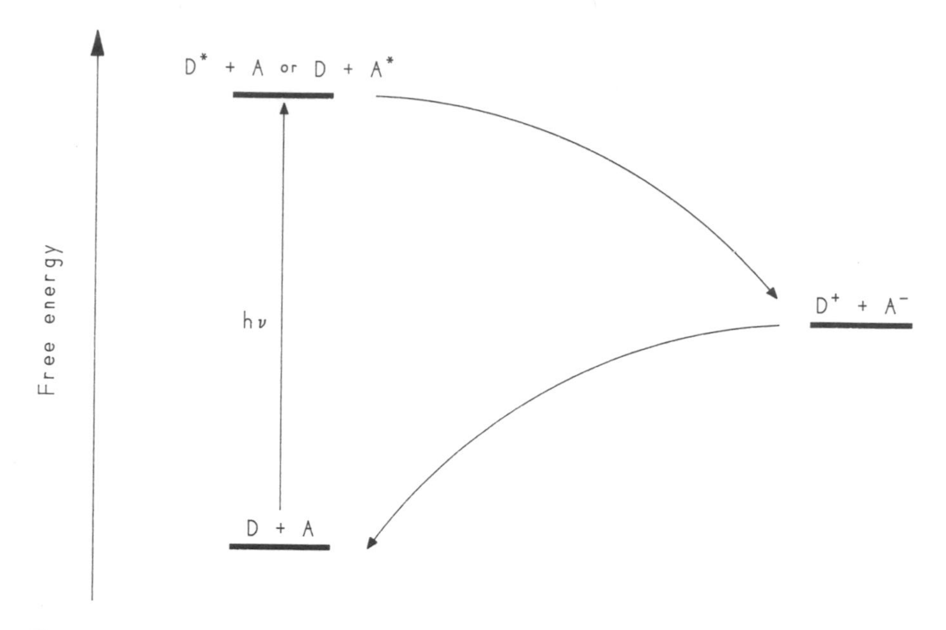

Figure 1.22 An archetypal energy diagram for photoinduced electron transfer.

sum of the free-energy changes for oxidation of the donor and reduction of the acceptor:

$$\Delta G_{el} = \Delta G_{D \to D^+} + \Delta G_{A \to A^-} \quad (1.46)$$

where ΔG_{el} specifies the standard free-energy change. Combining Eqs. 1.38 and 1.46 yields

$$\Delta G_{el}(\text{eV}) = nF(-E_{D \to D^+} - E_{A \to A^-}) \quad (1.47$$

or, in keeping with the convention of writing all half-reactions as reductions and using redox potentials:

$$\Delta G_{el}(\text{eV}) = nF[E^0(D^+/D) - E^0(A/A^-)] \quad (1.48)$$

The free-energy change for an electron transfer between an excited donor and ground-state acceptor—the thermodynamically downhill reaction—can be derived in a similar manner:

$$\Delta G_{el}(\text{eV}) = nF[E^0(D^+/D^*) - E^0(A/A^-)] \quad (1.49)$$

Substitution of Eq. 1.42 into 1.49 leads to

$$\Delta G_{el}(\text{eV}) = nF[E^0(D^+/D) - E^0(A/A^-) - \Delta G_{00}] \quad (1.50)$$

where ΔG_{00} is the free energy in electron volts corresponding to the equilibrium energy, E_{00}. For most one-electron transfers, $nF \sim 1$, so that we can write

$$\Delta G_{el}(\text{eV}) = E^0(D^+/D) - E^0(A/A^-) - \Delta G_{00} \quad (1.51)$$

If we measure the excited-state energy in kilocalories per mole and redox potentials in volts, as is customary, Eq. 1.51 can be written as

$$\Delta G_{el}(\text{kcal mol}^{-1}) = 23.06[E^0(D^+/D) - E^0(A/A^-)] - \Delta G_{00} \tag{1.52}$$

A similar treatment applied to the electron transfer between A* and D also leads to Eq. 1.52.

To apply Eq. 1.52, the only experimental parameters required are the excited-state energy of the photoexcited molecule and the redox potentials of the ground states of both the electron donor and acceptor. A point worth noting is that Eq. 1.52 is applicable to an equilibrated excited state and not to a Franck–Condon state. We assume that electron transfer occurs *after* the equilibration of the Franck–Condon excited state to the relaxed, equilibrated state. This should not be taken to mean that electron transfer cannot take place involving unrelaxed excited states. There are, in fact, examples in the literature describing electron transfer in these cases, but we will not consider them in this book.

There is an important refinement that can be introduced into Eq. 1.52. Suppose for now that the donor and acceptor are neutral molecules. Electron transfer would then generate two charged species—D^+ and A^-. When this ion pair is formed, attractive Coulombic forces will draw the two ions closer together and result in a release of energy. This attraction is given by a work term, w_p, derived from Coulomb's law.

$$w_p(\text{kcal mol}^{-1}) = \frac{(z_{D^+} z_{A^-})e^2}{d_{cc}\varepsilon_s} = \frac{332(z_{D^+} z_{A^-})}{d_{cc}\varepsilon_s} \tag{1.53}$$

where z_{D^+} and z_{A^-} are the charges on the molecules, ε_s is the static dielectric constant of the solvent, and d_{cc} is the center-to-center separation distance in Ångströms (Å) between the two ions. Combining Eqs. 1.50 and 1.53 gives

$$\Delta G_{el}(\text{kcal mol}^{-1}) = 23.06[E^0(D^+/D) - E^0(A/A^-)] - w_p - \Delta G_{00} \tag{1.54}$$

which is called the Rehm–Weller equation.[1] According to the work term in Eq. 1.54, the presence of electrostatic interactions between two ions should influence the free-energy change accompanying electron transfer. Under polar conditions where ε_s is large (e.g., $\varepsilon_s = 37$ for acetonitrile), $w_p \sim -1.3$ kcal/mol for the usual case where $z_{D^+} = +1$ and $z_{A^-} = -1$ and $d_{cc} \sim 7$ Å. When using Eq. 1.54 to calculate ΔG_{el} for a reaction in a given solvent, the choice of redox potentials is important. Preferably, they should be measured in the same solvent or in a solvent with a similar dielectric constant. Since most of the redox potentials given in the tables later in this chapter have been measured in polar solvents, the use of these redox potentials to calculate ΔG_{el} in nonpolar media may lead to erroneous values. Equation 1.54 can be modified to calculate free energies in nonpolar solvents using redox potentials measured in polar media but this treatment is covered in Chapter 2.

At the heart of electron-transfer photochemistry, Eq. 1.54 states concisely the fundamental thermodynamic condition for spontaneous electron transfer between neutral reactants: $\Delta G_{el} < 0$.

To exemplify the use of the Rehm–Weller equation, let us examine the quenching of singlet 2-cyanonaphthalene by a series of electron-donating quenchers. The reaction is typical of many found in the literature where the excited state is an electron acceptor:

$$^{1}[\text{2-cyanonaphthalene}]^{*} + D \rightarrow [\text{2-cyanonaphthalene}]^{\overline{\cdot}} + D^{\dot{+}} \qquad (1.55)$$

ΔG_{el}s, which are listed in Table 1.6, were calculated using Eq. 1.54 from values of $E^0(D^{\dot{+}}/D)$ and $E^0(A/A^{\overline{\cdot}})$ tabulated in Tables 1.5 and 1.6. The reaction is sensitive to $E^0(D^{\dot{+}}/D)$ for the series of quenchers and parallels the ability of the donor to give up an electron [i.e., the donor with the lowest $E^0(D^{\dot{+}}/D)$ is most effective], as illustrated in Fig. 1.23.

In the more general case where the reactants are charged species, we can employ another form of Eq. 1.54. Equation 1.56 includes a Coulombic term (w_r where $w_r = -w_p$) that expresses the work needed to bring charged reactants to a sufficiently close distance (this work will be negative if the charges are of opposite sign, and positive for charges of same sign):

$$\Delta G_{el}(\text{kcal mol}^{-1}) = 23.06[E^0(D^+/D) - E^0(A/A^-)] - w_p + w_r - \Delta G_{00} \qquad (1.56)$$

Equation 1.56 is intended to be used to calculate the ΔG_{el}s of electron transfers of charged transition metal complexes.

1.7. Photophysical and Electrochemical Properties of Electron Donors and Acceptors

A critical factor in photoinduced electron transfer lies in the successful matching of a donor and acceptor with suitable electrochemical and photophysical properties. The properties of these reactants must not only be compatible with exothermic

Table 1.6 FREE-ENERGY CHANGES FOR ELECTRON TRANSFER BETWEEN ELECTRON DONORS AND SINGLET 2-CYANONAPHTHALENE[a]

Donor	$E^0(D^{\dot{+}}/D)$ (V)	ΔG_{el} (kcal mol^{-1})
Anisol	1.76	+4.6
1,2,4-Trimethoxybenzene	1.12	−11.5
N,N-Dimethylaniline	0.78	−18.4
TPD	0.16	−32.3

[a] ΔG_{el} calculated with Eq. 1.54; $\varepsilon_s = 37$; $d_{cc} = 7$ Å; redox potentials vs. SCE in acetonitrile from Rehm, D., and Weller, A. *Isr. J. Chem.* **1970,** *8,* 259; TPD, *N,N,N′N′*-tetramethyl-*p*-phenylenediamine.

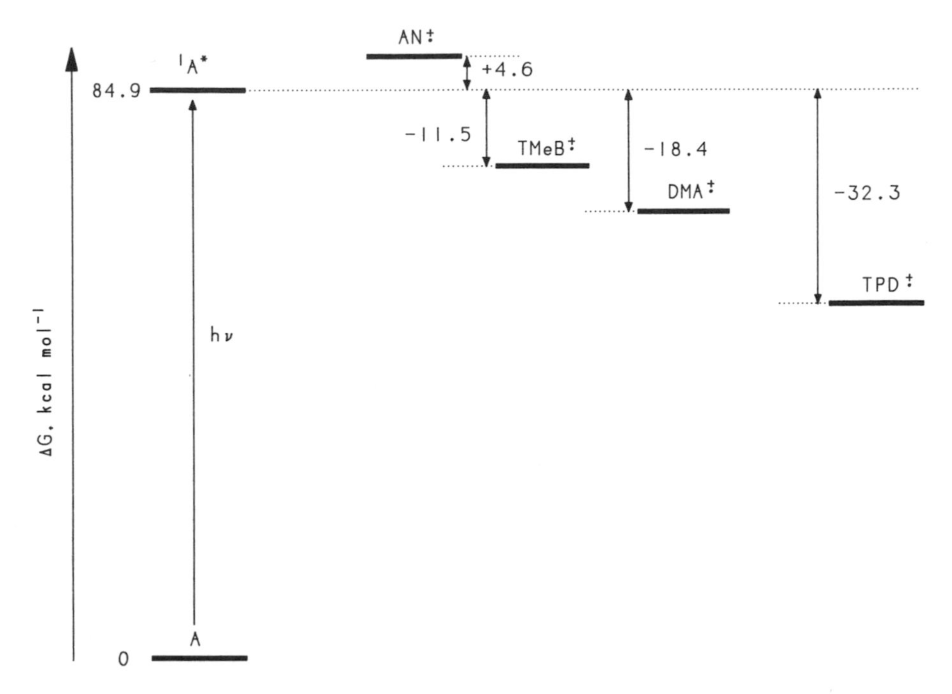

Figure 1.23 Energy diagram for the quenching of singlet 2-cyanonaphthalene by organic electron donors. Abbreviations: A = electron acceptor (2-cyanonaphthalene); AN = anisol; TMeB = 1,2,4-trimethoxybenzene; DMA = N,N-dimethylaniline; TPD = N,N,N′,N′-tetramethyl-*p*-phenylenediamine.

electron transfer but the lifetime of the excited sensitizer—whether it is an electron donor or acceptor—must be sufficiently long to allow quenching by electron transfer to take place [i.e., decay pathways should not compete with electron transfer (for our purposes, decay pathways include deactivation of the excited state by emissive and nonemissive processes)]. For studies dealing with certain aspects of photoinduced electron transfer, chemists try to select and synthesize compounds which meet certain photophysical, photochemical, and electrochemical criteria. Knowledge of the excited state energies of the sensitizers and their redox potentials, as well as the redox potentials of the electron donating or accepting quenchers is thus an essential requirement for investigating photoinduced electron-transfer processes. Tables 1.7 and Table 1.8 list the photophysical properties of organic and inorganic sensitizers. The data in these tables include the electronic excited state energies as well as relevant photophysical properties. The latter includes absorption wavelengths, quantum yields of the important radiative and nonradiative pathways, and excited-state lifetimes.

Obtaining the rate constant, k_q, when an electron donor or acceptor quenches an excited state by an electron-transfer mechanism is facilitated if the sensitizer displays luminescence. To estimate k_q from singlet excited-state lifetimes, for

Table 1.7 PHOTOPHYSICAL AND ELECTROCHEMICAL PROPERTIES OF ORGANIC SENSITIZERS[a,b]

Sensitizer	$E^0(D^{+}/D)$	$E^0(A/A^{-})$	λ_{00} (nm)	E_{00}(S) (kcal mol^{-1})	E_{00}(T) (kcal mol^{-1})		τ_f (ns)	τ_p (s)	Φ_f	Φ_p
Acetophenone	—	−1.85	363	78.7	73.8		—	0.6	0.00	—
4-Methoxyacetophenone	—	−1.50	353	81.0	71.5		—	0.26	—	0.72
N,N-Dimethyl aniline	—	0.81	—	88	77		—	—	—	—
Anthracene	1.16	−1.93	375	76.3	42.7		5.3	—	0.27	—
9,10-Dicyanoanthracene	—	−0.89	433	66	—		19.6	—	—	—
2,6,9,10-Tetracyanoanthracene	—	−0.45	440	65	—		15.2	—	—	—
p-Dicyanobenzene	—	−1.64	290	98.6	70.5		9.7	—	—	—
N,N,N',N'-Tetramethylbenzidine	0.32	—	345	83	63		10	—	0.30	—
Methyl *p*-cyanobenzoate	—	−1.76	301	95	72		—	—	—	—
Benzophenone	—	−1.68	384	74.4	69.2		0.005	0.006	0.0	0.7
2,4,6-Triphenylpyrillium tetrafluoroborate	—	−0.29	440	65	53		—	—	—	—
Chloranil	—	0.02	—	—	62.3		—	—	—	—
Naphthalene	1.62	−2.29	311	92	60.9	105	2.25	0.21	0.10	
1-Cyanonaphthalene	—	−1.98	320	89.4	57.4	8.9	—	—	—	
2-Cyanonaphthalene	—	−2.13	320	84.9	—	—		—	—	—
1,4-Dicyanonaphthalene	—	−1.28	359	86.4	55.5	10.1	—	—	—	
2-Methoxynaphthalene	1.42	—	318	90	59.9	15	—	—	—	
Perylene	0.85	—	435	65.8	35.1	6	—	0.87	—	
Phenanthrene	1.58	−2.20	345	82.9	62	61	3.7	0.13	0.13	
Pyrene	1.20	—	372	77	48.1	475	0.5	0.53	—	

[a] From Table IX in Kavarnos, G. J., and Turro, N. J. *Chem. Rev.* **1986,** *86,* 401, and references therein.

[b] τ_f and Φ_f refer to fluorescence lifetimes and quantum yields at room temperature; τ_p and Φ_p refer to triplet lifetimes and quantum yields at 77 K; all other values were obtained at room temperature in acetonitrile or other polar solvents.

Table 1.8 ELECTROCHEMICAL AND PHOTOPHYSICAL PROPERTIES OF INORGANIC SENSITIZERS[a,b]

Sensitizer	E^0 (D^+/D)	E^0 (A/A^+)	λ_{mx} (nm)	E_{00} (kcal mol^{-1})	Φ_{em}	τ (μs)
$Cr(bpy)_3^{3+}$	—	-0.26[c]	455	39.4	—	77
$Cu(dpp)^{2+}$	0.39[d]	—	439	41.5	0.0004	0.310
$MgPc^{4+}$	0.75[c]	—	620	41.3 (S)	0.72 (S)	0.0082 (S)
	0.75	—		25.1 (T)	—	280 (T)
$Pt_2(P_2O_5)_4H_8^{4-}$	<1	-1.4	367	57.7	0.52	<0.002
$ReCl_8^{2-}$	>1.25	-0.85	—	40.4	—	0.14
$Rh_2(dicp)_4^{2+}$	0.89	-1.4	553	39	—	<0.002
$Ru(bpy)_3^{2+}$	1.29	-1.35	454	48.9	0.042	0.85
UO_2^{2+}	—	0.06[c]	420	58.6	~1	6.6
ZnPc	0.92[c]	-0.65[c]	672	42.2 (S)	0.3 (S)	0.0038 (S)
				26.1 (T)	—	—
$ZnTMPyP^{4+}$	1.18[c]	—	—	47.3 (S)	0.025 (S)	0.0014 (S)
				36.2 (T)	0.9 (T)	655 (T)
ZnTPP	0.95[e]	-1.11[e]	560	47.3 (S)	0.04 (S)	0.0027 (S)
				36.7 (T)	0.88 (T)	1200 (T)

[a] From Table X in Kavarnos, G. J., and Turro, N. J. *Chem. Rev.* **1986,** *86,* 401, and references therein.

[b] All values were obtained at room temperature in acetonitrile or other polar solvents; bpy, 2,2′-bipyridine, dicp, 1,3-diisocyanopropane; dpp, 2,9-diphenyl-1,10-phenanthroline; Pc, phthalocyanine; Pc^{4+}, tetra(*N*-methylaza)phthalocyanine; $TMPyP^{4+}$, tetramethylpyridinium porphyrin; TPP, tetraphenylporphyrin.

[c] vs. NHE in water.

[d] vs. ferrocenium/ferrocene in methylene chloride.

[e] vs. organic solvent.

Table 1.9 REDOX POTENTIALS OF ORGANIC ELECTRON DONORS[a]

Quencher	$E^0(D^{\dot{+}}/D)^b$
N,N,N',N'-tetramethyl-*p*-phenylenediamine	0.16
N,N,N',N'-tetramethylbenzidine	0.32
1,4-diazabicyclo[2.2.2]octane (DABCO)	0.57
Phenothiazine	0.59[c]
N,N-Diethylaniline	0.76
N,N-Dimethylaniline	0.81
Quadricyclane	0.91
Triethylamine	0.96
Indene	1.52
2-Methoxynaphthalene	1.42
1,3,5-Trimethoxybenzene	1.49
1,1-Diphenylethylene	1.52
Norbornadiene	1.54

[a] From references in Table XI in Kavarnos, G. J. and Turro, N. J. *Chem. Rev.* **1986,** *86,* 401, unless otherwise noted.
[b] Redox potentials, V vs. SCE in acetonitrile.
[c] Nocera, D. G., and Gray, H. B. *J. Am. Chem. Soc.* **1981,** *103,* 7349.

Table 1.10 REDOX POTENTIALS OF ORGANIC ELECTRON ACCEPTORS[a]

Quencher	$E^0(A/A^{\dot{-}})^b$
Tetracyanoethylene	0.24
Chloranil	0.02
Tetracyanoanthracene	−0.45
Methylviologen	−0.45
p-Benzoquinone	−0.54
Oxygen	−0.33[c]
9,10-Dicyanoanthracene	−0.89
p-Dinitrobenzene	−1.18[d]
1,4-Dicyanonaphthalene	−1.28
Dimethylbicyclo[2.2.1]hepta-2,5-diene-2,3-dicarboxylate	−1.5
9-Cyanoanthracene	−1.58
p-Dicyanobenzene	−1.64
Benzophenone	−1.68
Nitrobenzene	−1.76[d]
Acetophenone	−1.85
Anthracene	−1.93
1-Cyanonaphthalene	−1.98
trans-Stilbene	−2.26

[a] From references in Table XI in Kavarnos, G. J., and Turro, N. J. *Chem. Rev.* **1986,** *86,* 401, unless otherwise noted.
[b] Redox potentials, V vs. SCE in acetonitrile, unless otherwise noted.
[c] Sawyer, D. T. *Oxygen Chemistry*. Oxford, New York, 1991, p. 21.
[d] vs. ferroceniuim/ferrocene in methylene chloride.

Table 1.11 REDOX PROPERTIES OF INORGANIC QUENCHERS[a]

Quencher	$E^0(D^+/D)^b$	$E^0(A/A^-)^b$
$Ru(bpy)_3^{2+}$	1.29[c]	−1.35[c]
$Ru(bpy)_3^{2+}$	1.26	−1.33
$Cr(bpy)_3^{2+}$	—	−0.26
$Cr(CN)_6^{3-}$	>1.6	−1.52
$Fe(CN)_6^{4-}$	0.45	≤1.7
Eu^{2+}	−0.43	—
Fe^{2+}	0.73	—
Fe^{3+}	—	0.73

[a] From Table XII in Kavarnos, G. J., and Turro, N. J. *Chem. Rev.* **1986,** *86,* 401, and references therein.
[b] Redox potentials, V vs. NHE in water, unless otherwise noted.
[c] vs. SCE in acetonitrile.

example, one may measure the singlet lifetimes at various quencher concentrations. This procedure is based on the *Stern–Volmer* equation:

$$\frac{1}{\tau_f} - \frac{1}{\tau_f^0} = 1 + k_q[Q] \tag{1.57}$$

where τ_f and τ_f^0 are fluorescence lifetimes in the presence and absence of quencher, respectively. As more quencher is added, τ_f progressively decreases. k_q can then be determined by plotting $1/\tau_f$ vs. [Q] and measuring the slope. When k_q is within the range of $\sim 10^9$ to $\sim 10^{10}$ M^{-1} s^{-1}, the reaction is so rapid that it is *diffusion controlled,* which is to say that the rate-determining step is diffusion of the sensitizer and quencher to a *quenching sphere*. In Chapter 6, the factors influencing k_q will be discussed.

The redox potentials listed in Tables 1.9 and 1.10 are applicable to both excited sensitizers and ground-state quenchers when using Eqs. 1.54 or 1.56. The parameters relevant to organic electron donors are given in Table 1.9. The compound listed at the top of the table is the best donor [i.e., its redox potential being the smallest (the least positive)]. The redox potentials for one-electron reduction of selected organic acceptors are presented in Table 1.10. The acceptors near the top of the table have the most positive redox potentials for one-electron reduction and therefore are the best oxidants. The redox potentials of selected transition metal complexes are given in Table 1.11.

1.8. Energy Transfer versus Electron Transfer

Before concluding this chapter, it will be instructive to compare the energetics of electron transfer and energy transfer. These pathways can be illustrated by the well-known dot and arrow symbolism familiar to chemists as shown in 1.58:

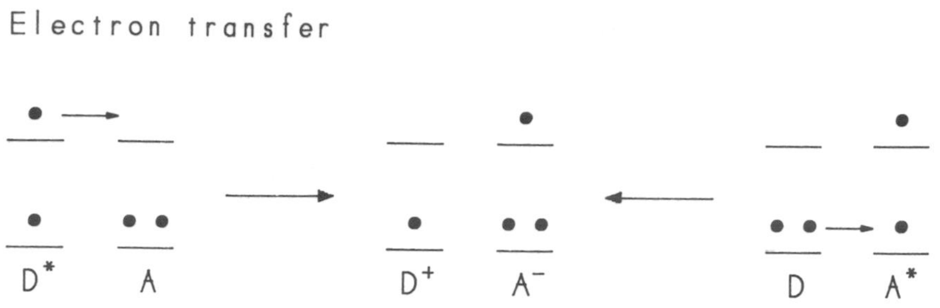

Electron transfer

D^* A → D^+ A^- ← D A^*

(1.58)

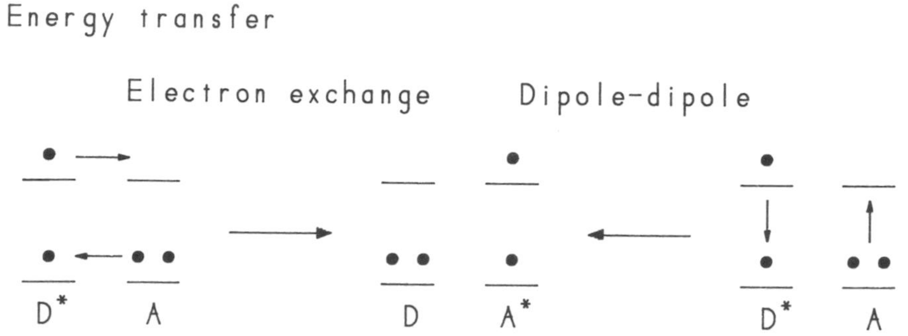

Energy transfer

Electron exchange Dipole–dipole

D^* A → D A^* ← D^* A

In this scheme, the horizontal lines represent orbitals populated by electrons (shown by dots). In energy transfer, there are two pathways for generating A's excited state, A*. First, there may be a mutual exchange of electrons, a mechanism known as *energy transfer by electron exchange*. The fact that D gives up an electron to A, while the latter in turn returns an electron to D, implies mutual or close proximity of D* and A. In other words, for the electrons to travel back and forth, there must be some overlap of the electron clouds on D* and A. In this respect, electron transfer is quite similar to energy transfer via electron exchange. In both of these pathways, overlap between the electron clouds of the reactants is an important feature.

Energy transfer can also take place via a long-range dipole–dipole interaction. In the dipole–dipole interaction, since there are no mutual electron exchanges, electronic transitions take place exclusively within D and A. D* acts like an oscillating dipole—analogous to a transmitting antenna—by inducing the formation of A*. In contrast to electron exchange, the dipole–dipole mechanism does not require D and A to be in close proximity. In fact, the dipole–dipole mechanism may be operative over many tens of angströms.

The free-energy changes accompanying energy transfer either by electron exchange or the dipole–dipole mechanism may be formulated as follows:

$$\Delta G_{en} = \Delta H_{en} - T\Delta S_{en} = -\Delta G_{D^* \rightarrow D} + \Delta G_{A^* \rightarrow A} \tag{1.59}$$

$$= -\Delta H_{D^* \rightarrow D} + T\Delta S_{D^* \rightarrow D} + \Delta H_{A^* \rightarrow A} - T\Delta S_{A^* \rightarrow A} \tag{1.60}$$

Table 1.12 PHOTOPHYSICAL PARAMETERS OF ORGANIC QUENCHERS[a]

	E_{00}(S) (kcal mol^{-1})	E_{00}(T)
Amines		
N,N-Diethylaniline	90	68
N,N-Dimethylaniline	88	68.4
Triethylamine	>90	>90
1,4-Diazabicyclo[2.2.2]octane (DABCO)	—	>90
N,N,N′,N′-Tetramethylbenzidine	83	63
N,N,N′,N′-Tetramethyl-*p*-phenylenediamine	—	60.9
Methoxy compounds		
Anisole	96.6	63
2-Methoxynaphthalene	85.3	62
Nitro compounds		
Nitrobenzene	>98	60
Cyano compounds		
p-Dicyanobenzene	98.6	70.5
1,4-Dicyanonaphthalene	79.6	—
Viologens		
Methyl viologen	—	71.5
Carbonyl compounds		
Acetophenone	78.7	73.8
Benzophenone	74.4	69.2
p-Benzoquinone	—	53
Olefins		
1,1-Diphenylethylene	97.9	—
Indene	>97	60
Norbornadiene	>100	70
Miscellaneous		
Quadricyclane	—	~80
Dimethylbicyclo[2.2.1]hepta-2,5-diene-2,3-dicarboxylate	—	53
Oxygen	22.5	—

[a] From Table XI in Kavarnos, G. J., and Turro, N. J. *Chem. Rev.* **1986,** *86,* 401, and references therein.

Equation 1.60 can be simplified given the reasonable assumption that the enthalpy difference is equal to the spectroscopic energy, E_{00}. Further, if it is assumed that the entropy differences between the ground and excited states are negligible (in general $T\Delta S < 0.1$ kcal mol^{-1}), Eq. 1.60 becomes

$$\Delta G_{en} = -E_{00}(D^*) + E_{00}(A^*) \tag{1.61}$$

When values of $E_{00}(D^*)$ and $E_{00}(A^*)$ are obtained from emission studies, one can estimate ΔG_{en}.

Frequently, when investigating a new quenching reaction, a chemist may want to establish whether electron or energy transfer is the predominant quenching pathway. The chemist will need to know the redox potentials of the reactants as well as their zero spectroscopic energies. It is quite desirable, therefore, to compare the excited-state energies of the quencher and the excited state. The excited-state energies of organic and inorganic quenchers are tabulated in Tables 1.12 and 1.13.

Table 1.13 PHOTOPHYSICAL PARAMETERS OF INORGANIC QUENCHERS[a]

Quencher	E_{00} (kcal mol^{-1})
Eu^{2+}	89.5
$Fe(CN)_6^{4-}$	67.8
$Cr(bpy)_2^{2+}$	39.4
$Cr(CN)_6^{3-}$	35.5
$Ru(bpy)_2^{2+}$	48.9
Fe^{3+}	36.9
Fe^{2+}	29.9

[a] From Table XII in Kavarnos, G. J., and Turro, N. J. *Chem. Rev.* **1986,** *86,* 401, and references therein.

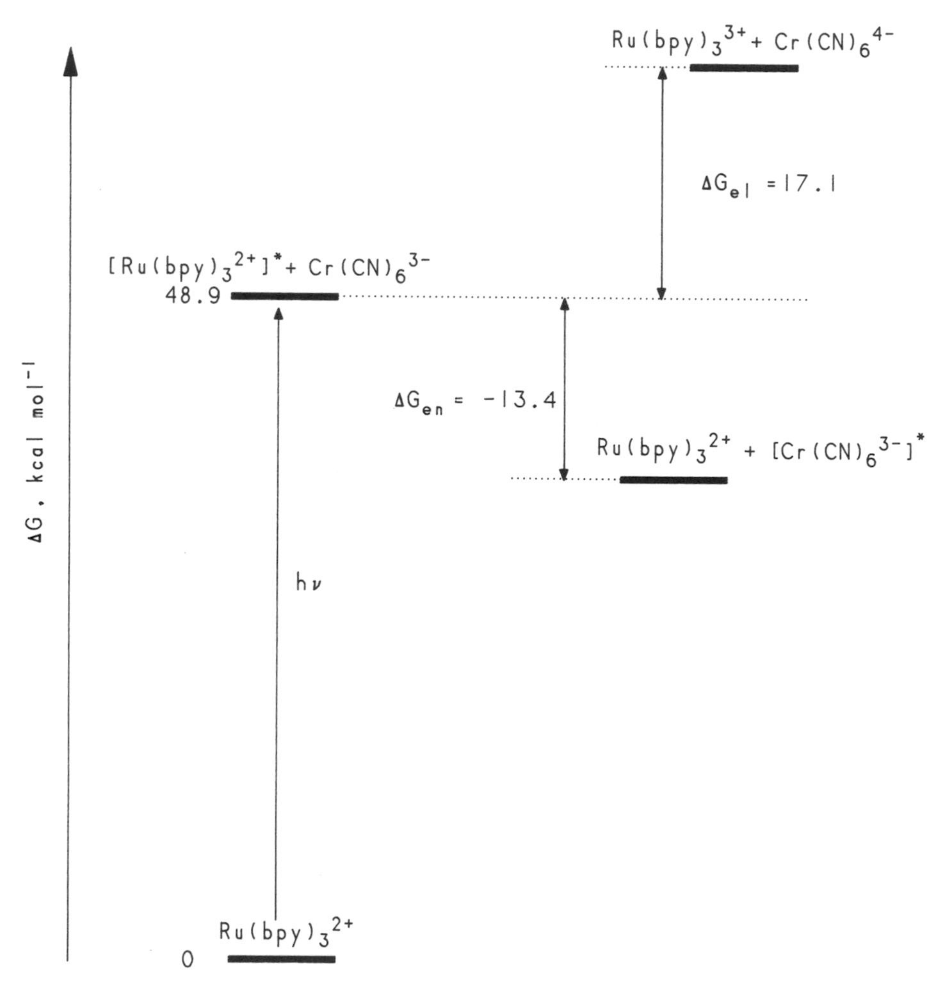

Figure 1.24 Energy diagram for the quenching of $[Ru(bpy)_3^{2+}]^*$ by $Cr(CN)_6^{3-}$. Energy transfer is the thermodynamically favored quenching pathway.

With the data in these tables, we can make a pretty good guess at the nature of the quenching mechanism. $[Ru(bpy)_3^{2+}]^*$, as we have noted earlier, can be quenched by a variety of electron donors and acceptors. These reactions proceed by electron transfer but this is not always true. For example, $Cr(CN)_6^{3-}$ is known to quench the emission of $[Ru(bpy)_3^{2+}]^*$.[2] However, using Eq. 1.56 we quickly find out that electron transfer is not thermodynamically allowed. The $E^0(D^+/D^*)$ of $Cr(CN)_6^{3-}$ (Table 1.11) suggests that this quencher is difficult to oxidize by $[Ru(bpy)_3^{2+}]^*$ ($\Delta G_{el} > 18$ kcal mol^{-1}). Moreover, since $E^0(A^*/A^-)$ is only -1.52 V, a relatively low value, reduction of $Cr(CN)_6^{3-}$ can also be ruled out. In this case, $\Delta G_{el} \sim 17$ kcal mol^{-1}. We conclude that both oxidation and reduction of $Cr(CN)_6^{3-}$ by $[Ru(bpy)_3^{2+}]^*$ are not likely mechanisms. Since $[Cr(CN)_6^{3-}]^*$ lies ~13.4 kcal mol^{-1} lower than the excited state of $Ru(bpy)_3^{2+}$, an energy-transfer pathway is more likely. As Fig. 1.24 shows, energetics favors energy transfer, not electron transfer. That energy transfer is indeed a major mechanism has been confirmed by the observation of sensitized emission from $[Cr(CN)_6^{3-}]^*$.

Suggestions for Further Reading

Scope of Photoinduced Electron Transfer

Roth, H. D., *Top. Curr. Chem.* **1990,** *156,* 1. A brief history of photoinduced electron transfer.

Review of Photochemical Principles and Definitions

Ballhausen C. J., and Gray, H. B. *Molecular Electronic Structures: An Introduction.* Benjamin/Cummings, Reading, MA, 1980.

Cotton, F. A., and Wilkinson, G. *Advanced Inorganic Chemistry.* Wiley, New York, 1980, Chapter 20, Pertinent and useful descriptions of molecular orbital theory and transition metal complexes.

Dewar, M. J. S., and Dougherty, R. C. *The PMO Theory of Organic Chemistry.* Plenum/Rosetta, New York, 1975. PMO stands for pertubation molecular theory. Chapters 1 and 6 contain pertinent information and background.

Turro, N. J. *Modern Molecular Photochemistry.* Benjamin/Cummings, Menlo Park, CA, 1978. Chapters 2 to 4 provide an excellent introduction to excited states, electronic configurations, and electronic transitions.

Wayne, R. P. *Principles and Applications of Photochemistry.* Oxford Science Publications, Oxford, 1988.

Photophysical and Electrochemical Properties of Electron Donors and Acceptors

Balzani, V., Bolletta, F., Gandolfi, M. T., and Maestri, M. *Top. Curr. Chem.* **1978,** *75,* 1. A good introduction to the role of transition metal complexes in photoinduced electron transfer.

Fox, M. A., and Chanon, M. (eds.). *Photoinduced Electron Transfer. Part A. Conceptual Basis.* Elsevier, Amsterdam, 1988. Chapter 1 is a good introduction to the evaluation of redox properties of molecules.

Julliard, M., and Chanon, M. *Chem. Br.* **1982,** 558. Redox properties of excited states.

Energy Transfer versus Electron Transfer

Balzani, V., Moggi, L., Manfrin, M. F., Bolletta, F., and Laurence, G. S. *Coord. Chem. Rev.* **1975,** *15,* 321. An introduction to sensitization processes of coordination compounds.

Scandola, F., and Balzani, V. *J. Chem. Ed.* **1983,** *60,* 814. Energy transfer involving metal complexes.

References

1. Rehm, D., and Weller, A. *Isr. J. Chem.* **1970,** *8,* 259.
2. Juris, A., Gandolfi, M. T., Manfrin, M. F., and Balzani, V. *J. Am. Chem. Soc.* **1976,** *98,* 1047.

Problems

1. Calculate ΔG_{el} for the reactions in Table 1.6 in acetonitrile assuming $d_{cc} \sim 4$ Å. Discuss the effect of separation distance on the reactions.
2. Calculate ΔG_{el} for the quenching of the singlet and triplet states of 1-cyanonaphthalene by the electron donors in Table 1.6, assuming polar conditions and a center-to-center separation distance of ~7 Å. What do your calculated values say about the ability of the triplet state to perform as an electron acceptor?
3. Considering only energetics, predict what quenching pathway(s)—electron or energy transfer—should dominate for each of the following pairs in acetonitrile:

   ^{1}[*p*-dicyanobenzene]* and indene
   ^{3}[*p*-dicyanobenzene]* and 1,3,5-trimethoxybenzene
   ^{3}[benzophenone]* and *p*-dicyanobenzene
   ^{3}[benzophenone]* and phenothiazine
   [$Ru(bpy)_3^{2+}$]* and phenothiazine
   ^{1}[9-cyanoanthracene]* and anisole
   ^{3}[acetophenone]* and *N,N*-dimethylaniline
   [$Cr(bpy)_3^{3+}$]* and Fe^{2+}
   [UO_2^{2+}]* and *N,N*-diethylaniline

4. Using their redox potentials, arrange the singlet and triplet states of the cyanoaromatic molecules in Table 1.7 in increasing order to act as electron acceptors.
5. Which molecule should have the *lower* ionization potential, methylbenzene or hexamethylbenzene? Can you offer an explanation for your choice? (Kochi, J. K. *Adv. Free Radical Chem.* **1990,** *1,* 53.)
6. Classify the reactants in each pair below as an electron acceptor or electron donor:

   [$Pt_2(P_2O_5)_4H_8^{4-}$]* and *N,N,N′,N′*-tetramethyl-*p*-phenylenediamine
   [$Rh_2(dicp)_4^{2+}$]* and MV^{2+}
   ^{1}[$MgPc^{4+}$]* and MV^{2+}
   [$Ru(bpy)_3^{2+}$]* and $Fe(CN)_6^{4-}$

Determine w_r, w_p, and ΔG_{el} in acetonitrile. What is the effect of the work terms on the electron transfers?

7. Anions are known to quench triplets of some carbonyl compounds. Given that the redox potentials of I^-, SCN^-, OH^-, and Cl^- are 1.40, 1.50, 2.29, and 2.55 V, respectively, what is a likely mechanism for the quenching of triplet benzophenone by these anions? [Hoshino, M., and Shizuka, H. In Fox, M. A., and Chanon, M. (eds.), *Photoinduced Electron Transfer. Part C. Photoinduced Electron Transfer Reactions: Organic Substrates*. Elsevier, Amsterdam, 1988, Chapter 4.5.]
8. Suggest a mechanism which explains why methyl viologen (paraquat) is such a strong electron acceptor:

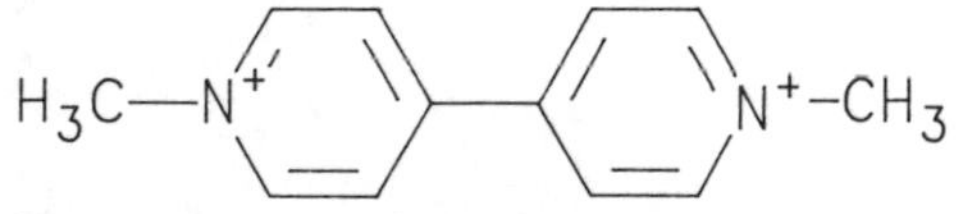

H_3C-N^+ N^+-CH_3

methyl viologen

9. Derive Eq. 1.52 for electron transfer between an excited acceptor and ground-state donor.

CHAPTER

2

Properties of Charge-Transfer Intermediates in Photoinduced Electron Transfer

We now begin to examine the nature of the charge-transfer intermediates in photoinduced electron transfer and the techniques used to study them. In Chapter 1, we derived thermodynamic expressions that chemists use to predict that electron transfer is or is not a dominant pathway. However, thermodynamic arguments should be supplemented with direct identification of the products of electron transfer. For this purpose, a number of experimental approaches have found an important place. These include spectroscopic detection by flash spectroscopy, photoconductivity, photoacoustic calorimetry, pulse radiolysis, and electron-spin studies. Although based on widely different principles, these experimental techniques allow direct study of the charge-transfer intermediates. The aim of this chapter will be to show how these laboratory techniques can be used to establish an electron-transfer mechanism as well as to clarify some of the subtle properties of charge-transfer intermediates. The discussion in this chapter will be largely restricted to photoinduced electron transfer taking place in solution at encounter distances. We will not be concerned with electron transfer in the solid state. This topic is reserved for later chapters.

2.1. The Nature of Charge-Transfer Intermediates

At the outset, let us review the main features of the solvent cage model insofar as it plays a vital role in our attempt to conceptualize solution-phase photoinduced electron transfer. The pathways shown below represent one way of classifying photoinduced electron-transfer reactions in solution:

Electron transfer via a solvent-separated ion pair (SSIP):

$$\underset{}{D^* + A} \rightleftharpoons \underset{\text{encounter complex}}{D^* \cdots A} \rightleftharpoons \underset{\text{SSIP}}{[D^+ \cdots A^-]} \rightleftharpoons \underset{\text{free ions}}{D^+ + A^-}$$

Electron transfer via an exciplex (EX): (2.1)

$$\underset{}{D^* + A} \rightleftharpoons \underset{\text{encounter complex}}{D^* \cdots A} \rightleftharpoons \underset{\text{EX}}{[D^+ A^-]^*} \rightleftharpoons \underset{\text{free ions}}{D^+ + A^-}$$

Scheme 2.1 shows a free ion pair being generated via a solvent-separated ion pair or an exciplex. Both mechanisms involve initial formation of an encounter complex. In the idealized representation in Fig. 2.1, photoinduced transfer between two neutral spherical organic molecules takes place within the encounter complex after numerous collisions. During one of these collisions, electron transfer may prevail (if allowed by thermodynamics!) resulting in formation of two ions, each with opposite charge. When an ion pair is initially formed in a solvent cage, we refer to the two ions in the pair as *geminate ions* (the Latin *geminus* means "twin-born"). Immediately, Coulombic forces draw these ions into close proximity, perhaps as close as 3–4 Å depending on their structures. These ions are called contact ions. The binding forces holding the ions in close proximity are determined for the most part by Coulombic attractions. Since the ions are so close, the electron may be "returned" to the donor very rapidly resulting in no net change. At the same time, however, polar solvent molecules can rearrange and subsequently stabilize the charge on the contact ions before the return of the electron. As the solvent molecules surround the ions, they begin to penetrate the space between the ions to form a solvent-separated ion pair. As the ions are "pried" apart, the ions may eventually separate to such large distances that the ions are no longer "correlated." At this stage, they are free to enter the cages of other ions, or free to participate in chemical transformations. This process is *ion dissociation or charge separation*.

In effect, there is competition between electron return and dissociation. Our task is to understand the factors that infuence this competition.

2.1.1. The Role of Solvent and Coulombic Effects

We now direct our attention to the interplay of solvent polarity and Coulombic interactions, as this has important consequences on the yield of long-lived free ions. To develop a working model of charge separation, we return to the dielectric constant, ε_s, which was introduced in Chapter 1. As was mentioned in that chapter, one of the terms that affects the free-energy change is $e^2/\varepsilon_s d_{cc}$. This term tells us

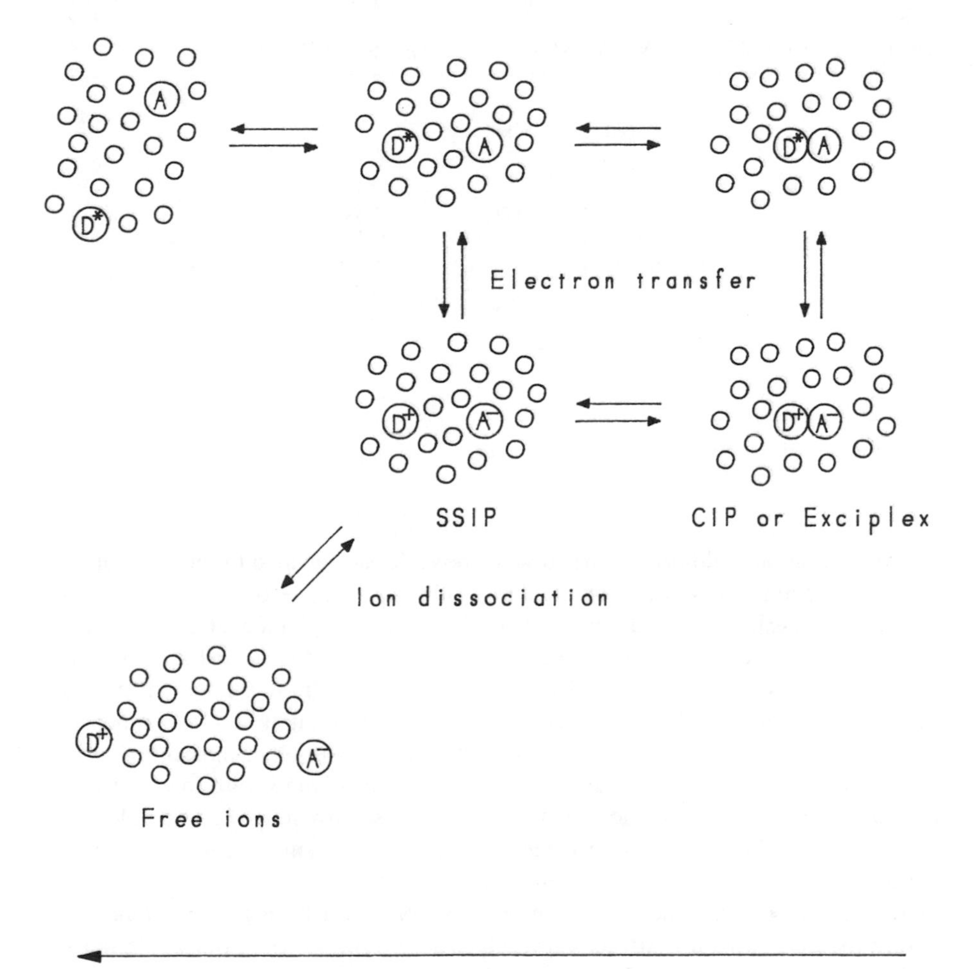

Figure 2.1 This figure summarizes the pathways in photoinduced electron transfer. Following photoexcitation, the donor and acceptor diffuse to encounter distances where electron transfer becomes favorable. Subsequent to formation of the ion-pair state, the ion pair can "return" to ground-state reactants or dissociate to free ions. (Adapted with permission from Kavarnos, G. J. *Top. Curr. Chem.* **1990,** *156,* 21. Copyright 1990 Springer-Verlag.)

that small separation distances and solvents with low dielectric constants [e.g., nonpolar solvents (Table 2.1)] should favor photoinduced electron transfers. But is a nonpolar medium the ideal environment? Not really, and here is the reason. The reciprocal dielectric constant, $1/\varepsilon_s$ in $e^2/\varepsilon_s d_{cc}$, can be regarded as a "screening" term. When ions are formed from neutral reactants, polar solvent molecules rapidly surround each ion. As they surround the ions and penetrate the space between the ions, polar solvent molecules "screen" the electrostatic interactions. This blocking

Table 2.1 DIELECTRIC CONSTANTS OF SELECTED SOLVENTS[a]

Solvent	ε_s
Water	80.2
Dimethyl sulfoxide	46.68
Glycerol	42.5
Acetonitrile	37.5
Dimethylformamide	36.71
Methanol	33.7
Benzonitrile	25.70
Ethanol	24.45
n-Propanol	20.6
Tetrahydrofuran	7.58
Benzene	2.275
Cyclohexane	2.023

[a] ε_s, static dielectric constant; data from Murov, S. L. *Handbook of Photochemistry*. Marcel Dekker, New York, 1973.

of the Coulombic field allows the ions to move farther apart into the bulk of the solvent until in the limit of complete dissociation, $e^2/\varepsilon_s d_{cc} \sim 0$.

Thus, we desire polar conditions where the ion pair is given a slight reprieve to escape from the solvent cage. In the polar environment, the radical ions become well separated. This will block electron return. Escape from the solvent cage is indeed a key feature of photoinduced electron transfer in solution. It is this propensity to escape from the cage that allows ion pairs to react with other molecules.

In contrast, the partners in an ion pair formed in solvents with low dielectric constants (Table 2.1) have a tendency to stay in close proximity, favoring electron return. Under nonpolar conditions, the screening term is small, and consequently the ion pair will have a shorter lifetime.

In the solvent cage model, geminate ions engage in primary *recombinations* before the ions leave the solvent cage. Electron return, which is often referred to as *homogeneous recombination,* occurs usually on a microsecond timescale.

Before concluding this section, we should note the distinction between electron return and reversible electron transfer, k_{-el}, as used in this book. Electron return means "return" to the ground state by ion recombination; reversible electron transfer is return *to the excited state*. Reverse electron transfer to the excited state is usually endothermic; electron return to the ground state is exothermic. The reader should note that this practice is not universally observed in the literature. Occasionally, the terms "reverse" or "reversible" electron transfer are used to denote ion recombination to ground-state reactants. It may not always be strictly correct to ignore the importance of k_{-el} in some reactions.

2.1.2. Ion Pairs

We can use Fig. 2.2 to develop simple expressions for the quantum efficiencies and lifetimes of ion-pair formation in solution. The types of intermediates shown in Fig. 2.2 include the excited state, the encounter complex, solvent-separated ions, and free ions. The quantum efficiency for the production of free ions can be shown by a simple kinetic derivation to be

$$\phi_{IP} = \frac{k_{dis}}{k_{dis} + k_{ret}} \tag{2.2}$$

which says that the efficiency of generating charge-separation ions is determined by the competition between the rates of ion dissociation k_{dis}, and electron return, k_{ret}. To understand the underlying factors that mediate this competition, we need to review kinetic expressions for k_{dis} and k_{ret}. Starting with ion dissociation, we write Eq. 2.3:

$$k_{dis} = \frac{D}{d_{cc}^2}\left[\frac{w_{IP}}{1 - \exp(-w_{IP})}\right] \tag{2.3}$$

D is the diffusion constant, which is discussed in Chapter 6. Equation 2.3, which is known as the Eigen equation, expresses the fact that solvent polarity and viscosity, Coulombic work terms, and separation distance affect the kinetics of charge separation. An expression for w_{IP}, which is the work term for two ions separated by at least one solvent molecule, can be written as

$$w_{IP}\ (\text{kcal mol}^{-1}) = \frac{332 q_A q_D e^2}{\varepsilon_s d_{cc} k_B T} \tag{2.4}$$

where k_B is the Boltzmann constant and T is the temperature. Again, in Eq. 2.4, we note the presence of the screening term. As before, this factor can be regarded as a correction to the rate of charge separation for charged ionic species. From Eqs. 2.3 and 2.4, we can calculate k_{dis} keeping in mind that we want k_{dis} to be much greater than k_{ret} (if the electron returns to its origin, no useful purpose is served; the reaction is inefficient). For the typical case where $q_A = 1$ and $q_D = -1$, D

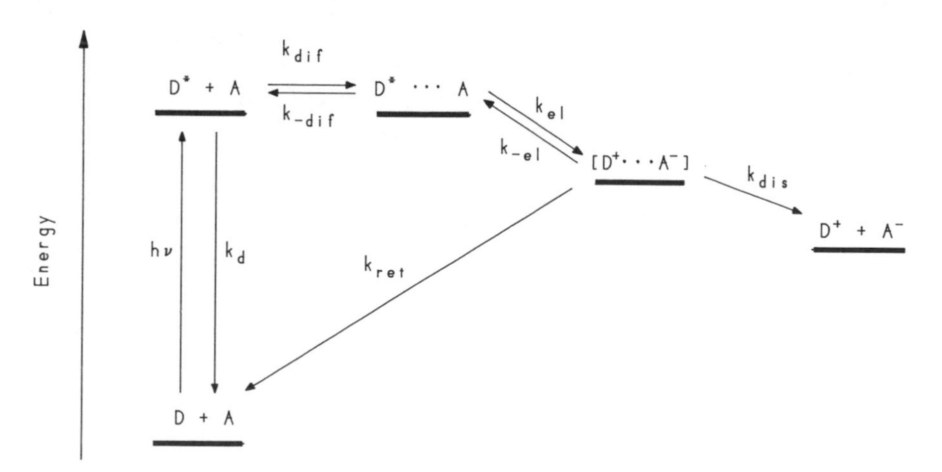

Figure 2.2 Energy diagram showing the pathways to free-ion formation in solution *via* a solvent-separated ion pair.

$= 2 \times 10^{-5}$ cm^2 s^{-1}, and $d_{cc} = 7$ Å, $k_{dis} = 7.8 \times 10^8$ s^{-1} in acetonitrile (ε_s = 37). Contrast this value with the rate constant calculated in tetrahydrofuran (ε_s = 7.6), where $k_{dis} \sim 10^5$ s^{-1}. These calculated rate constants compare very well with the measured values obtained in studies dealing with radical ion pairs formed by quenching of singlet pyrene by *N,N*-dimethylaniline ($k_{dis} \sim 5 \times 10^8$ s^{-1} in acetonitrile and $\sim 10^6$ s^{-1} in tetrahydrofuran).[1] These measured rate constants are consistent by the model presented in the previous section that polar solvent molecules can lessen the barrier to charge separation by assisting in the formation of solvent-penetrated ion pairs.

Let us examine further the effects of Coulombic interactions on ion dissociation. For the ion dissociation of charged species, let us consider the quenching of $[Ru(bpy)_3{}^{2+}]^*$ first by an electron donor (2.5) and then by an electron acceptor (2.6):

$$[Ru(bpy)_3{}^{2+}]^* + D \xrightarrow{k_{el}} [Ru(bpy)_3{}^{+} + D^{+}] \xrightarrow{k_{dis}} Ru(bpy)_3{}^{+} + D^{+}$$
$$\downarrow k_{ret} \qquad (2.5)$$
$$Ru(bpy)_3{}^{2+} + D$$

$$[Ru(bpy)_3{}^{2+}]^* + A \xrightarrow{k_{el}} [Ru(bpy)_3{}^{3+} + A^{-}] \xrightarrow{k_{dis}} Ru(bpy)_3{}^{3+} + A^{-}$$
$$\downarrow k_{ret} \qquad (2.6)$$
$$Ru(bpy)_3{}^{2+} + A$$

It should be evident that ion dissociation should be more favorable in 2.5 where there is mutual repulsion between ions. But separation of ions is less favorable in 2.6 where Coulombic attractive forces hold the ions together. In fact, for 2.5, using Eqs. 2.3 and 2.4, we calculate $k_{dis} \sim 3 \times 10^9$ s^{-1}, and for 2.6, $k_{dis} \sim 1 \times 10^8$ s^{-1} in acetonitrile (assuming that $d_{cc} \sim 10.9$ Å).[2]

To now we have omitted other medium effects on ion dissociation. These, too, can exert a powerful effect. For example, salt effects have been shown to enhance charge separation. Salt effects increase the ionic strength of a solution and are due to specific stabilizing interactions between the ion pair and added salts. There is now substantial evidence that underscores the role of salt effects in photoinduced electron transfer. Much of this evidence has been obtained in studies dealing with the effect of added salts on product yield and distribution. Several examples are presented in Chapter 3.

Thus, the rate of electron return, k_{ret}, has important consequences on ion dissociation. This is a crucial problem in natural and artificial photosynthesis. In attempting to optimize ϕ_{IP}, chemists have sought ways to ensure that $k_{ret} \ll k_{dis}$. As noted above, Coulombic interactions and solvent polarity are among the factors which control k_{ret}. Later in this chapter, we shall observe that electronic spin can also influence the rate of electron return.

2.1.3. Exciplexes

Suppose a pair of neutral but *planar* organic molecules react (Scheme 2.1, second pathway). Since they may be able to adopt a sandwich-like orientation, these reactants may form a exciplex. As we noted in Chapter 1, an exciplex is a charge-transfer species that displays emission—a distinct property of exciplex. As in the case of contact ions, electron return to ground state reactants is an important pathway because of the close proximity of the charge-transfer intermediates. The Coulombic attraction between two ions of opposite charge keeps them in close proximity and prevents the ions from diffusing out of the cage. To the extent that the ions do escape the cage, as they tend to do in highly polar solvents, the exciplex can be very difficult to detect.

Exciplex formation is dependent on similar thermodynamic arguments we developed for the formation of ion pairs in Chapter 1. The greater the difference between IP of the donor and EA of the acceptor, the more favorable is exciplex formation. The dipole moment of such an exciplex is predicted to be smaller than the dipole of an ion pair because of the presence of a smaller amount of charge on each reactant. (For two charges separated by a certain distance, the dipole moment has dimensions of charge $\times$ distance. One debye unit is equal to 10^{-18} electrostatic units. For reference, the moment of a dipole of an electron separated from a proton by 1 Å is 4.8 debye units. When there are significant differences between the IP and EA of both reactants, the dipole moments will be large, sometimes exceeding 14 debye units.)

The possibility of exciplex formation depends on the structure of the reacting partners. Planar organic molecules are ideal choices for exciplex formation since these reactants can adapt face-to-face geometries at distances of 3–4 Å. Exciplexes have also been identified in intramolecular systems linked by flexible chains. In these molecules, exciplex formation is preceded by numerous chain motions, which eventually bring the donor and acceptor into a favorable orientation. A variety of intramolecular exciplex structures is possible. These exciplex structures, which may differ only slightly from one another, can undergo interconversions in solution.

To gain a better understanding of exciplexes, it is useful to consider the energy diagram shown Fig. 2.3. This figure shows the change in energy (vertical axis) that accompanies a reaction (horizontal axis). Let us consider two ground-state species in a fluid medium engaging in a series of collisions at a range of distances. The nature of these interactions is dictated by both electronic and structural features of the reactants. However, there are also forces opposing these attractive interactions. These opposing forces may be electronic in origin. For example, repulsion between the bonding electrons in the ground states of both donor and accept may prevent the molecules from approaching to close distances. Steric repulsion between the molecules may also prevent a close approach. The net result is that at close separation distances, there may be repulsion in the ground state. In the excited state, intermolecular repulsive forces may be relaxed. Thus, after photoexcitation, the reactants may approach on the excited surface to form an encounter complex and finally an exciplex. As long as the reactants remain well separated, emission for the excited reactant (called *monomer emission*) will be observable. If the complex equilibrates to a minimum on the excited-state surface corresponding to the equilibrium separation distance of the exciplex, the exciplex may then undergo emission or radiationless decay to the ground-state surface. Deactivation occurs as a vertical, Franck–Condon transition from a minimum on a reaction surface to a position of high

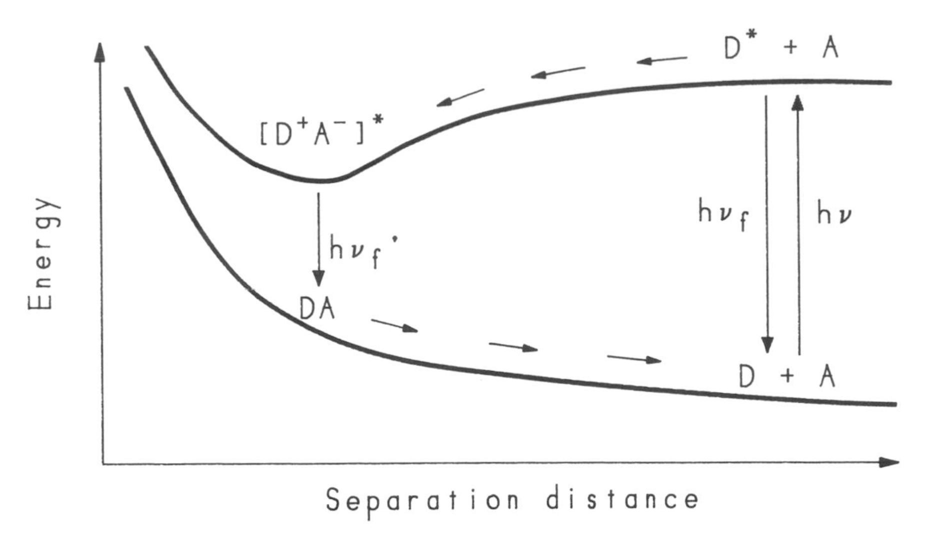

Figure 2.3 An energy curve showing the energy changes accompanying formation of an exciplex. The exciplex is formed following collisions between reactants. This places the pair at an energy minimum. At this stage, the reactant partners are in close union sharing electronic excitation and charge. Decay to the ground-state surface places the complex into the ground state. The donor and acceptor then pull apart as shown on the lower energy curve.

energy on the ground state, so that the reactants will separate rapidly until their energy reaches a minimum point on the surface.

Let us take a closer look at the electronic characteristics of exciplexes. In general, exciplexes in photoinduced electron transfer have strong charge-transfer character and negligible "excited-state" character. The nature of these exciplexes is perhaps better appreciated if we resort to a valence-bond formulation. The wavefunction (Ψ) of an exciplex can be regarded as the sum of the separate wavefunctions (ψ) of locally excited and charge-transfer states:

$$\Psi_{EX} = c_1\psi(D^*A) + c_2\psi(DA^*) + c_3\psi(D^+A^-) \quad (2.7)$$

The wavefunctions describe the "system," in this case the exciplex, and for our purposes tell us on which reactant partner the electronic excitation energy and electronic density are localized. The coefficients correspond to the contribution of each state to the entire wavefunction. The weighted values of each coefficient depend partly on the energetics of energy and electron transfer. If energy transfer is thermodynamically the *only* allowed quenching pathway, the $c_3 \sim 0$. When electron transfer is the *exclusive* pathway, $c_2 \sim 0$. Thus, for a predominantly charge-transfer exciplex, we write

$$\Psi_{EX} \sim c_3\psi(D^+A^-) \quad (2.8)$$

Exciplexes, say those formed between two identical aromatic hydrocarbons, are likely to form strongly nonpolar "exciplexes" since their IPs and EAs are identical (these exciplexes are often called *excimers*). They too will display strong emission.

On the other hand, strongly polar exciplexes (which are formed between "dissimilar" reactants) usually display little emission.

The very nature and character of exciplexes in photoinduced electron transfer are strongly dependent on the polarity of the medium. Thus, since an exciplex is a dipolar species with a dipole moment, the exciplex solvation enthalpy in a solvent of known polarity can be calculated from Eq. 2.9:

$$\Delta H_{D^+A^-} = -\frac{\mu^2}{\rho^3}\left(\frac{\varepsilon_s - 1}{2\varepsilon_s + 1}\right) \tag{2.9}$$

ρ is the sphere radius and, for most exciplexes, is estimated to range between ~4 and ~8 Å. μ^2/ρ^3 is the solvation free energy and can be calculated from the shift to the red of the maximum of exciplex emission as the solvent's polarity increases:

$$\bar{\nu}_{EX} = \bar{\nu}^0 - \frac{2\mu^2}{hc\rho^3}\left(\frac{\varepsilon_s - 1}{2\varepsilon_s + 1} - \frac{n^2 - 1}{4n^2 + 2}\right) \tag{2.10}$$

where $\bar{\nu}_{EX}$ is the wave number corresponding to the fluorescence maximum, $\bar{\nu}^0$ is the fluorescence maximum in a vacuum, h is Planck's constant, n is the refractive index, and c is the velocity of light. In the calculation of μ^2/ρ^3, $\bar{\nu}_{EX}$ is plotted vs. the second term on the right of Eq. 2.10 for a series of solvents. For the anthracene-*N,N*-diethylaniline pair, for instance, this procedure yields a value of $2\mu^2/hc\rho^3 = 8.3 \times 10^3\ \text{cm}^{-1}$. Assuming that ρ has the same value as the radius of the solvent cavity (i.e., $\rho = 5$ Å), Weller obtained a value of $\mu = 10$ debye.[3]

The fluorescence emission of exciplexes is usually studied by *fluorescence emission spectroscopy*. This technique is carried out in a fluorescence spectrometer where a beam of photons excites the sensitizer, which subsequently reacts with the quencher to form the exciplex. Exciplex emission is monitored with a monochromator and photomultiplier. Details of the procedure are described in standard photochemistry textbooks.

2.2 The Energetics of Exciplex and Ion-Pair Formation

In this section we present expressions that can be used to calculate the free-energy changes accompanying photoinduced electron transfer in polar and nonpolar solvents. Since the nature of the pathway of an electron transfer is dependent on the dielectric constant of the solvent, these equations can be used to show the dependence of exciplex and ion-pair formation on the solvent polarity. Thus, the free energy of exciplex formation is

$$\Delta G_{EX}\ (\text{kcal mol}^{-1}) = 23.06[E^0(D^+/D) - E^0(A/A^-)]^{37} - \Delta G_{00} - \frac{\mu^2}{\rho^3}\left(\frac{\varepsilon_s - 1}{2\varepsilon_s + 1} - 4.3\right) + 8.8 \tag{2.11}$$

where the superscript 37 indicates that the redox potentials were measured in a polar solvent with a dielectric constant of $\varepsilon_s = 37$ (acetonitrile).

An expression for the free energy involved in formation of a solvent-separated ion pair can also be derived:

$$\Delta G_{SSIP}\ (\text{kcal mol}^{-1}) = 23.06\,[E^0(D^+/D) - E^0(A/A^-)]^{37} - \Delta G_{00} - 166\left(\frac{1}{r_D} + \frac{1}{r_A}\right)\left(\frac{1}{37} - \frac{1}{\varepsilon_s}\right) - \frac{332}{\varepsilon_s d_{cc}} \tag{2.12}$$

(Further discussion of Eqs. 2.11 and 2.12 is given in Appendices II and III.) Note the difference between Eq. 2.12. and Eq. 1.54. In Eq. 2.12, the free-energy change in any solvent polarity can be calculated using the redox potentials measured in polar solvents. This is the equation we want to use when attempting to evaluate the free energy in nonpolar solvents.

With Eqs. 2.11 and 2.12 now at hand, we can now determine quantitatively under what solvent-polarity conditions ion-pair formation is favored over exciplexes. Plots of ΔG_{EX} and ΔG_{SSIP} vs. ε_s demonstrate the dependence of the energies of exciplexes and solvent-separated ions on solvent polarity (Fig. 2.4).[4] These plots were constructed for the photoinduced electron transfer between triplet chlorophyll *a* and *p*-benzoquinone. They show graphically the dependence of the energies of exciplexes and solvent-separated ion pairs on solvent polarity. For this reaction system, exciplex formation dominates in nonpolar media, but in polar media, solvent-separated ion-pair formation prevails. By using the expressions given in this

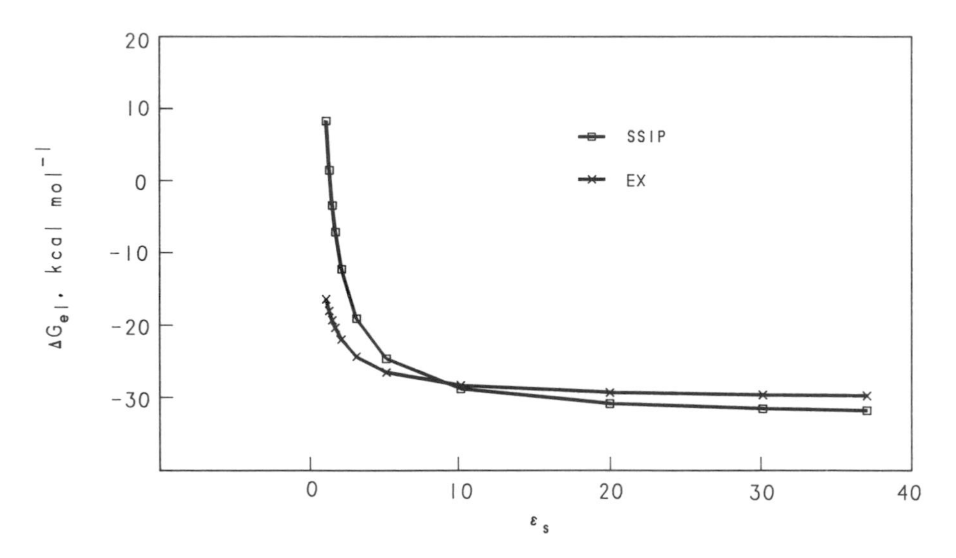

Figure 2.4 Plots of ΔG_{EX} and ΔG_{SSIP} *vs.* ε_S showing the variations of the free energies of exciplexes and solvent-separated ion-pair formation on the dielectric constant of the solvent. The example given here is for the quenching of triplet chlorophyll *a* by the electron acceptor, *p*-benzoquinone. These plots are based on Eqs. 2.11 and 2.12. The molecular parameters are: 5 Å for the radius (r_D) of triplet chlorophyll *a*; 3 Å for the radius (r_A) of *p*-benzoquinone; and 8 Å for the center-to-center separation distance ($d_{cc} = r_D + r_A$); $\rho = 8$ Å; and $\mu \simeq$ 10 debye.

section, the chemist can judiciously direct the "course" of a photoinduced electron transfer by choice of solvent.

2.3. Flash Spectroscopy

2.3.1. Detection of Transient Ion Pairs

In the previous section, we discussed how solvent polarity, charges on the reactants and products, and salt media can be manipulated to improve the yield of dissociated ions. Now, we may ask, How does one detect and identify dissociated ion pairs? One of the most powerful and universally used techniques to study transient intermediates is flash photolysis. In this technique, a short-lived pulse of light initiates the reaction by exciting the sensitizer. The pulse may be emitted from an conventional light source or a more sophisticated laser (light amplification by stimulated emission). After excitation of the sensitizer, the appearance of transient intermediates can be monitored in several ways. One of these approaches involves *flash spectroscopy* where the initial light pulse is normally of very high intensity and of short duration so as to generate a sufficiently large concentration of transient species. Shortly after the first pulse, a second pulse from a broad-band light source is directed into the sample compartment. The absorption spectra of transient intermediates, which may be radical ions and excited states, can be recorded immediately following decay of the initial flash by a suitable arrangement consisting of a monochromator to monitor the absorption of the unknown species. An absorption spectrum over a specified range of wavelengths or a recording of the absorbance vs. time (time-resolved spectroscopy) is thus obtained. The transient absorption spectrum can be compared with absorption spectra obtained by other means. The flash system must be capable of monitoring and resolving the change over time of the short-lived transient absorption.

To study the kinetic processes of short-lived charge-transfer transients, one may employ a continuous light source in place of the second light pulse. The wavelengths from the continuous light source are then collimated and directed through the reaction vessel. An absorption or emission band due to the transient is monitored at a single wavelength over time in order to obtain kinetic data.

Since the transient intermediates generated by the flash may decay quite rapidly by electron return, the signals due to decay of the initial flash must be sufficiently resolved and separated from the signals due to the transient species of interest. For this reason, conventional microsecond flash photolysis may lack sufficient resolution for detection of short-lived transients. Modifications can be introduced to improve the signal-to-noise ratio. For example, laser pulsed spectroscopy provides a powerful probe of the time evolution of the short-lived, ion-pair intermediates. Lasers operate by *stimulated emission* from an excited state. In stimulated emission, a photon of given frequency encounters an excited state and releases a second photon (Fig. 2.5). Since the two photons are in phase and propagate in the same direction, they may excite other ground-state species or induce the emission from excited states they may encounter. This leads to an amplification of the "exciting" photons. Stimulated emission occurs only when the population of the excited state exceeds that of the ground state. This is known as a population inversion and can occur by a process known as "optical pumping" (Fig. 2.6). Optical pumping can be achieved by photoexcitation of a dye (usually by another laser) or by electrical excitation. A laser is generated in a cavity consisting of a reflecting mirror at one end and a partially transmitting plate at the other end. The size of the cavity is "tuned" to the wavelength of emission from the upper energy state. After a generation of a photon, the formation of additional photons is stimulated and the intensity of photons is amplified with reflection of the photons back and forth within the cavity. The buildup of photons takes place until there is no longer any population inversion. Once a pulse is formed, the laser light emits as a train of photons.

Laser pulses are well defined in terms of energy and spectral output and have durations ranging from 10^{-6} to 10^{-13} s. In nanosecond transient spectroscopy, a continuous light source is used to produce a transient absorption or emission signal at a single wavelength. In nanosecond flash photolysis, two lasers are sometimes used in combination. The first laser excites formation of the transient. After a short interval, the second laser pulse excites a fluorescent material to induce emission of broad-band fluorescence, which in turn can function as the spectroscopic probe. This allows a wide range of wavelengths to be monitored at one time. In the technique called *Q-switching,* oscillations within the laser cavity are blocked. This results in a very large inverse population whose energy can be released by a Q-switch. The result of this modification is a very short nanosecond pulse of high power.

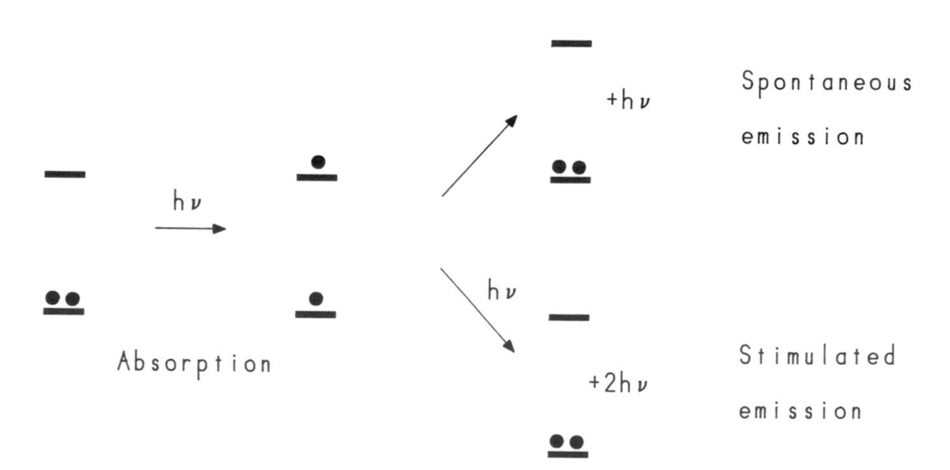

Figure 2.5 A laser operates by stimulated emission. A photon can interact with an excited state to stimulate the release of a second photon. These photons are in phase and propagate in the same direction, eventually colliding with and being absorbed by ground states or stimulating the release of additional photons from excited states. This phenomenon results in amplification of the laser light signal.

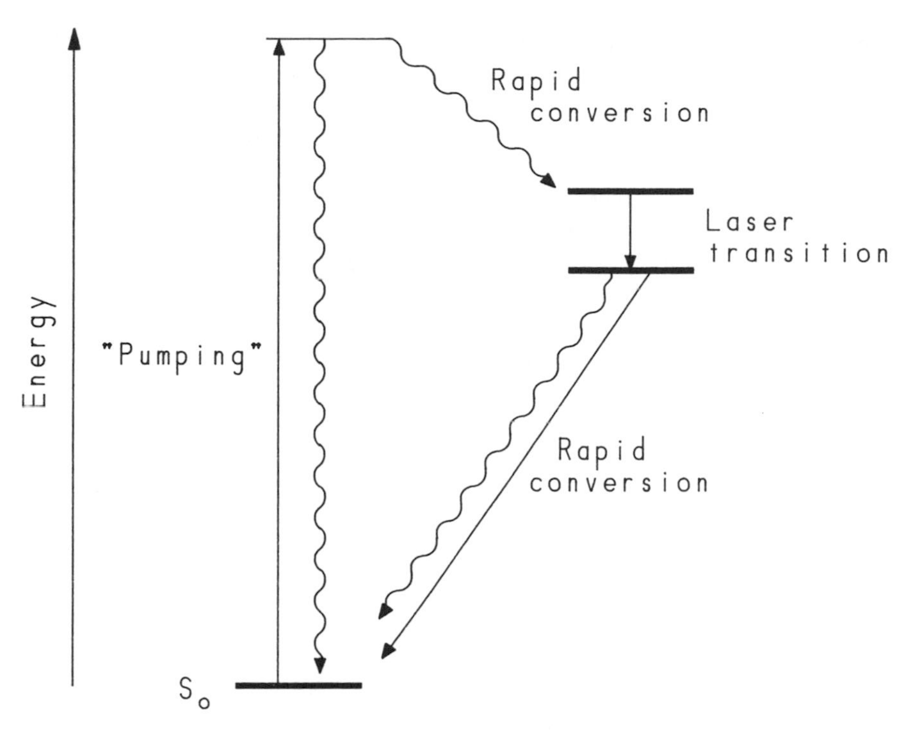

Figure 2.6 One configuration of a laser. Optical pumping results in a population inversion which favors stimulated emission. See text for explanation.

In picosecond flash spectroscopy, short picosecond pulses equally spaced in time are generated by a technique called *mode-locking*. The laser cavity contains longitudinal and transverse waves or "modes." The temporal behavior of laser output depends on the relative phases and amplitudes of these modes in a randomed manner. In mode-locking, the laser modes oscillate with fixed phases and amplitudes. The laser pulses produced by employing the technique of mode-locking have large bandwidths but very short lifetimes. Dye lasers, which have large bandwidths, are therefore useful for the generation of picosecond pulses. In picosecond flash spectroscopy, the probe beam is delivered by the photolytic flash and spectrally modified for broad-band transient absorption experiments. Because of the extremely short duration of the exciting pulses, picosecond laser spectroscopy extends the scope of spectroscopy enormously and allows insight into the ultrafast dynamic and temporal behavior of reactants and radical ions on a time scale between $\sim 10^{-9}$ and 10^{-14}. The rapid motions and orientation of the donor and acceptor prior to quenching, the subsequent dynamic motions of the radical ions, and ion recombination can be studied effectively by picosecond spectroscopic techniques.

It is important to recognize that unless the time scale of the flash is short enough, ions can easily escape spectroscopic detection. The lifetime of the flash becomes an even more critical factor if electron return or chemical transformations are much faster than ion separation. It is desirable, therefore, to use conditions that are most favorable for electron transfer such as the use of at least moderately polar solvents.

In applying flash spectroscopy, it should also be noted that, in general, ions having the same charge can be detected with greater ease than the "conventional" ion pair with one positive and one negative charge. Electron return between ions of opposite charge is usually so fast so as to elude spectroscopic detection. However, ionic species possessing the identical charge repel one another and consequently dissociate rapidly from the solvent cage. The quenching of metal complexes by electron acceptors and donors affords us an example to examine the role of Coulombic interactions, since these complexes are usually charged species. The reduction of $Ru(bpy)_3^{2+}$ by certain donors turns out to be an easy reaction to track

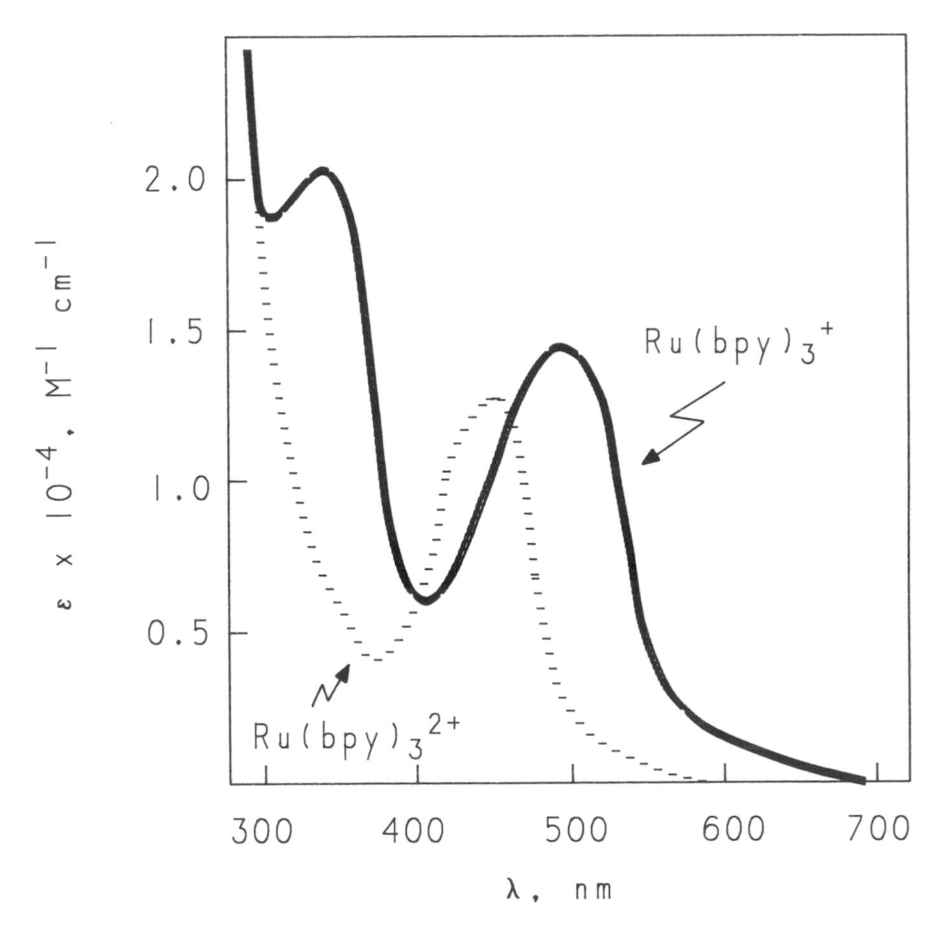

Figure 2.7 The absorption spectra of $Ru(bpy)_3^+$ (solid line) and $Ru(bpy)_3^{2+}$ (dotted line). (Adapted with permission from Mulazzani, Q. G., Emmi, S., Fuochi, P. G., Hoffman, M. Z., and Venturi, M. *J. Am. Chem. Soc.* **1978,** *100,* 981. Copyright 1978 American Chemical Society.)

by flash spectroscopy since two positive ions are formed in the electron transfer. The spectrum of $Ru(bpy)_3^{+}$, which is formed when $[Ru(bpy)_3^{2+}]^*$ reacts with electron donors, is particularly prominent in the visible with a maximum at 510 nm and an extinction coefficient of 14,000 M^{-1} cm^{-1} (Fig. 2.7).[5] In the quenching of $[Ru(bpy)_3^{2+}]^*$ by Eu^{2+}, flash spectroscopy with a 530 nm pulse generated with a neodymium laser verified the presence of $Ru(bpy)_3^{+}$:[6,7]

$$[Ru(bpy)_3^{2+}]^* + Eu^{2+} \rightarrow Ru(bpy)_3^{+} + Eu^{3+} \qquad (2.13)$$

The free-energy change for this reaction is calculated as $\Delta G_{el} = -28.7$ kcal mol^{-1}), so we predict this reaction to be thermodynamically feasible. A transient absorption spectrum of $Ru(bpy)_3^{+}$ was observed from 490 to 510 nm (a region where Eu^{2+} and Eu^{3+} do not absorb). The transient absorption is probably due to $Ru(bpy)_3^{+}$. The disappearance of the transient absorption takes place on a long time scale, as would be expected because of slow electron return between two positive ions.

The reaction between $[Ru(bpy)_3^{2+}]^*$ and MV^{2+} involves initial electron transfer to produce a strongly blue colored intermediate, $MV^{\dot{+}}$:[8]

$$[Ru(bpy)_3^{2+}]^* + MV^{2+} \rightarrow Ru(bpy)_3^{3+} + MV^{\dot{+}} \qquad (2.14)$$

Reaction 2.14 is also thermodynamically allowed ($\Delta G_{el} = -10$ kcal mol^{-1}). Energy transfer from $[Ru(bpy)_3^{2+}]^*$ to MV^{2+}, as a competitive mechanism, was ruled out because both the singlet and triplet excited states of MV^{2+} lie higher than the excited state energy of $[Ru(bpy)_3^{2+}]^*$. Since the extinction coefficients of $Ru(bpy)_3^{3+}$ at 420 and 680 nm are only 680 and 3,300 M^{-1},[9] the presence of $MV^{\dot{+}}$ was monitored by a buildup of absorption at 600 nm, which parallels a concomitant decrease of the emission of $[Ru(bpy)_3^{2+}]^*$ at 615 nm. Eventually, on a longer time scale, the absorption due to $MV^{\dot{+}}$ was found to disappear because of electron return:

$$Ru(bpy)_3^{3+} + MV^{\dot{+}} \rightarrow Ru(bpy)_3^{2+} + MV^{2+} \qquad (2.15)$$

There is also sufficient spectroscopic evidence of quenching of $[Ru(bpy)_3^{2+}]^*$ by strong electron donors such as amines.[10] This is consistent with thermodynamic arguments. The radical ion yield in these reactions approaches unity. Energy transfer is excluded because of the high-lying energy states of the amines. Transient absorption spectra of $Ru(bpy)_3^{2+}$–amine systems are shown in Fig. 2.8.

In contrast to the above examples, it has been difficult to detect the ions formed when $[Ru(bpy)_3^{2+}]^*$ is quenched by an electron attractor such as an nitroaromatic compound. The following reaction sequence can be written for a one-electron reduction of nitrobenzene (NB):

$$[Ru(bpy)_3^{2+}]^* + NB \xrightarrow{CH_2Cl_2} Ru(bpy)_3^{3+} + NB^{\dot{-}} \qquad (2.16)$$

$$Ru(bpy)_3^{3+} + NB^{\dot{-}} \rightarrow Ru(bpy)_3^{2+} + NB \qquad (2.17)$$

Although reaction 2.16 is exothermic in methylene chloride ($\varepsilon_s = 4.8$) with $\Delta G_{el} = -10.5$ kcal mol^{-1}, spectroscopic detection of transient nitrobenzene radical anion is difficult using microsecond flash spectroscopy. The two ions within the solvent

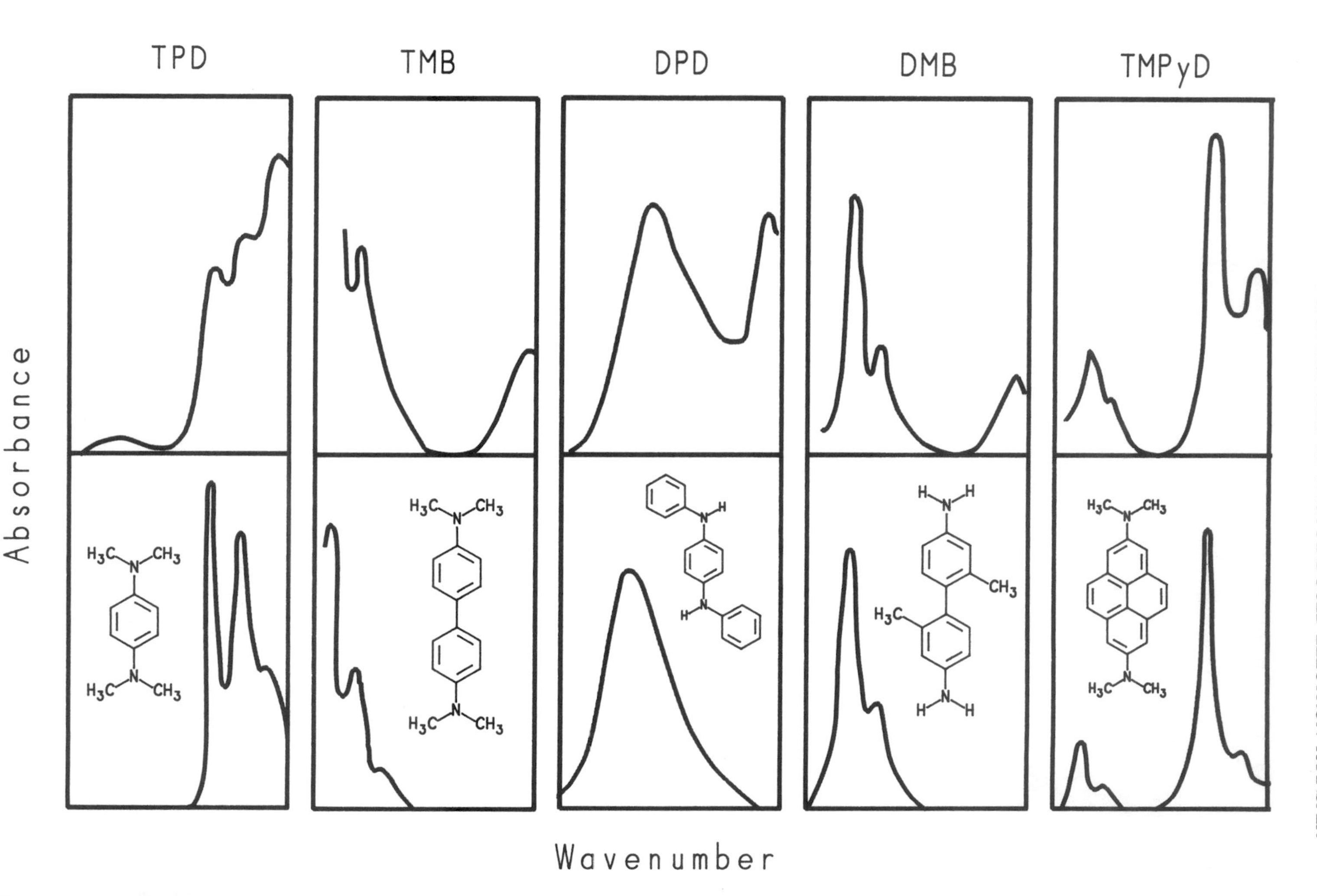
TPD
TMB
DPD
DMB
TMPyD
Absorbance
Wavenumber
H_3C
CH_3
N
H

cage remain in close proximity because of Coulombic attractive forces and because of nonpolar conditions. Electron return rapidly reduces the lifetimes of the ion pair making direct spectroscopic observation of the nitrobenzene anions difficult. This example underscores the following point: simply because electron-transfer products are not observed in a spectroscopic analysis does not in itself rule out an electron-transfer pathway. The lifetime of the intermediates may be quite small because of rapid decay processes such as electron return. The resolution of the spectroscopic technique must be selected for the particular reaction system under investigation. Conversely, spectroscopic evidence alone does not rule out competing pathways. An electron-transfer pathway may be only a small part of the quenching mechanism. Spectroscopic methods are intended to complement kinetic and thermodynamic studies.

Coulombic effects in substituted porphyrins or phthalocyanines also play an important role in the reactions of these sensitizers. Positive or negative functional groups can be placed around the periphery of the porphyrin or phthalocyanine ring, e.g.,

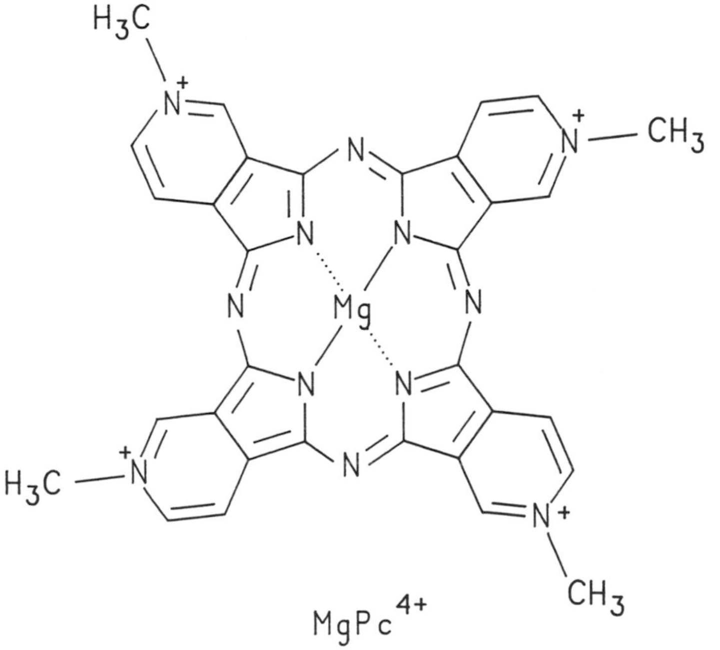

$MgPc^{4+}$

I

Figure 2.8 Transient absorption of Ru(bpy)$_3$2+-amine systems. This work was performed with a dye laser photolysis system. Lower part of diagram displays absorption spectra of amine cations produced by γ-radiation of glass solutions of known amines at 77 K. TPD = N,N,N′,N′-tetramethyl-*p*-phenylenediamine; TMB = N,N,N′,N′-teteramethylbenzidine; DPD = N,N′-diphenyl-*p*-phenylenediamine; DMB = 3,3′-dimethylbenzidine; TMPyD = N,N,N′,N′-tetramethyl-1,6-pyrenediamine. (Reprinted with permission from Shioyama, H., Masuraha, H., and Mataga, N. *Chem. Phys. Lett.* **1982,** *88,* 161. Copyright 1982 Elsevier Science Publishers.)

For example, flash spectroscopy has revealed the presence of long-lived radical ions in the quenching of singlet and triplet states of a substituted magnesium phthalocyanine ($MgPc^{4+}$, **1**) in the presence of high concentrations of MV^{2+}.[11] Electrostatic repulsion between the positive species is expected to help to prevent rapid electron return.

2.3.2. Dynamic Motion of Charge-Transfer Species

During a chemical reaction, the reactants engage in rapid motions on a very short timescale. These are called *dynamic motions* and may include the positional changes with time of the reacting molecules, changes in the orientations of the reactants, and conformational motions. Normally, these motions are so fast that they are quite difficult to trace. However, with the advent of time-resolved spectroscopy using laser excitation with very short pulses, it became possible to study dynamic motion. For example, picosecond flash photolysis has revealed dynamic molecular motion in the quenching of triplet benzophenone by triethylamine in both inter- and intramolecular pairs (**2** and **3**)[12]:

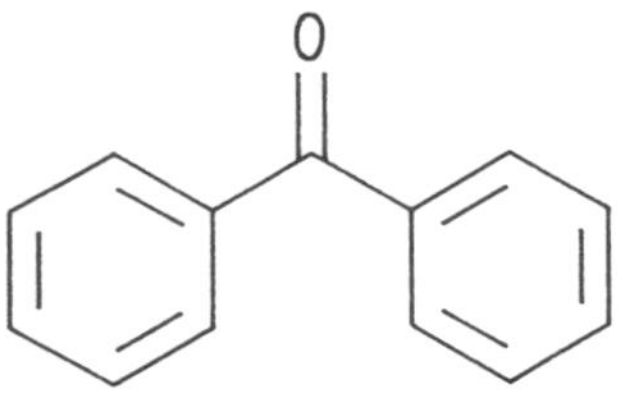

$:N(CH_2CH_3)_3$

2 3

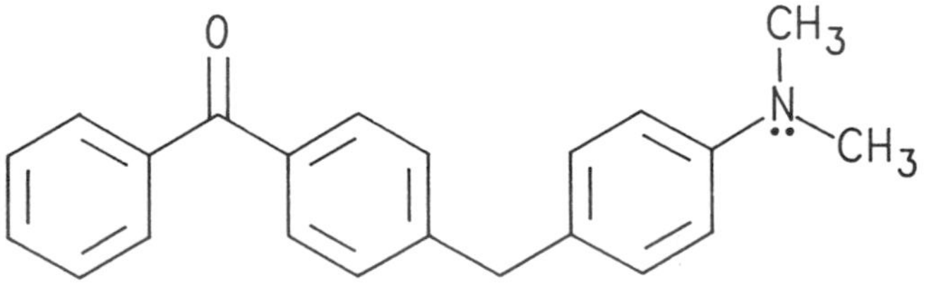

4

Within 25 ps after quenching, the spectrum of the radical anion of benzophenone appears, displaying its maximum at 720 nm. After 25 ps, the maximum peak begins to shift reaching 690 nm in about 400 ps. It was interesting to compare this reaction with the dynamic motion of an intramolecular donor–acceptor pair. When the time evolution of photoexcited 4-(*p*-dimethylamino)benzylbenzophenone (an intramolecular donor–acceptor pair) was followed under similar conditions, again there was the appearance of the 720 nm peak but no subsequent spectral shift. In both the inter- and intramolecular systems, the appearance of solvent-separated ion pairs is consistent with an electron-transfer mechanism. In the case of the intermolecular benzophenone–amine system, the ions can return to contact distances to form a contact ion pair. On the other hand, the chain in the intramolecular system prevents the formation of a contact ion pair between the carbonyl and amino groups, so that no hypsochromic shift is observed. As we shall show in Chapter 4, donors and acceptors connected by spacer molecules often display much different behavior than when they are free reactants. The spacer group can have a dramatic effect on the freedom of the reactants to move toward or away from one another. Because of its extraordinary resolution, picosecond time-resolved spectroscopy helps us to obtain an improved picture of the dynamics of charge separation in these systems.

2.3.3 Resonance Raman Spectroscopy

A major drawback to studying reactions by transient absorption spectrometry lies in lack of detail usually displayed by the spectral bands of transients. This may impose a limit on spectral identification of an unknown transient. In transient resonance Raman (RR) spectroscopy, however, considerable definition of the spectral bands can be achieved. When used in conjunction with nanosecond or picosecond laser spectroscopy, the method is highly sensititive and provides detailed information on the vibrational structures of short-lived intermediates.

What is resonance Raman spectroscopy? The Raman effect is due to the inelastic scattering of a photon. When samples are excited by light, most of the light usually passes through the sample. However, a small percentage of photons exchange energy with molecules and are thereby scattered, leaving the molecules in higher energy states. Subsequently, these molecules may undergo relaxations. The energy of the scattered photons depends on whether the photons have lost or gained energy during collision. The energy of the scattered light corresponds to certain vibrational modes in the molecule and is therefore quantized. *Stokes lines* are defined as scattered emissions at frequencies less than the frequency of exciting radiation. *Anti-Stokes lines* correspond to higher frequencies.

In resonance Raman spectroscopy, the energy of the exciting photon nearly equals the absorption band of a transient. If the energy of the incident photons corresponds to the energy of an active Raman mode, a striking enhancement of the Stokes lines is observed. A laser beam provides the degree of monochromacity and intensity required to obtain the desired resonance Raman effect. Because of the enhanced sensitivity of the Raman method, time-resolved investigations can be carried out. Typically, the sample is exposed to two laser beams. One short pulse excites the sample to generate transients and the other beam is used to detect the Raman spectra. Nanosecond or picosecond pulses can be used.

Resonance Raman spectroscopy has proven to be a powerful technique to study various transients which accompany a photoinduced electron-transfer reaction. For example, we have already observed the $[Ru(bpy)_3{}^{2+}]^*$ can be quenched by MV^{2+} by an electron-transfer mechanism (reaction 2.14). This was suggested by the energetics of the reaction and by detection of $MV^{\dot{+}}$ by time-resolved absorption spectrophotometry. Additional support for electron transfer has been provided by a resonance Raman study. When mixtures of $Ru(bpy)_3{}^{2+}$ and MV^{2+} were exposed to two laser beams, highly detailed and resolved vibrational spectra were obtained.[13] The laser beams were simultaneously focused onto a sample cell and the resonance Raman spectrum detected at $\lambda = 350.6$ nm. The resonance Raman spectra obtained

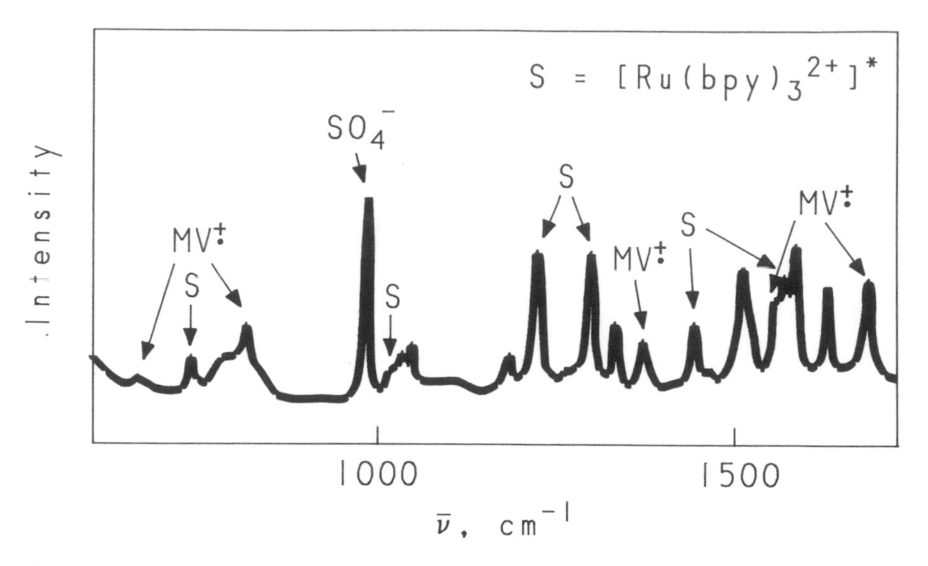

Figure 2.9 Resonance Raman spectra of $Ru(bpy)_3^{2+}/MV^{2+}/K_2SO_4$ solutions taken after excitation with the 350.6 nm wavelength of a focused laser beam. Peak assignments were made by comparison with standards. (Adapted with permission from Forster, M. and Hester, R. E. *Chem. Phys. Lett.* **1982,** *85*, 287. Copyright 1982 Elsevier Science Publishers.)

in this fashion were compared to Raman spectra of pure mixtures of the intermediates. In Fig. 2.9 are shown the resonance Raman spectra of aqueous solutions of $Ru(bpy)_3^{2+}$ and MV^{2+} exposed to light. The remarkable detail of the resonance Raman spectra of the photoexcited solutions corresponds closely to the spectra of the standards. This observation supports an electron-transfer pathway. There was no evidence of $[MV^{2+}]^*$, which would have been formed in energy transfer.

Time-resolved Raman spectroscopy has also provided valuable mechanistic information on the quenching reactions of excited *trans*-stilbene.[14] The fluorescence of *trans*-stilbene (**5**) undergoes quenching in the presence of tertiary amines and diamines such as 1,4-diazabicyclo[2.2.2]octane (DABCO, **6**):

$$[\mathbf{5}]^{*} + \mathbf{6} \longrightarrow \mathbf{5}^{\cdot -} + \mathbf{6}^{\cdot +} \tag{2.18}$$

Using Eq. 1.54, we calculate the free-energy of oxidation of DABCO to its radical cation by singlet *trans*-stilbene as $\Delta G_{el} = -10$ kcal mol^{-1}. Energy transfer from singlet *trans*-stilbene to populate singlet DABCO is ruled out since the singlet

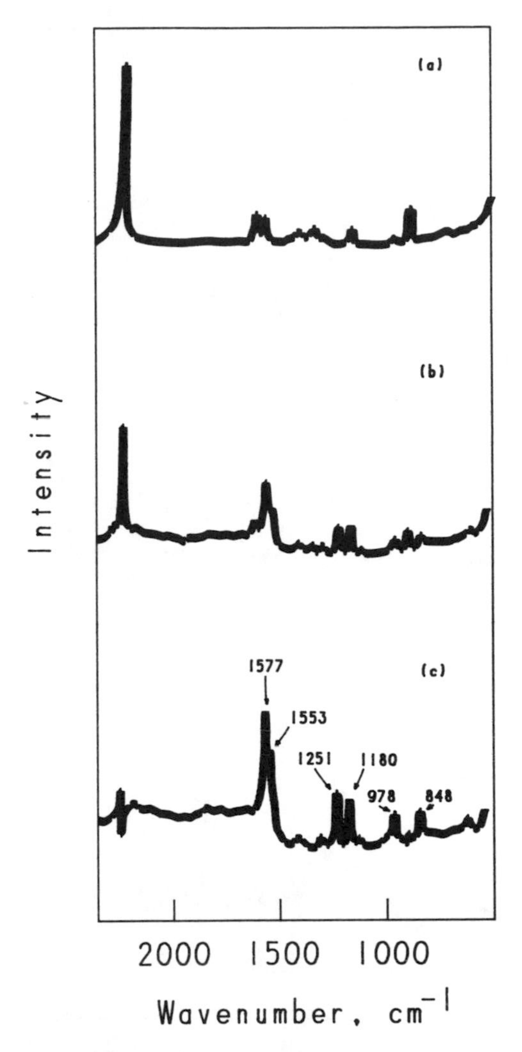

Figure 2.10 Raman spectra of *trans*-stilbene anion obtained by excitation of solutions of *trans*-stilbene and ethyldiisopropylamine in CH_3CN. Solvent peaks are labeled S. (a) "normal" Raman spectrum of *trans*-stilbene and the amine on excitation with a probe wavelength of 480 nm (the reactants are in their ground state); (b) the spectrum obtained 60 ns after excitation with a laser pulse. (c) a difference spectrum between (a) and (b). The difference spectrum contains the peaks corresponding to the transient anion radical of *trans*-stilbene produced as a result of photoinduced electron transfer. (Adapted with permission from Hub, W., Schneider, S., Dörr, F., Oxman, J. D., and Lewis, F. D. *J. Am. Chem. Soc.* **1984,** *106,* 708. Copyright 1984 American Chemical Society.)

energy of the amine is much higher than that of the sensitizer's singlet. Confirmatory evidence of electron transfer was provided by the direct evidence of the *trans*-stilbene radical anion by resonance Raman spectroscopy (Fig. 2.10). The Raman spectrum was obtained at ~60 ns after excitation with a pulsed nitrogen laser. The vibrational frequencies of this transient correlate well with literature values.

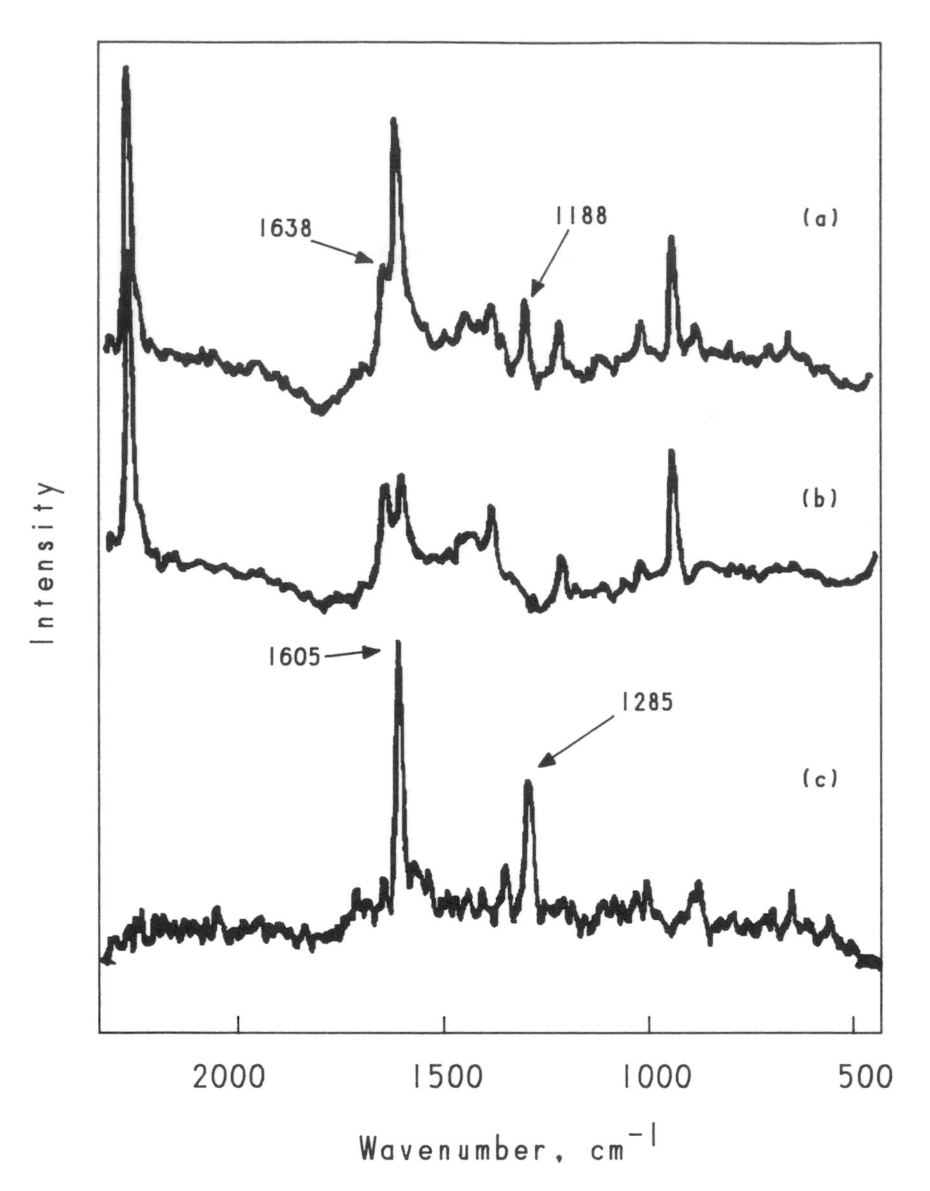

Figure 2.11 (a) The resonance Raman spectrum of a solution of *trans*-stilbene and fumaronitrile 60 ns after excitation with light; (b) Raman spectrum of *trans*-stilbene and fumaronitrile with a probe wavelength of 480 nm; and (c) the difference spectrum with additional peaks due to radical cation of *trans*-stilbene. Identification of the radical cation of *trans*-stilbene was based on the disappearance of a C=C stretching frequency at 1638 cm^{-1} and the appearance of a peak at 1285 cm^{-1}. This new peak is probably due to stretching of the central bond in the radical cation. Note the substantial difference between the resonance Raman spectra of the radical cation and of the radical anion shown in Fig. 2.10. (Adapted with permission from Hub, W., Klüter, U., Schneider, S., Dörr, F., Oxman, J. D., and Lewis, F. D. *J. Phys. Chem.* **1984,** *88,* 2308. Copyright 1984 American Chemical Society.)

Singlet *trans*-stilbene is one of those sensitizers that can be quenched by both electron donors and acceptors. In the presence of a strong acceptor such as fumaronitrile (**7**), time-resolved resonance Raman reveals the vibrational spectrum of the *trans*-stilbene cation.[15] This spectrum correlates well with the spectrum obtained in the presence of other electron acceptors. As shown in Fig. 2.11, the resonance Raman spectrum of the *trans*-stilbene cation can be resolved from the absorption spectrum of the anion. The features of this spectrum support the following mechanism:

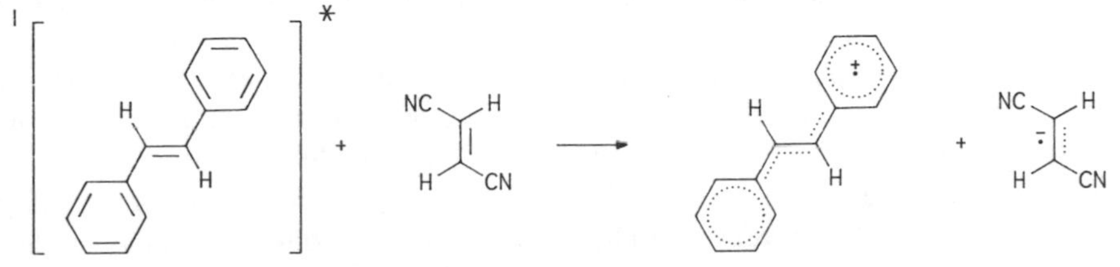

5 + 7 → (2.19)

2.4. Photoconductivity

So far, we have implicitly assumed that the ion-pair intermediates generated in photoinduced electron transfer are equilibrated charge-transfer complexes. In fact, the generation of equilibrated charge-transfer intermediates by excited-state electron transfer is the result of a series of steps that starts out with the formation of a high-energy, short-lived ion pair. As a relaxation proceeds, intermediates lower in energy content are successively formed. After the first charge-transfer intermediate is formed, solvation and Coulombic interactions play a critical role in stabilizing each successive intermediate. Initially, short-lived and nonequilibrated ions pairs may be formed but they rapidly relax into stabilized ion pairs. The application of time-resolved fluorescence spectroscopy, nano- and picosecond flash photolysis combined with photoconductivity measurements has provided evidence of the involvement of unrelaxed charge-transfer complexes. This work has provided a more complete picture of the dynamic events occurring during photoinduced electron transfer.

Photoconductivity is the ability of certain substances to conduct electricity on exposure to light. Transient photoconductivity is the rise and fall of photocurrent that results after the absorption of a pulse of light. In photoinduced electron transfer, the transient photocurrent is due to the appearance of conductive ion pairs. The rise in photocurrent signals the generation of these species by electron transfer. This is followed by the fall in the photocurrent, which marks their disappearance by pathways such as electron return.

For a photocurrent experiment to be successful, free ions must escape from the solvent cage. These dissociated ions can operate somewhat like "charge carriers" in the sense that they are capable of conducting electric current. The escape of ionic species from a solvent cage is expressed as a ratio of rate constants for ion dissociation and recombination processes (Eq. 2.2). In effect, the yield of ions depends on a competition between in-cage recombination (followed by electron return) and ionic dissociation. However, when the ions do escape from their cages, their concentrations must not be so high that the ions immediately cross one another's paths to undergo electron return. Practically speaking, the concentration of reactants should be $\ll 0.1$ M in acetonitrile or $\ll 0.3 \times 10^{-5}$ M in benzene. As was demonstrated earlier in this chapter, the competition between electron return and ion separation depends on the solvent's dielectric constant (i.e., the ability to solvate ions) and the molecular structures of the reactants. Photoconductivity is thus favored in polar solvents.

When transient photoconductivity measurements are taken after excitation with nanosecond or picosecond laser pulses, a picture of the dynamical behavior of ion radicals emerges. For example, laser-induced photoconductivity studies have revealed such behavior in the quenching of pyrene by *N,N*-dimethylaniline (Fig. 2.12).[16] In nonpolar solvents, no photoconductivity was observed, although there was observable exciplex emission. In moderately polar solvents, the rise in photocurrent was resolved into rapid and slow components (not shown in Fig. 2.12).

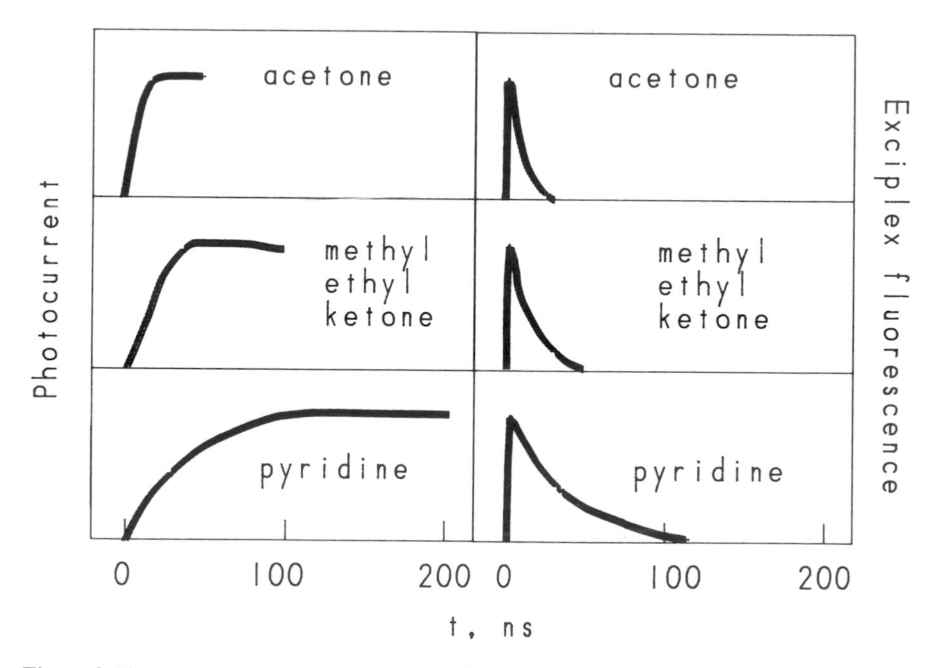

Figure 2.12 The curves on the left represent the rise curves of photocurrent in the quenching of singlet pyrene by N,N-dimethylaniline. For comparison, the decay of exciplex fluorescence is shown on the right. Studies were performed in acetone (1), methyl ethyl ketone (2), and pyridine (3) solvents. (Reprinted with permission from Hirata, Y., Kanda, Y., and Mataga, N. *J. Phys. Chem.* **1983,** *87,* 1659. Copyright 1983 American Chemical Society.)

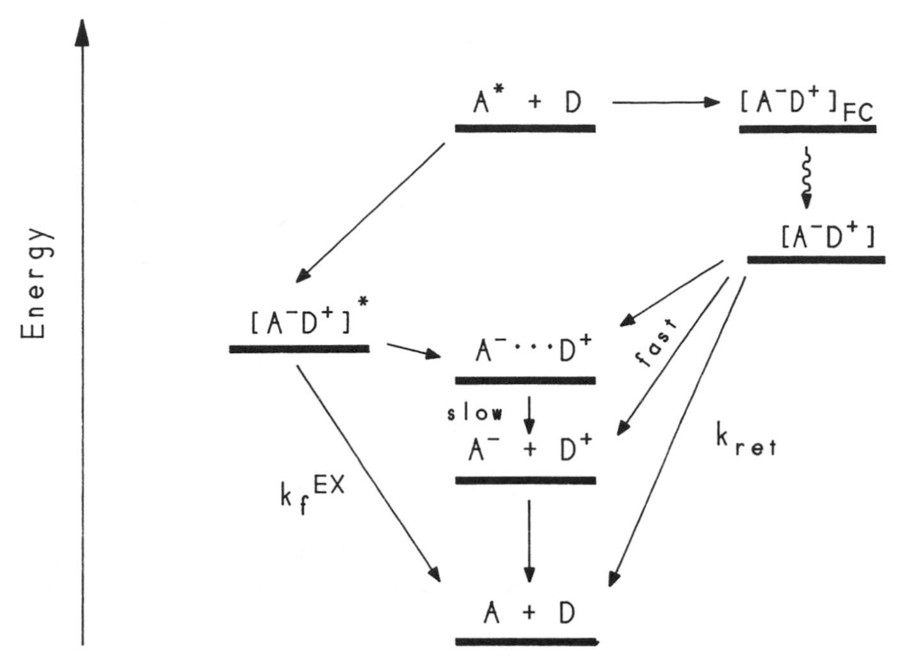

Figure 2.13 The nature of an unrelaxed, unequilibrated charge-transfer species is shown in this diagram. This conceptualization is based on extensive work by Mataga (see text).

The rapid rise occurs before the completion of exciplex decay. Thus, dissociated ions do not result from the exciplex. According to Mataga, the rapid rise in photoconductivity might be the result of ion formation from an "unrelaxed" precursor state. Ion dissociation from the solvent-separated ion pair may account for the slow rise. Mataga's conceptualization of a non-relaxed charge-transfer state is that of a short-lived transient polar species nonequilibrated with respect to solvent orientation.[17,18] This species, which is depicted in 2.20 as a contact ion pair, arises directly from a Franck-Condon state:

$$^{1}A + D \rightarrow [A^{\overline{\cdot}}\, D^{\dot{+}}]_{FC} \rightarrow [A^{\overline{\cdot}}\, D^{\dot{+}}] \rightarrow A^{\overline{\cdot}} + D^{\dot{+}} \qquad (2.20)$$

The energy diagram for the reaction in 2.20 is shown in Fig. 2.13. In Mataga's model, electron transfer initially forms a high-energy ion-pair species that subsequently relaxes to a lower energy, equilibrated structure. Relaxation to the equilibrated structure may be accompanied by a partial realignment of solvent molecules surrounding the ion pair. The formation of free ions from this state accounts for the rapid rise in photocurrent. The important conclusion from Mataga's work is that in the pyrene-*N,N*-dimethylaniline system as well as other reaction systems, the experimental photocurrent and fluorescence measurements can best be explained by invoking several types of charge-transfer intermediates.

2.5. Pulse Radiolysis

To study the properties and behavior of radical ions formed during photoinduced electron transfer, it is often desirable to generate them by an independent route. This approach can provide additional support for an electron-transfer mechanism. By comparing spectroscopic and other properties of the radical ions formed as a result of photoinduced electron transfer with those of radical ions generated by the independent route, we can study and compare the chemical behavior of these intermediates. A particularly powerful approach for the generation of radical ions involves the use of pulsed, high-energy electrons with energies ranging up to about 30 MeV. After generation of the radical ion intermediates by bombardment with a high-energy pulse, they can be studied by time-resolved flash spectroscopy, resonance Raman spectroscopy, and transient conductivity. For an example, the reader may want to review Fig. 2.8, which compares the transient absorption spectra of amines generated by photoinduced electron transfer with an excited electron acceptor and by γ-radiation.

The primary step in pulse radiolysis usually involves radiation of a dilute sample with a high-energy pulse. The energy absorbed by the solvent gives rise to the generation of reactive radical intermediates. In the radiolysis of water, for example, a hydrated electron (e_{aq}^-), hydrogen atom, and hydroxyl radical ($OH^{\cdot}$) can be generated. As highly reactive intermediates, they are capable of functioning as electron donors or acceptors. The hydrated electron is an excellent reducing agent, whereas the hydroxyl radical performs as an electron acceptor. The concentration of a radical species formed as a result of pulse radiolysis can be manipulated by a judicious choice of solvent or the addition of electron scavengers. For example, radical anions of aromatic molecules can be generated by radiolysis of alcohol or ether solutions:

$$Ar + e_{aq}^- \rightarrow Ar^{\overline{\cdot}} \tag{2.21}$$

Similarly, the radical anions of carbonyl compounds, nitro compounds, and quinones can be formed by exposure to solvated electrons. Aromatic radical cations can be formed by the radiolysis of halogenated solutions:

$$RX \rightarrow RX^+ + e^- \tag{2.22}$$

$$RX + e^- \rightarrow R^{\cdot} + X^- \tag{2.23}$$

$$RX^{\dot{+}} + Ar \rightarrow RX + Ar^{\dot{+}} \tag{2.24}$$

where X is a halogen atom. One-electron reduction and oxidation products of porphyrins and related compounds can be formed by pulse radiolysis. For example, the 18-membered π-electron system of the porphyrin molecule is amenable to electron loss and addition.

2.6. Electron and Nuclear Spin

2.6.1. Electron Spin

The intrinsic magnetic moment of an electron is referred to as its "spin." Electron spin plays a crucial role in both the dynamics of ion pairs and methods used to study them. With an operational model of electron spin, we will examine the effects of magnetic fields on excited-state electron transfer. For this purpose, we will trace the spin changes which take place when a radical ion pair is generated by the quenching of an excited state (Fig. 2.14).

The *spin angular momentum* vector of an electron can adopt one of two orientations along an axis, say the z-axis. Since the overall spin of the ion pair reflects the combined spin of the donor and acceptor pair because of spin conservation, the spin of each electron in the radical ions partner can adopt one of the two orientations, so that either the combined spins of each ion cancel to give a singlet state or add up to one to give a triplet state. The overall spin state of the radical ion pair is classified as singlet or triplet, although the spin of each individual radical ion is a doublet. This statement holds true for *correlated* spin pairs by which it is meant that the spins of each ion are coupled with one another. When the ions are separated by an infinite distance, the spin of each unpaired electron adopts a random orientation with respect to the spin of the other (i.e., the spins are now *uncorrelated*).

Figure 2.14 also shows the formation of a triplet state of one of the reactants directly from the triplet ion-pair species. This pathway is called *triplet recombination* and is an important feature of many photoinduced electron-transfer reactions. Triplet recombination is possible when the energy of a triplet state of one of the reactants lies just below the energy of the radical ion pair. Triplet recombination occurs during the return of the electron but because of spin conservation, a triplet state rather than singlet ground-state reactants is generated.

When a spin correlated ion pair is formed within a solvent cage, the spin state may become altered by spin rephasing (i.e., a singlet radical ion pair may evolve into a triplet state or a triplet state may convert into a singlet state). A radical ion pair experiences this spin evolution because the spin multiplicity of a radical ion pair is a dynamic property. To model changes in spin, it is useful to consider Fig. 1.11 and then the vector diagram of a singlet and triplet state shown in Fig. 2.15. There are three possible spin orientations of the triplet state corresponding to the three sublevels of the triplet state designated as T_+, T_-, and T_0. For the singlet state, there is only one possible spin orientation. In the pure singlet state, the electron spin vectors are oriented in opposite directions and cancel out (i.e., $S = 0$ and $M_s = 0$); in the pure triplet state, $S = 1$ and $M_s = 0, 1$, or -1. There are three possible spin orientations for the triplet. In the T_+ and T_- states, the spin vectors point in the same direction (i.e., $M_s = 1$ and -1). In the T_0 state, the spin vectors point in opposite directions, but $M_s = 0$.

The spin vector of an electron can be visualized as revolving around the z-axis with a precessional frequency, ω. This precessional frequency, which is related to spin rephasing, is proportional to the strength of the magnetic field that a spinning

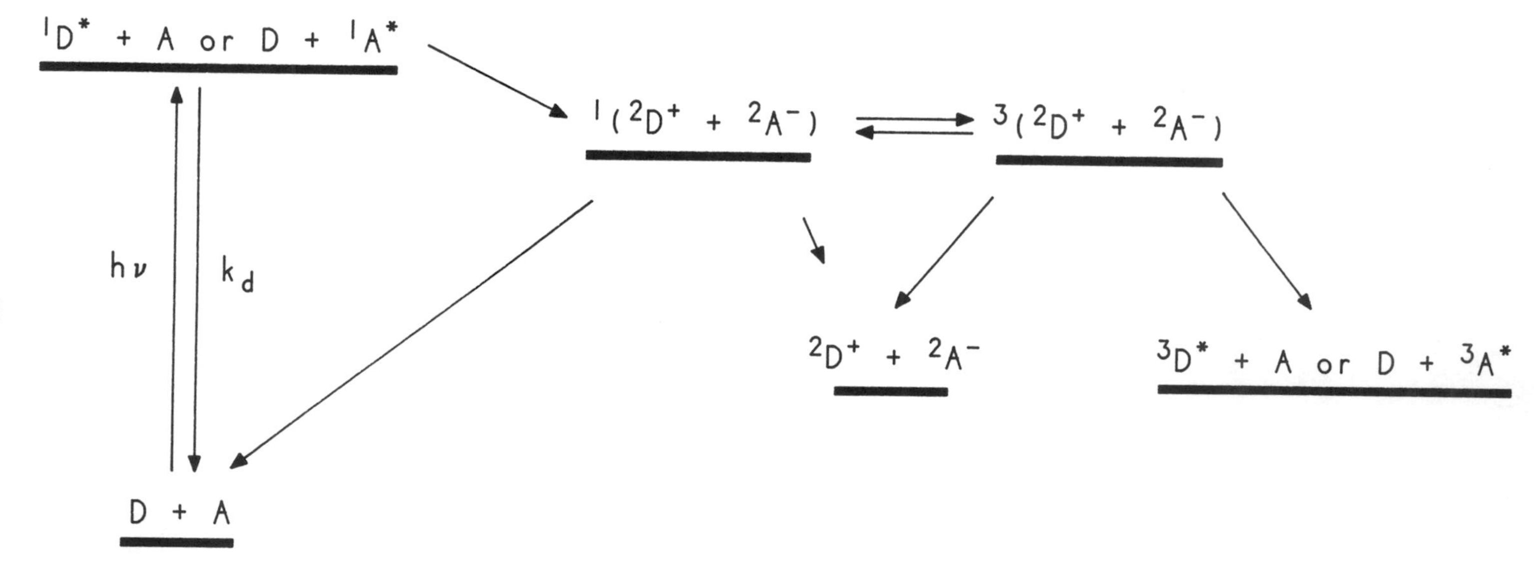

Figure 2.14 Electronic spin changes occurring during photoinduced electron transfer.

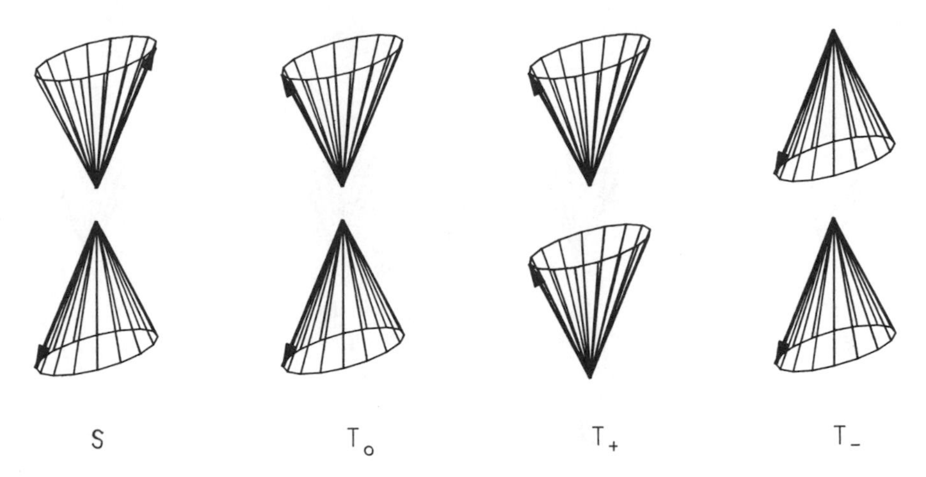

Figure 2.15 This vector model represents the coupled angular momenta of two electron spins.

electron may experience [i.e., $\omega \propto \mu_m \times H_0$, where μ_m is the magnetic moment of the spinning electron and H_0 is the strength of a magnetic field in Gauss (G)]. The source of the magnetic field may be a nearby electron, nucleus, or an external magnetic field. Depending on the strengths of these magnetic fields, a pure singlet state may rephase into a pure triplet state or vice versa.

2.6.2. Nuclear Spin

Atomic nuclei also possess an intrinsic spin and magnetic moment. As in the case of electrons, there are two possible projections of the spin vector along the z-axis. We shall now consider the interactions, or coupling, between the electron and nuclear spin.

Normally, spin coupling between nearby electrons is so weak that these interactions can be neglected. However, the magnetic moments of nearby nuclei can couple with electron spin and thereby induce singlet–triplet interconversions. This interaction is termed nuclear-electronic *hyperfine coupling*. Two mechanisms for an interconversion between a singlet and triplet state (intersystem crossing) are represented in Fig. 2.16. The first mechanism, a spin flip, depends on the exchange of spin angular momentum between an electron coupled with a nucleus. A magnetic field operating in the plane perpendicular to the z-axis induces the spin flip. Thus, the transitions between T_0 and T_+ or T_- states are accompanied by spin flips. The second mechanism involves a *spin rephasing* or *spin evolution*. The spin rephasing mechanism depends on a difference in precessional frequencies ($\Delta\omega$) of the spinning electrons.

In the absence of an external magnetic field, the major source for spin flipping and spin evolution may be hyperfine coupling arising from the interaction between the spins of unpaired electrons with the spins of nearby nuclei possessing magnetic

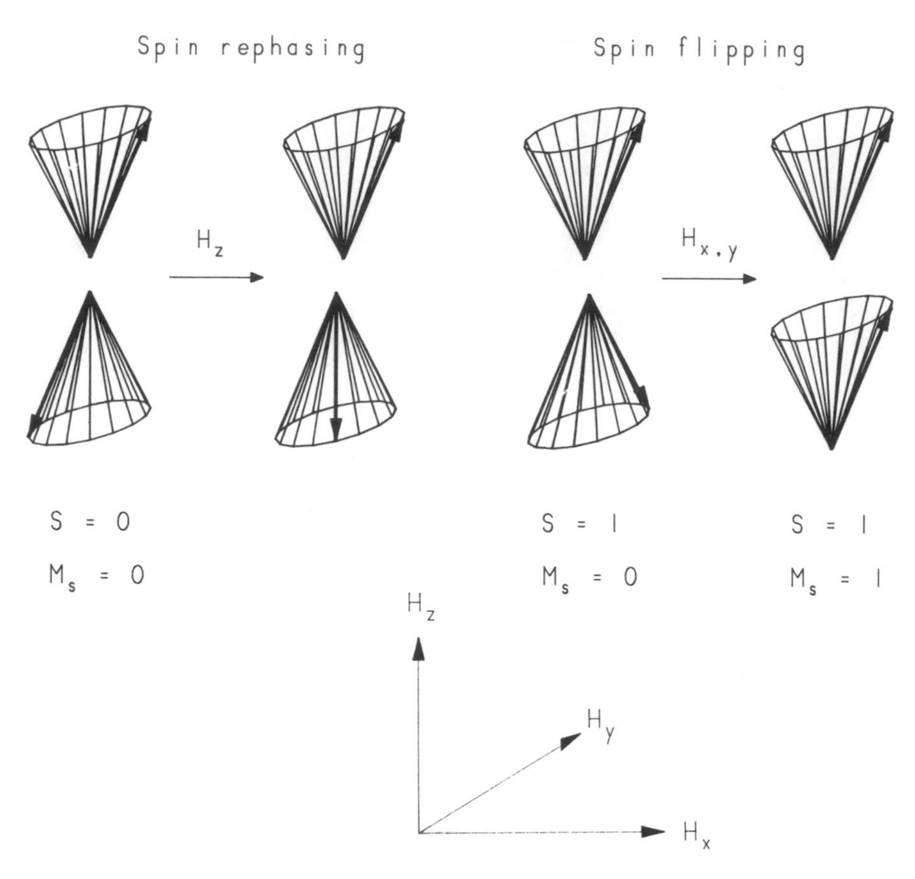

Figure 2.16 The spin rephasing and flipping mechanisms which bring about singlet-triplet interconversions. (Adapted from Fig. 8.12 in Turro, N.J. *Modern Molecular Photochemistry;* Benjamin/Cummings: Menlo Park, CA, 1978 with permission of the publisher. Copyright 1978 Benjamin/Cummings.)

moments (paramagnetic nuclei). The rates of precession of the two electrons differ by $\Delta\omega = \Delta a_i M_I$ where a_i is defined as the hyperfine coupling constant of the *i*th nucleus and M_I is its magnetic quantum number. Examples of paramagnetic nuclei (those with magnetic moments) are ^{1}H ($I = 1/2$), ^{13}C ($I = 1/2$), and ^{17}O ($I = 5/2$). a_i usually may range from a few to several hundred gauss. The energy due to hyperfine coupling is designated as ΔE_{hc}.

When we apply these concepts to organic radical ion pairs, we can regard spin flips and spin evolution via hyperfine interactions as the major mechanisms for intersystem crossing between singlet and the three isoenergetic triplet ion pairs (Fig. 2.16). The spin of the radical ion pair will not change until each ion experiences a different magnetic field set up by neighboring nuclei. Spin rephasing is determined by magnetic fields along the *x*-, *y*-, and *z*-axes; spin flips are largely influenced by fields within the *x*- and *y*-planes.

When the force of an external magnetic field is imposed on two spinning electrons coupled with the magnetic field of neighboring nuclei, we must take into account the fact that the external magnetic field may destroy the degeneracy between the energies of the three triplet states. The splitting of the energies of the triplet sublevels by the external magnetic field arises because of the *Zeeman interaction*. The energy difference between the sublevels is equal to $\beta_m \times g \times H_0$, where β_m, the Bohr magneton, has a value of 0.927×10^{-20} erg G^{-1}. In an unpaired electron, for example, $g = 2.003$. In a radical ion, g differs from the value for a free electron because of spin-orbital coupling. For organic radical ions, the difference in g values typically fall between 10^{-2} and 10^{-3}. It is the difference in g values, Δg, which accounts for the changes in spin multiplicity in a radical ion pair. This coupling, which arises from the Zeeman interaction, is known as the Δg mechanism. Hence, the splitting of the triplet sublevels is related to the strength of an applied magnetic field and to the magnitude of Δg.

We shall now show that singlet–triplet interconversions in radical ion pairs can be modulated by the Zeeman interaction. In developing a picture of singlet–triplet transitions, we must, however, include the effects of electron exchange, given by the exchange integral, J. It must be noted that singlet and triplet states should lie within ΔE_{hc} for transitions to occur. J depends on the distance separating the two ions and consequently may affect singlet–triplet transitions. The exponential dependence of J can be written as

$$2J = 2J^0 \exp(-\alpha d_{hc}) \tag{2.25}$$

J^0 is the exchange interaction at 7 Å, and d_{hc} is the distance where the radical ions must remain briefly to allow hyperfine-induced spin alignment to take place. Typically, $2J^0 = 1.89 \times 10^{10}$ G, and $1/\alpha = 0.468$ Å. For successful spin realignment, ΔE_{hc} should be greater than $2J$. Because J depends on separation distance (Table 2.2), we predict that exchange interactions should be more important for closely held ions such as exciplexes and contact ion pairs than for solvent-separated ions. J values must be evaluated in relationship to distance and the energy due to hyperfine-induced coupling. For $\Delta E_{hc} = 10$ G ($\sim 2.7 \times 10^{-6}$ kcal mol^{-1}), spin alignment will be favored only at distances where $d_{cc} \gtrsim 10$ Å ($J \sim 1.3 \times 10^{-6}$ kcal mol^{-1}).

Table 2.2 THE DEPENDENCE OF J ON SEPARATION DISTANCE[a]

d_{cc} (Å)	J (kcal mol^{-1})
4	0.48
5	0.069
6	0.007
7	8.1×10^{-4}
8	9.7×10^{-5}
9	1.1×10^{-5}
10	1.3×10^{-6}

[a] J calculated with Eq. 2.25; $2J^0 = 1.89 \times 10^{10}$ G; $1/\alpha = 0.468$ Å; 1 G = 2.67×10^{-7} kcal mol^{-1}.

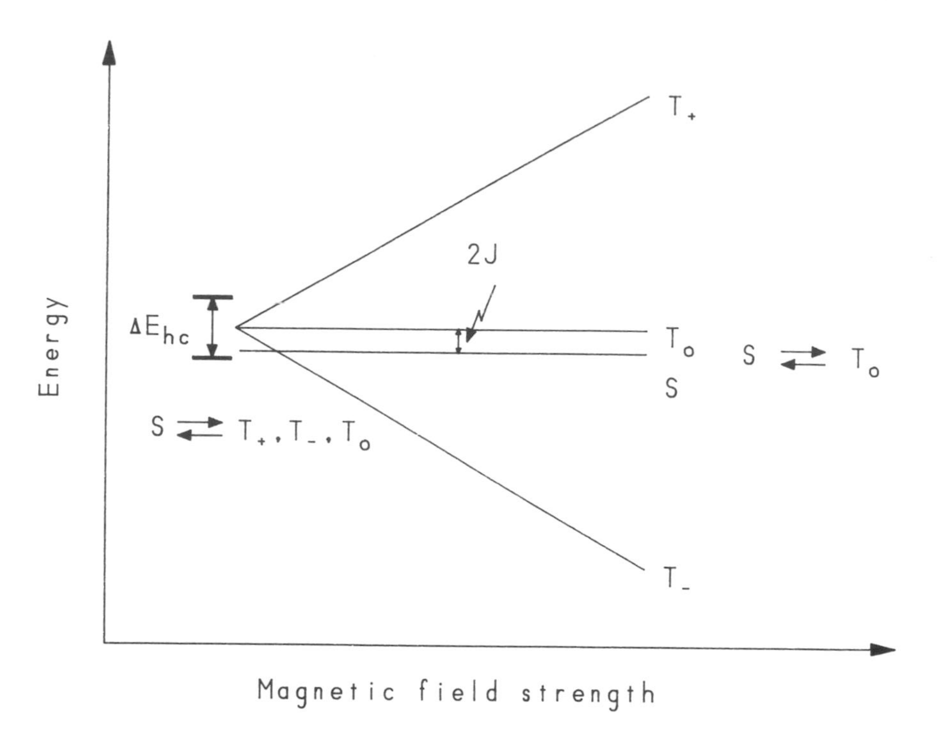

Figure 2.17 Singlet-triplet transitions when J is small. (Adapted with permission from Kavarnos, G. J. *Top. Curr. Chem.* **1990,** *156,* 21. Copyright 1990 Springer-Verlag.)

At closer distances, electronic interaction "overwhelms" hyperfine-induced coupling. With larger hyperfine coupling, say $\Delta E_{hc} \sim 200$ G ($\sim 5.4 \times 10^{-5}$ kcal mol^{-1}), spin alignment can occur at closer distances, say $d_{cc} \sim 9$ Å.

Let us now consider the effects of an external magnetic field on singlet–triplet interconversions. We will assume an extreme case where there is virtually no electron exchange between the radical ions ($J \sim 0$). Let us assume, momentarily, that there is no external magnetic field (Fig. 2.17), so that the Zeeman interaction is also negligible. Under these conditions, S, T_+, T_- and T_0 states are almost degenerate. Consequently, interconversions between singlet and triplet states are energetically allowed. Now if we supply a weak magnetic field ($H_0 \sim 100$ to 500 G), the states begin to split. The frequency of spin flips required for transitions between T_+, T_-, and S states starts to fall because of the increasing energy barriers between these states. Spin rephasing due to hyperfine interactions begins to assume more importance for transitions between the T_0 and S states. In stronger magnetic fields, say $H_0 \sim 10{,}000$ G, the differences in g values may be more pronounced ($\Delta g > 10^{-2}$). The spins now precess at different rates as determined by the strength of the imposed magnetic field and the differences in g values:

$$\omega = \left(\frac{\beta_m}{\hbar}\right) \Delta g H_0 \pm \Delta a_i M_I \tag{2.26}$$

Equation 2.26 states that in the presence of an external magnetic field, the precessional frequency is determined by the *g*-factor difference and differences in hyperfine coupling constants due to the magnetic fields of nearby nuclei ($\hbar = h/2\pi$). Hyperfine coupling due to nearby nuclei spins can be additive or subtractive relative to the external magnetic field. In the absence of hyperfine coupling, the second term on the right drops out, so that the precessional rate is due only to differences in the *g*-factors. Although in strong magnetic fields, the frequencies of spin flips are reduced because of the increased energy separation, transitions between T_0 and S states are enhanced because these large fields favor spin evolution.

In the case where there is substantial electron exchange between the radical ions, $2J > \Delta E_{hc}$ (Fig. 2.18). At zero or low magnetic fields, there are no singlet–triplet transitions since the hyperfine mechanism is not possible. At some higher field, these transitions may become possible, provided that crossings between these states exist, as shown in Fig. 2.18.

We conclude that the fate of ions generated within a solvent cage is related to spin flip and spin evolution mechanisms. These mechanisms, in turn, depend on the separation distance and the strength of an external magnetic field. When singlet or triplet states are quenched by ground states, the overall spin of the radical ion pair will be singlet or triplet, respectively. Since the pathways available to the radical ion pair must be consistent with spin conservation, singlet radical ion pairs

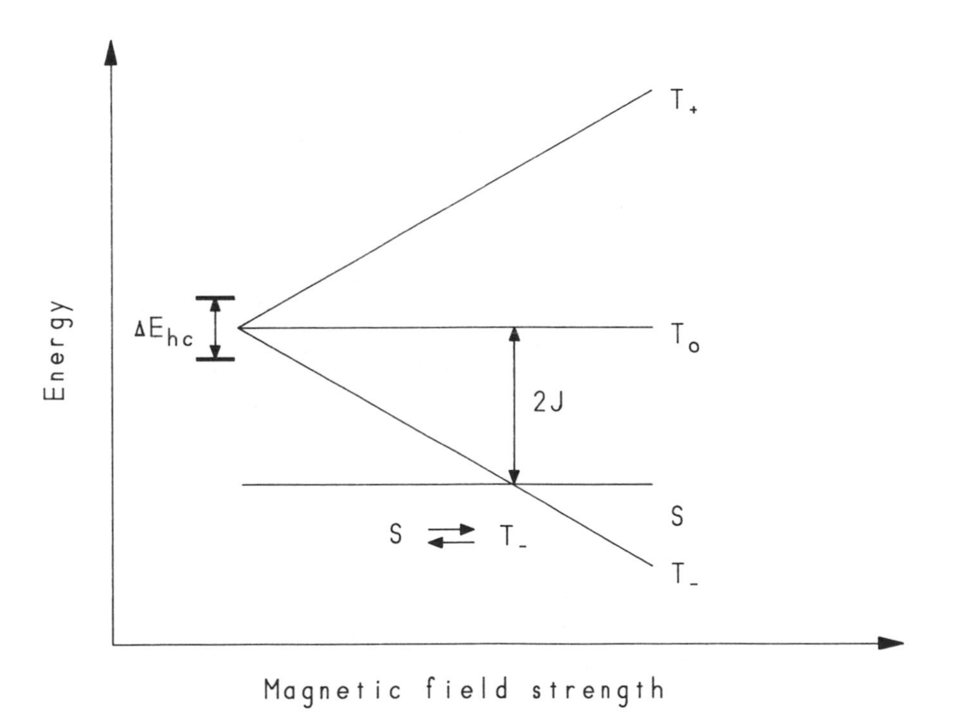

Figure 2.18 Singlet-triplet transitions when $2J > \Delta E_{hc}$.

should undergo electron return, whereas a triplet radical ion pair is likely to recombine to form a triplet state species or to dissociate from the solvent cage. Once formed, the two radical ions may also separate slightly and experience spin evolution. When the radicals reencounter, they may be in a different spin state. Thus, in the presence of an external magnetic field, quenching of singlet states may give triplet products, and likewise quenching of triplet states lead to singlet products. Accordingly, the ratio of electron return to ion dissociation should be related to the strength of an external magnetic field, which induces spin evolution.

Electron return between ions within an ion pair formed by reencounters occurs on a microsecond timescale (homogeneous recombination as distinct from the more rapid geminate recombination). Thus, on the basis of the splitting of a triplet state into three sublevels, one predicts that the statistical probability of generating triplet and singlet ion pairs should be 75 and 25%, respectively. The physical picture of two time domains in triplet recombination is one that has been developed by Weller.[19] For example, the quenching of singlet pyrene by *N,N*-dimethylaniline in polar solvents has been investigated in the absence and presence of a magnetic field of about 500 G. This system can be described as shown in 2.27:

$$\begin{array}{ccccccc} {}^1A^* + D & \leftrightharpoons & {}^1[{}^1A^* + D] & \rightarrow & {}^1[{}^2A^{\cdot -} + {}^2D^{\cdot +}] & \leftrightharpoons & {}^3[{}^2A^{\cdot -} + {}^2D^{\cdot +}] \rightarrow {}^3A^* + D \\ & & & & \downarrow \qquad \searrow\!\!\nwarrow & & \swarrow\!\!\nearrow \\ & & & & A + D \qquad {}^2A^{\cdot -} & + & {}^2D^{\cdot +} \end{array} \tag{2.27}$$

A = aromatic hydrocarbon, D = aromatic amine. In the absence of a magnetic field, time-resolved spectroscopy reveals the decay of the radical ion pair within two time domains corresponding to the primary and secondary geminate recombination processes to form triplet pyrene ($^3A^*$). Whereas electron return is favored for the singlet state, ion dissociation and triplet recombination are the dominant pathways for the triplet ion pair. Recall that the singlet and three triplet spin states are approximately at the same energy levels in the absence of the magnetic field, and that the triplet radical ion pair is predominantly formed by hyperfine-induced intersystem crossing within the ion pair. Thus, when an external magnetic field of 500 G is applied, the yield of triplet pyrene is reduced to 50%. The magnetic field splits the energy levels of the three triplet sublevels to reduce intersystem crossing to the triplet (the Zeeman interaction).

2.6.3. Electron Spin Resonance (ESR) and Electron Nuclear Double Resonance (ENDOR)

In electron spin resonance (ESR), advantage is taken of the fact that the magnetic moment of a spinning electron corresponds to two different energy states in a magnetic field of strength H_0. A transition between the two states can occur if the system is exposed to electromagnetic radiation with a frequency, ω:

$$\Delta E = \hbar\omega = g\beta H_0 \tag{2.28}$$

If a free electron is placed within a strong magnetic field and the electromagnetic radiation is "tuned," a single absorption of the radiation will take place when its

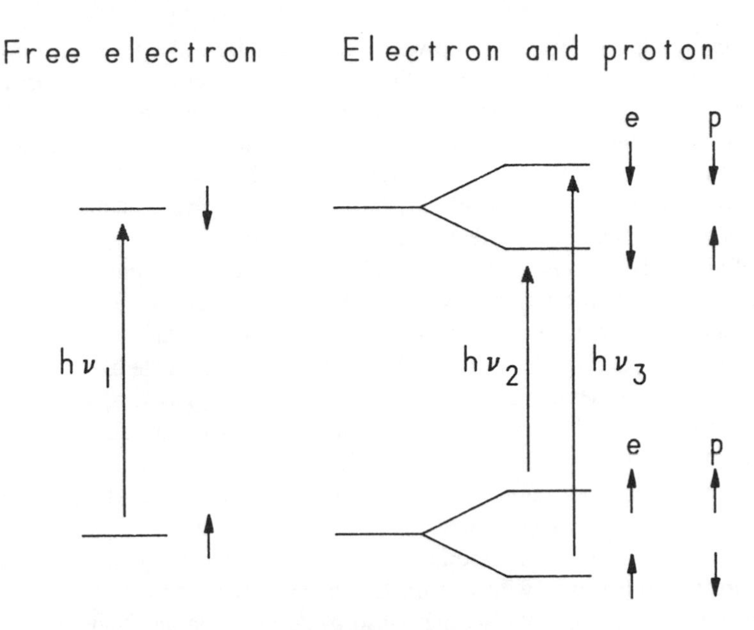

Figure 2.19 An energy diagram is shown depicting the interaction of a radical with one unpaired spin (S = 1/2) and one nucleus with spin M_I = 1/2.

energy is equal to the energy difference between the states. If the electron is associated with a proton, two transitions are possible, each corresponding to two spin states of the proton (Fig. 2.19). Thus, in a hydrogen atom, two absorption signals are observed, and the difference between them is related to the magnitude of the hyperfine coupling constant, a_H. In organic molecules, the electron-nuclear spin coupling is more complex, leading to a multiplet of absorption patterns. In molecules with delocalized π-orbitals, the unpaired electron may be delocalized throughout the molecule allowing for more electron–proton coupling.

The energy levels associated with an ESR transition are about equally populated with a slightly greater population in the lower level. Transitions between these levels can thus be stimulated by the application of microwave energy. A transition from the lower to upper level and from the upper to lower involves absorption and emission, respectively. The ESR signal results because of the slightly greater population in the lower level. Under continuous radiation, we would predict that eventually the population of the two energy levels should equalize with the result that there would be no ESR signal. But this does not occur because of spin relaxation processes from the higher energy state. These pathways are radiationless transitions between the electron–nucleus system and the thermal motion of the surrounding lattice. Spin relaxation may entail electron spin, nuclear spin, and cross-relaxation processes (cross-relaxation refers to simultaneous electron and nuclear spin relaxations). During an ESR experiment with low microwave power, spin relaxation processes from the upper to lower state can normally maintain a thermally equilibrated population of the two levels. This allows for an observable ESR signal.

However, at higher microwave powers, transitions from the lower to upper level become competitive with spin relaxation. The result is an equalization of the population of both levels (this is called saturation). Consequently, the ESR signal decreases. For this reason, the microwave power should be sufficiently low.

In ESR, nuclei with magnetic moments (e.g., H, F, and ^{13}C) induce hyperfine coupling. The usefulness of ESR lies in its ability to give valuable structural information on radical ion transients and excited states.

Unfortunately, because of the short lifetime of radical ion intermediates in photoinduced electron transfer, relatively few ESR experiments have been undertaken in this area. Typically, one requires a radical ion steady-state concentration of at least $\sim 10^{-8}$ M. Because the yields and lifetimes of ion products in solution at room temperature are generally quite low because of electron return, the steady-state concentrations needed to observe radical–ion intermediates are difficult to achieve.

A modification of ESR has successfully been introduced to increase its resolution and sensitivity. This technique, which is called *electron nuclear double resonance,* or ENDOR, has been used to investigate charge-transfer intermediates. In ENDOR, nuclear spin transitions are introduced by using a radio frequency (rf) field. On exposure to radio waves with certain frequencies, the magnetic moments of the nuclei in the parallel orientation may absorb the radiation and undergo a spin flip to the antiparallel alignment. After absorption, the nucleus in the higher energy state relaxes and returns to the lower state by spin relaxation. According to Fig. 2.20, the interaction of a radical with one unpaired electron spin ($S = 1/2$) and one nucleus with $M_I = 1/2$ leads to splitting to give four energy levels. A microwave-induced transition—the ESR transition—and the electron spin-lattice relaxation are shown (there are other relaxation processes—nuclear spin and cross-relaxation—but these are assumed to occur on a longer time scale). In ESR, since increasing the microwave power eventually populates the upper and lower levels about equally *(saturation),* there is a substantially reduced ESR signal. In ENDOR, the nuclear spin transitions are excited with a strong rf field (Fig. 2.20). Both of the upper nuclear spin levels should then be populated equally (since nuclear spin relaxation is not competitive with the transition). The population of the upper level of the ESR transition then becomes depleted with respect to the lower level. The net result is a dramatic increase in the ESR signal.

In ENDOR, resolution is also significantly increased over ESR. In ESR, the linewidth is typically $\sim$10 G, as compared to only $\sim$0.01 G in ENDOR. Another important advantage of ENDOR is that the spectra are considerably simpler than in ESR.

The following example is given to illustrate the combined use of ESR and ENDOR to detect charge-transfer intermediates in photoinduced electron transfer. This reaction is the photoirradiation of magnesium or zinc tetrakis (4-sulfonatophenyl)-porphyrin (MgTPPS and ZnTPPS) in a glassy matrix in the presence of the electron acceptor, $K_3Fe(CN)_6$.[20] These metalloporphyrins, like many porphyrins, are electron-rich molecules with relatively low redox potentials. In the presence of an appropriate electron acceptor, they readily undergo one-electron oxidation, e.g.,

$$^{1,3}[\mathrm{MgTPPS}]^* + \mathrm{A} \rightarrow \mathrm{MgTPPS}^{\dot{+}} + \mathrm{A}^{\dot{-}} \qquad (2.29)$$

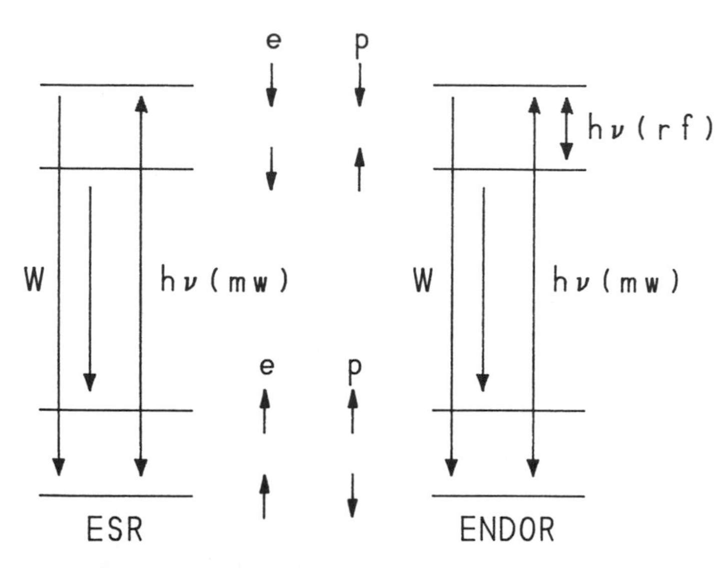

Figure 2.20 Energy diagram comparing the transitions in ESR and ENDOR. In ESR, transitions occur with sufficiently low microwave power to avoid saturation. Electron spin-lattice relaxation processes, denoted by W, sustain a thermal equilibrium of spin states. With increasing microwave power, the populations of the lower and upper states become equal, the result of which is a decrease in the microwave signal. In ENDOR, a radiofrequency field induces transitions between the upper nuclear spin states. As the populations of these nuclear spin states become equal, the upper level associated with the ESR transition becomes depopulated, with a resulting enhancement of the ESR signal. The ENDOR response depends on the relative magnitudes of cross relaxation pathways where both electron and nuclei experience spin flipping.

Electron transfer from singlet and triplet states of the porphyrins is exothermic: $\Delta G_{el} = \sim -35$ kcal mol^{-1} and $\Delta G_{el} = \sim -23$ kcal mol^{-1}, respectively. Since the lifetime of ion products in solution is usually too small for ESR detection because of rapid electron return, the objective in this work was to attempt to detect the intermediates in a solid matrix. Figure 2.21 shows the effects of photoillumination of MgTPPS and $K_3Fe(CN)_6$ with an argon ion laser in the cavity of an ESR spectrometer. The experiment was carried out by irradiating the sample during cooling to 100 K. The solid line in Fig. 2.21 represents the ESR signal due to ^{3}[MgTPPS]* in the absence of the acceptor. Irradiation of the porphyrin in the presence of the acceptor while the reaction vessel was being cooled led to the disappearance of the triplet porphyrin ESR signal and the appearance of a doublet signal due to the radical cation of MgTPPS (dotted line).

That quenching by electron transfer is the principal mechanism was further supported by an ENDOR experiment involving the photoillumination of ZnTPPS in the presence of $K_3Fe(CN)_6$. With microwave power of 2 mw and an rf power of 100 w, the spectra shown in Fig. 2.22 were obtained. The doublet signal of the radical cation of the porphyrin arising from the oxidation of ZnTPPS by I_2 is shown together with the signal arising from the photoillumination of ZnTPPS with $K_3Fe(CN)_6$.

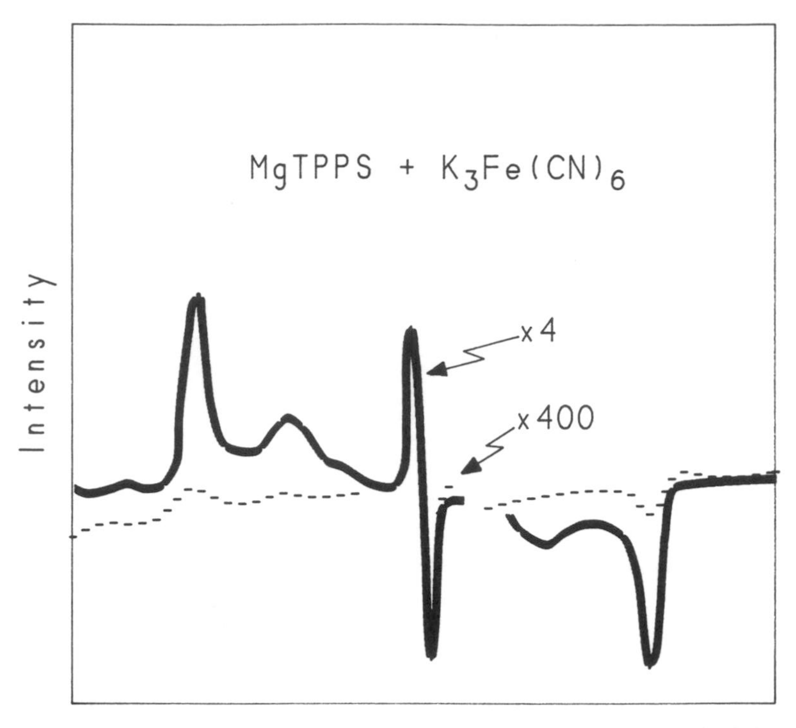

Figure 2.21 ESR signals in the photoillumination of MgTPPS and $K_3Fe(CN)_6$ with a laser in the cavity of an ESR spectrometer. See text for explanation. (Adapted with permission from van Willigen, H. and Ebersole, M. H. *J. Am. Chem. Soc.* **1987,** *109,* 2299.)

The appearance of the radical cation of ZnTPPS supports a quenching mechanism involving electron transfer. Note the improved resolution and detail of the ENDOR spectra. The spectra were similar for MgTPPS. These experiments illustrate the ability of ESR and ENDOR experiments to assist in identifying the products resulting from photoinduced electron transfer.

2.6.4. Chemically Induced Nuclear Polarization

The observation of enhanced or polarized signals in the nuclear magnetic resonance (NMR) spectra of free-radical reactions in the late 1960s was a significant development in photochemistry—one that would eventually prove to be highly useful in elucidating the complex mechanisms in photoinduced electron transfer. In conventional NMR spectroscopy, a substance is placed within a strong magnetic field, H_0, and subsequently exposed to an rf field. As we mentioned in previous sections, nuclei with spin may align in a parallel or antiparallel orientation with respect to the external magnetic field. As a result, there is a slight energy difference between the nuclear spin states: nuclei with spins in the parallel alignment are slightly more stable. The energy required to induce a nuclear spin flip is related to the strength

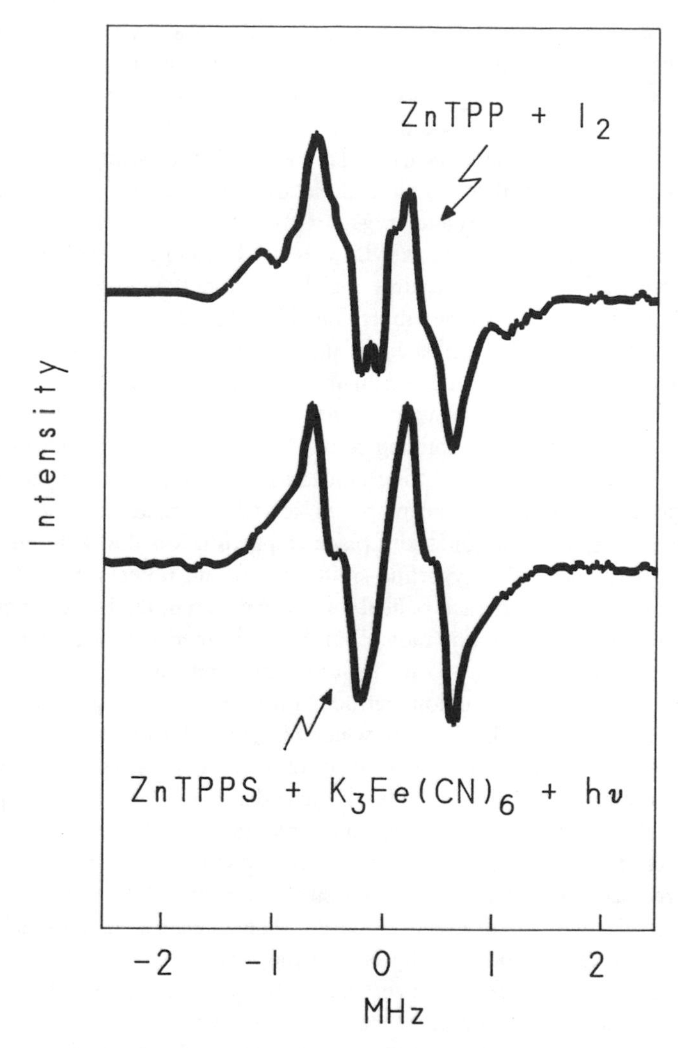

Figure 2.22 ENDOR experiment involving the photoillumination of ZnTPPS in the presence of $K_3Fe(CN)_6$. See text for discussion. (Adapted with permission from van Willigen, H. and Ebersole, M. H. *J. Am. Chem. Soc.* **1987,** *109,* 2299.)

of H_0 (the Zeeman interaction). Thus, at higher magnetic fields, the energy difference between the spin states is larger, and the exciting radio frequency must be adjusted to induce the transition. The term "resonance" implies that the external magnetic field and the rf field are matched to induce a spin flip. Since the magnetic field to which a nucleus is exposed is also affected by the spins of neighboring nuclei, the nuclei in a molecule may flip at different combinations of H_0 and rf field, giving rise to the complex spectra characteristic of NMR spectrometry.

If one photolyzes a sensitizer and quencher within the cavity of an NMR spectrometer, the NMR signals may display enhanced absorption or emission. The direction of these polarized signals can be used not only to confirm the electron-transfer pathway but also to probe the fine details of the mechanism.

How do these polarized signals arise? To answer this question, let us review the simple radical ion model of photoinduced electron transfer [i.e., the generation of radical ions within a solvent cage (Fig. 2.14)]. We assume that when a singlet electron donor or acceptor interacts with a neutral organic quencher, the overall spin of the radical ion pair immediately formed is singlet. Similarly, when a triplet sensitizer is quenched by another molecule, a triplet radical ion pair is formed. After its formation, a radical ion pair may undergo intersystem crossing to generate a pair with a different spin state. From the discussion in Section 2.6.1, we are reminded that the strength of a neighboring magnetic field (H_0), the g-factors of each electron, and the hyperfine splitting constants can induce electronic spin changes. If the g-factors and hyperfine splitting constants of each radical ion are different, we can expect to have a mechanism for singlet–triplet interconversion. With different g-factors, spin evolution takes place depending on the differences in the precessional frequency. The hyperfine splitting constant reflects the effects of nuclear spin on the electron. If each radical ion has a different nuclear configuration—let us say, for example, that one radical ion is substituted with a proton—then the precessional rate of the electron on the substituted radical should differ in phase from the unsubstituted radical ion partner. From Fig. 2.16, we recall the spin rephasing and spin flip mechanisms by which singlet and triplet states interconvert.

For there to be effective coupling of electronic and nuclear spins, the separation distance between the radicals should be of the order of molecular dimensions (i.e., 6–10 Å). This is so because J, the exchange integral, is distance dependent, as we noted in Section 2.6.2. When the radicals are separated by distances less than 6 Å, as they may be within an exciplex, J assumes a large value, and, consequently, the singlet–triplet energy difference is large. The nuclear-spin induced coupling mechanism for singlet–triplet mixing is then inoperative.

In practice, CIDNP is performed with an applied magnetic field sufficiently large so that the Zeeman interaction can separate the T_+ and T_- states from S_0 and T_0 (Fig. 2.17). At these energies, mixing occurs between S_0 and T_0 and not between S_0 and T_+ and T_-. The polarization of NMR spectra arises directly from the shifts in the Boltzmann distribution of nuclear spin populations when singlet–triplet interconversion takes place. These represent changes in the equilibrium position from a Boltzmann to a non-Boltzmann distribution of nuclear spin states. Non-Boltzmann distributions represent polarized nuclear spin populations, which are directly responsible for the enhanced absorption of emission observed in CIDNP.

To get an intuitive idea how spin sorting can take place by the shifts in Boltzmann populations, consider intersystem crossing taking place within a singlet radical ion pair formed by electron transfer (Fig. 2.23). Assume that the g-factors and hyperfine splitting constants on each radical ion are different. This will be the case, for instance, if one of the radical ions is substituted with a proton and the other is not. These conditions will favor intersystem crossing from the singlet to a triplet radical

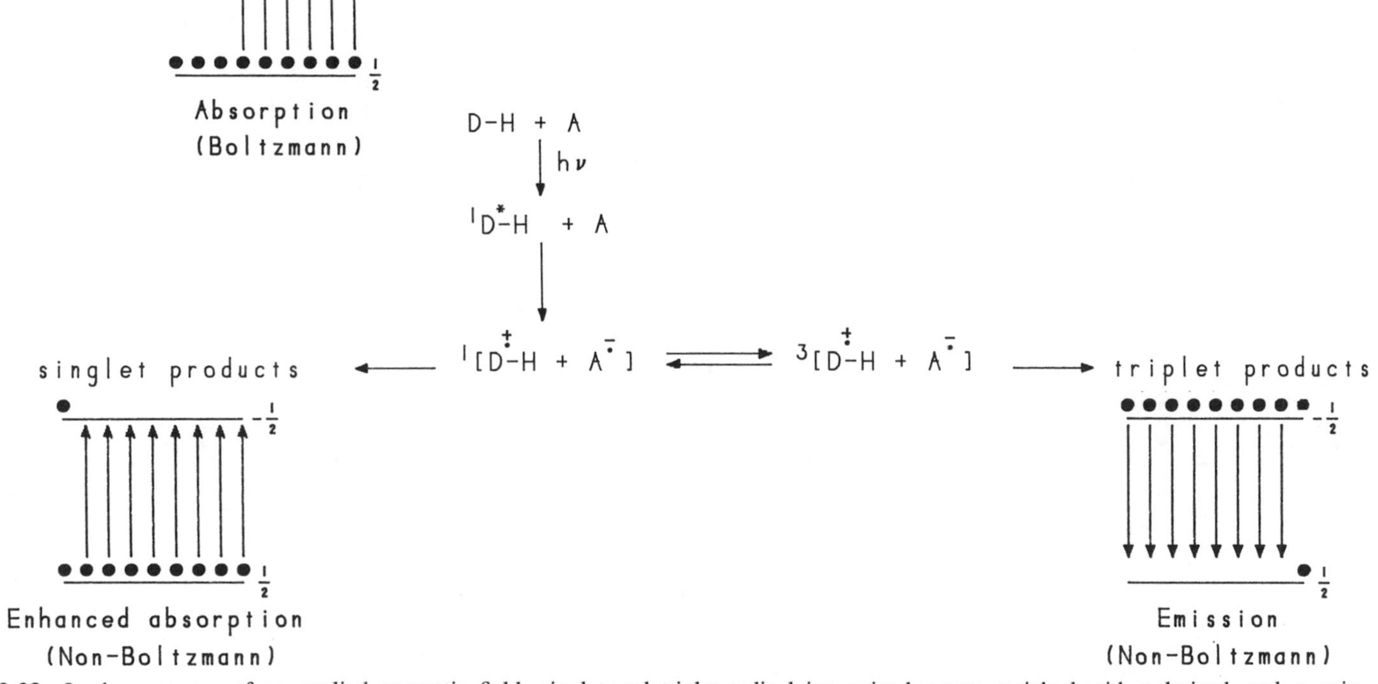

Figure 2.23 In the presence of an applied magnetic field, singlet and triplet radical ion pairs become enriched with polarized nuclear spin populations, which are directly responsible of the enhanced absorption or emission observed in CIDNP. This leads to spin sorting where "polarized" singlet and triplet "products" are formed. In this example, the nuclear spin population of the singlet state leads to enhanced NMR absorption in singlet products and enhanced emission in triplet products.

ion pair brought on by differences in the precessional frequencies of the electron spins. Initially, there is a Boltzmann distribution of nuclear spins ($M_I = \pm 1/2$) with a slight predominance of $M_I = +1/2$. In an NMR spectrometer, this would lead to a normal NMR signal. Assume that the precessional rate of the electron in the proton-substituted radical in the $-1/2$ state is faster than the rate of the $+1/2$ state. Then the radical ion pair containing the proton with $M_I = -1/2$ should reach a triplet spin configuration faster than the radical with the proton having spin $M_I = +1/2$. Accordingly, the triplet states will be enriched with spins of $M_I = -1/2$, whereas the singlet states will be enriched with $M_I = +1/2$. The triplet states now enriched with $M_I = -1/2$ will have a greater tendency to escape from the radical cage and undergo chemical transformation or eventual homogeneous recombination. But the singlet radical pairs, now enriched with $M_I = +1/2$, will have greater tendency to participate in electron return. As electron return takes place in competition with ion dissociation, the products arising from triplet state reactions will have an overabundance of $M_I = -1/2$ nuclear spins. The singlet reactants arising from electron return will have an abundance of $M_I = +1/2$ states. This simplified description assumes that spin relaxation is much slower than the radical ion reactions. In fact, spin relaxation from one nuclear spin state to the other lasts at least 1 s. This is much longer than typical radical ion reactions.

A similar argument can be applied to a triplet ion pair formed directly from the quenching of a triplet excited state.

From a theoretical viewpoint, the polarization of the products formed from escaped radicals should have a different sign than the polarization of the products or reactants formed from processes taking place within the solvent cage. In practice, the normal NMR signals will be enhanced (absorption) or suppressed (emission). Since spin polarization may last several tens of seconds, and radical ion processes are typically much faster, the spin polarization of the products can be detected by NMR and can be used to great advantage to study these fast radical ion reactions. The polarized absorption and emission signals observed in CIDNP experiments often display complex splitting patterns owing to nuclear spin–spin coupling. The resulting multiplets can be used for structural information regarding the radical ion intermediates.

The reactions shown below constitute a possible pathway by which CIDNP signals can arise when a donor and acceptor are irradiated within the cavity of an NMR spectrometer. It is assumed here that A is a singlet sensitizer, that the triplet state of D is energetically accessible from the triplet ion state, and that polarization is observed in D and in a product P: The right side up dagger represents nuclear spin polarization; the upside down dagger represents polarization of opposite sign:

$$A \xrightarrow{h\nu} {}^1A^* \tag{2.30}$$

$${}^1A^* + D \rightarrow {}^1[A^{\overline{\cdot}} + D^{\overset{+}{\cdot}}] \rightleftharpoons {}^3[A^{\overline{\cdot}} + D^{\overset{+}{\cdot}}] \tag{2.31}$$

$${}^1[A^{\overline{\cdot}} + D^{\overset{+}{\cdot}}] \rightarrow A + D\dagger \tag{2.32}$$

$$^3[\mathrm{A}^{\overline{\cdot}} + \mathrm{D}^{\dot{+}}] \rightarrow \mathrm{A} + {}^3\mathrm{D}^* \tag{2.33}$$

$$^3\mathrm{D}^* \rightarrow \mathrm{P}\!\uparrow\!\downarrow \tag{2.34}$$

The polarization that arises in D will be opposite in sign to the polarization in the product. The reaction pathway shown above requires that the triplet of D lie at a lower level than the energy of the ion-pair state. This feature of the pathway underscores the importance of energetics in determining the character of the nuclear polarization. When analyzing CIDNP patterns, it is important that all intermediates be accessible in thermodynamically allowed pathways.

For an example of an application of CIDNP, consider the photoconversion of the anthracene dimer (**8**) to anthracene (**9**) in the presence of the triplet sensitizer chloranil (CA)[20]:

hν, CA
CH_2Cl_2
2
9
8
O
Cl
Cl
Cl
Cl
O
CA
(2.35)

This reaction is an example of a [4+4] cycloreversion (Chapter 3). Photoexcited chloranil is known to form a singlet state that rapidly undergoes efficient intersystem crossing to a triplet. A good oxidizer, triplet chloranil can oxidize **8** to its radical cation, which subsequently fragments into anthracene. Calculations show that $\Delta G_{el} = -32$ kcal mol^{-1}. Energy transfer from singlet or triplet chloranil to the dimer is ruled out as being too endothermic. When the dimer and chloranil are irradiated in an NMR cavity, polarized NMR signals are observed: The NMR signals of protons substituted on the dimer are enhanced, whereas for the protons on anthracene, emission is observed (Fig. 2.24). CIDNP signals *are not* observed when chloranil and anthracene are irradiated in the absence of the dimer. These

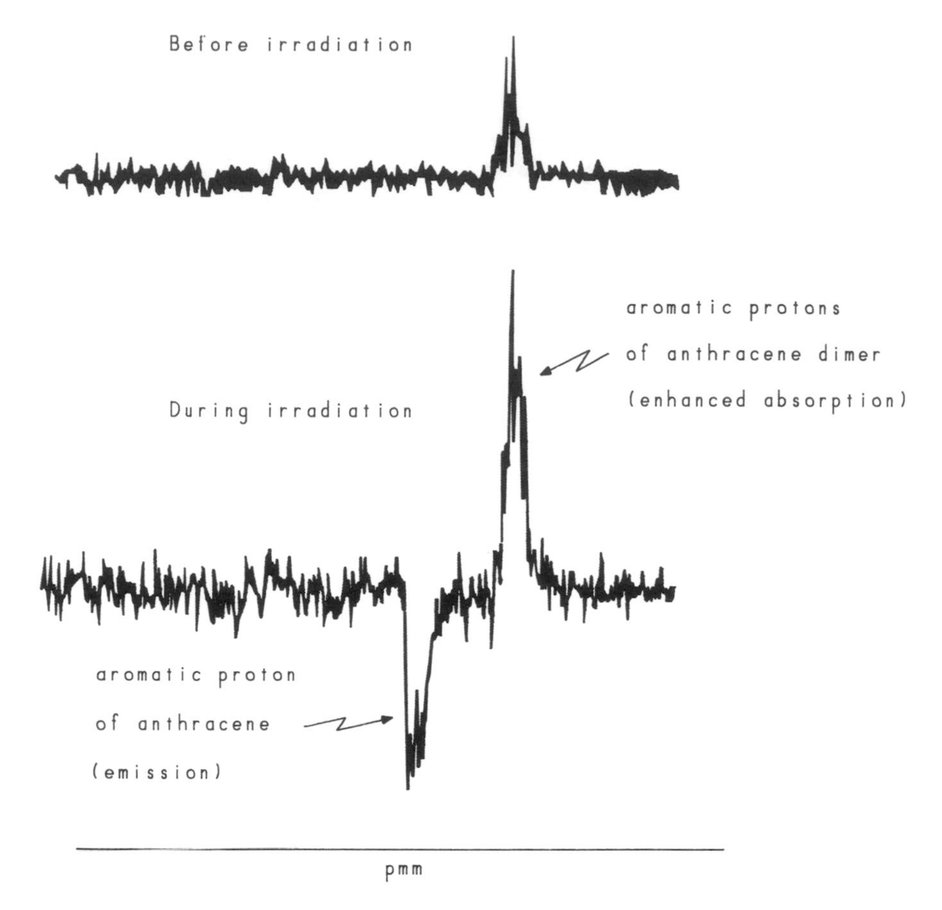

Figure 2.24 CIDNP signals of anthracene dimer (**8**) and anthracene (**9**) in the presence of chloranil preceding and during irradiation with light. Proton NMR signals are due to the aromatic hydrogens of **8** and **9**. (Adapted with permission from Barber, R. A., de Mayo, P., Okada, K., and Wong, S. K. *J. Am. Chem. Soc.* **1982,** *104*, 4995. Copyright 1982 American Chemical Society.)

observations are consistent with a mechanism involving hyperfine-induced intersystem crossing as shown below:

$$^{3}[CA^{\cdot-} + 8^{\cdot+}] \rightleftharpoons {}^{1}[CA^{\cdot-} + 8^{\cdot+}]$$

$$^{3}[CA^{\cdot-} + 8^{\cdot+}] \rightarrow CA + 9^{\dagger}$$

$$^{3}[CA^{\cdot-} + 8^{\cdot+}] \xrightarrow{\text{triplet recombination}} CA^{\cdot-} + {}^{3}9^{*} + 9^{\cdot+}$$

$$^{1}[CA^{\cdot-} + 8^{\cdot+}] \xrightarrow{\text{singlet recombination}} CA + 8^{\dagger} \qquad (2.36)$$

Note that the dimer is formed by electron return within the singlet ion pair. However, the opposite polarizations on anthracene can be due to (1) bond cleavage after ion dissociation, (2) fragmentation within the triplet ion pair, or (3) triplet recombination. The triplet recombination is thermodynamically allowed since the triplet energy of anthracene is ~4 kcal mol^{-1} below the energy of the ion-pair state. The observation of CIDNP signals supports an electron-transfer mechanism.

CIDNP has proven to be an invaluable tool in the study of triplet recombination processes. In the quenching of high-lying singlet states, it is not uncommon to observe CIDNP signals suggestive of triplet recombination. This is because the low-lying energy levels of triplet states can readily be populated from higher-lying singlet states. When this occurs, triplet reactions of the newly formed state give rise to CIDNP signals with opposite polarization with respect to the signals from the singlet manifold. A problem at the end of this chapter is given to illustrate the analysis of triplet recombination by CIDNP.

Suggestions for Further Reading

The Nature of Charge-Transfer Intermediates

Suppan, P. *Chimia* **1988,** *42,* 320. Presents an interesting viewpoint of the role of solvent polarity and Coulombic effects.

Ion Pairs

Bagdasar'yan, K. S. *Russ. Chem. Rev.* **1984,** *53,* 623. A thorough look at the solvent-cage model.

Weller, A. *Z. Phys. Chem. N. F.* **1982,** *130,* 93. Weller discusses the role of solvent in stabilizing ion pairs and exciplexes.

Exciplexes

Davidson, R. S. In *Advances in Physical Organic Chemistry,* Gold, V., and Bethell, D., eds. Academic Press, London, 1983, Vol. 19, p. 1. A review.

Forster, T. In *The Exciplex,* Gordon, M., and Ware, W. R. eds. Academic Press, New York, 1975, p. 1.

Kavarnos, G. J., and Turro, N. J. *Chem. Rev.* **1986,** *86,* 401.

Lewis, F. D. *Acc. Chem. Res.* **1979,** *12,* 152. A review of stilbene exciplexes.

Mataga, N. In *The Exciplex,* Gordon, M., and Ware, W. R., eds. Academic Press, New York, 1975, p. 113.

Mattes, S. L., and Farid, S. *Science* **1984,** *226,* 917.

Turro, N. J. In *The Exciplex,* Gordon, M., and Ware, W. R., eds. Academic Press, New York, 1975, p. 165. An interesting look at excimers and exciplexes.

Weller, A. In *The Exciplex,* Gordon, M., and Ware, W. R., eds. Academic Press, New York, 1975, p. 23.

The Energetics of Exciplex and Ion-Pair Formation

Weller A. *Z. Phys. Chem. N. F.* **1982,** *133,* 93.

Flash Spectroscopy

Arimitsu, S., Masuhara, H., Mataga, N., and Tsubomura, H. *J. Phys. Chem.* **1975,** *79,* 1255.

Darwent, J. R., Douglas, P., Harriman, A., Porter, G., and Richoux, M.-C. *Coord. Chem. Rev.* **1982,** 83.

Delaire, J. A., and Faure, J. In *Photoinduced Electron Transfer. Part B. Experimental Techniques and Medium Effects,* Chanon, M., and Fox, M. A., eds. Elsevier, Amsterdam, 1988, Chapter 2.1.

Fouassier, J. P., Lougnot, D. J., Paverne, A., and Wieder, F. *Chem. Phys. Lett.* **1987,** *135,* 30.

Hilinski, E. F., and Rentzepis, P. M. *Acc. Chem. Res.* **1983,** *16,* 224.

Kleinermanns, K., and Wolfrum, J. *Angew. Chem. Int. Ed. Engl.* **1987,** *26,* 38.

Rentzepis. P. M. *Science* **1982,** *218,* 1183.

Simon, J. D., and Peters, K. S. *Acc. Chem. Res.* **1984,** *17,* 277.

Zewail, A. H. *Science* **1988,** *242,* 1645. This paper discusses femtosecond spectroscopy, which was not covered in this chapter.

Resonance Raman Spectroscopy

Hutchinson, J. A., DiBenedetto, J., Hilinski, E. F., Hopkins, J. B., and Rentzepis, P. M. *SPIE, Laser Appl. Chem. Biophys.* **1986,** *620,* 73.

Photoconductivity

Hino, T., Akazawa, H., Masuhara, H., and Mataga, N. *J. Phys. Chem.* **1976,** *80,* 33.

Okada, T., Matsui, H., Oohari, H., Matsumoto, and Mataga, N. *J. Chem. Phys.* **1968,** *49,* 4717.

Ottolenghi, M. *Acc. Chem. Res.* **1973,** *6,* 153.

Taniguchi, Y., and Mataga, N. *Chem. Phys. Lett.* **1972,** *13,* 596.

Pulse Radiolysis

Neta, P., and Harriman, A. In *Photoinduced Electron Transfer. Part B. Experimental Techniques and Medium Effects,* Chanon, M., and Fox, M. A., eds. Elsevier, Amsterdam, 1988, Chapter 2.3.

Waltz, W. L. In *Photoinduced Electron Transfer. Part B. Experimental Techniques and Medium Effects,* Chanon, M., and Fox, M. A., eds. Elsevier, Amsterdam, 1988, Chapter 2.2. Gives a background on pulse radiolysis.

Electron and Nuclear Spin

De Kanter, F. J. J., Sageev, R. Z., and Kaptein, R. *Chem. Phys. Lett.* **1978,** *58,* 334.

Nolting, F., Staerk, H., and Weller, A. *Chem. Phys. Lett.* **1982,** *88,* 523.

Sagdeev, R. Z., Salikhov, K. M., and Molin, Y. M. *Russ. Chem. Rev.* **1977,** *46,* 297.

Schulten, K. and Weller, A. *Biophys. J.* **1978,** *24,* 295.

Steiner, U. E., and Ulrich, T., *Chem. Rev.* **1989,** *89,* 51. A thorough review of magnetic effects.

Turro, N. J. *Proc. Natl. Acad. Sci. U.S.A.* **1983,** *80,* 609.

Weller, A. *Pure Appl. Chem.* **1982,** *54,* 1885. Some of Weller's contributions are summarized in this article.

Electron Spin Resonance (ESR) and Electron Nuclear Double Resonance (ENDOR)

Kevan, L. In *Photoinduced Electron Transfer. Part B. Experimental Techniques and Medium Effects,* Chanon, M., and Fox, M. A., eds. Elsevier, Amsterdam, 1988, Chapter 2.7.

Kurreck, H., Kirste, B., and Lubitz, W. *Angew. Chem.* **1984,** *23,* 173. A review of the ENDOR technique.

Plüschau, M., Zahl, A., Dinse, K. P., and van Willigen, H. *J. Chem. Phys.* **1989,** *90,* 3153.

van Willigen, H., Vuolle, M., and Dinse, K. P. *J. Phys. Chem.* **1989,** *93,* 2.

Chemically Induced Nuclear Polarization

Boxer, S. G., Goldstein, R. A., and Franzen, S. In *Photoinduced Electron Transfer. Part B. Experimental Techniques and Medium Effects,* Chanon, M., and Fox, M. A., eds. Elsevier, Amsterdam, 1988, Chapter 2.4.

Closs, G. L., and Czeropski, M. S. *J. Am. Chem. Soc.* **1977,** *99,* 6127.

Turro, N. J. *Modern Molecular Photochemistry,* Benjamin Cummings, Menlo Park, CA, 1978, Chapter 8. This chapter contains an excellent discussion of magnetic effects and CIDNP in photochemistry. Much of the discussion in Section 2.6 is based on Turro's contributions to this field.

References

1. Weller, A. *Z. Phys. Chem. N. F.* **1982,** *130,* 129.
2. Tazuke, S., Kitamura, N., and Kim, H.-B. In *Supramolecular Photochemistry,* Balzani, V., ed. Reidel, Dordrecht, 1987, p. 87.
3. Beens, H., Knibbe, H., and Weller A. *J. Chem. Phys.* **1967,** *47,* 1183.
4. Chibisov, A. K. *Russ. Chem. Rev.* **1981,** *50,* 615.
5. Anderson, C. P., Salmon, D. J., Meyer, T. J., and Young, R. C. *J. Am. Chem. Soc.* **1977,** *99,* 1980.
6. Lin, C-T., Böttcher, W., Chou, M., Creutz, C., and Sutin, N. *J. Am. Chem. Soc.* **1976,** *98,* 6536.
7. Sutin, N. and Creutz, C. *Adv. Chem. Ser.* **1978,** *168,* 1.
8. Bock, C. R., Connor, J. A., Gutierrez, A. R., Meyer, T. J., Whitten, D. G., Sullivan, B. P., and Nagle, J. K., *J. Am. Chem. Soc.* **1979,** *101,* 4815.
9. Mulazzani, Q. G., Emmi, S., Fuochi, P. G., Hoffman, M. Z., and Venturi, M. *J. Am. Chem. Soc.* **1978,** *100,* 981.
10. Maestri, M., and Grätzel, M. *Ber. Bunsenges. Phys. Chem.* **1977,** 504.
11. Harriman, A., Porter, G., and Richoux, M.-C. *J. Chem. Soc., Faraday Trans. 2* **1981,** *77,* 1175.
12. Simon, J. D., and Peters, K. S. *J. Am. Chem. Soc.* **1982,** *104,* 6142.
13. Forster, M., and Hester, R. E. *Chem Phys. Lett.* **1982,** *85,* 287.

14. Hub, W., Schneider, S., Dörr, F., Oxman, J. D., and Lewis, F. D. *J. Am. Chem. Soc.* **1984,** *106,* 708.

15. Hub, W., Klüter, U., Schneider, S., Dörr F., Oxman, J. D., and Lewis, F. D. *J. Phys. Chem.* **1984,** *88,* 2308.

16. Hirata, Y., Kanda, Y., and Mataga, N. *J. Phys. Chem.* **1983,** *87,* 1659.

17. Masuhara, H. and Mataga, N. *Acc. Chem. Res.* **1981,** *14,* 312.

18. Hirata, Y., Kanda, Y., and Mataga, N. *J. Phys. Chem.* **1983,** *87,* 1659.

19. Nolting, F., Staerk, H., and Weller, A. *Chem. Phys. Lett.* **1982,** *88,* 523.

20. van Willigen, H., and Ebersole, M. H. *J. Am. Chem. Soc.* **1987,** *109,* 2299.

21. Barber, R. A., de Mayo, P., Okada, K., and Wong, S. K. *J. Am. Chem. Soc.* **1982,** *104,* 4995.

Problems

1. During flash spectroscopy of mixtures of perylene and *N,N*-diethylaniline in acetonitrile, emission of triplet perylene was observed. Using thermodynamic arguments, what is a possible mechanism for the appearance of the signal due to triplet perylene, given the fact that high quencher concentrations were used in this experiment? (Knibbe, H., Rehm, D., and Weller, A. *Ber. Bunsenges. Phys. Chem.* **1968,** *72,* 257.)
2. On sensitization with singlet 1-cyanonaphthalene, norbornadiene is observed to isomerize to quadricyclane:

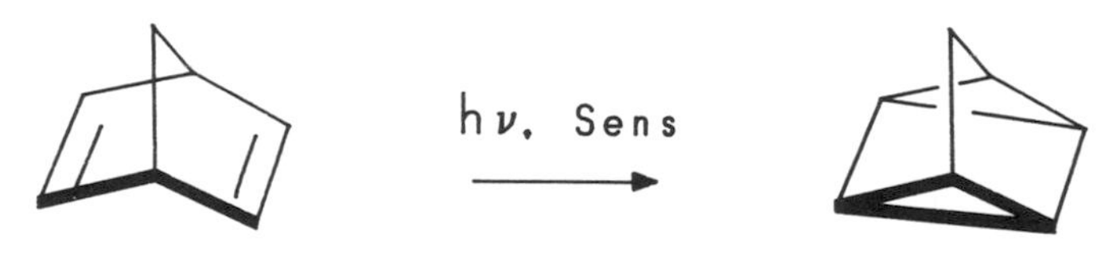

Sens = 1-cyanonaphthalene or chloranil

This isomerization is accompanied by polarized signals of quadricyclane and norbornadiene. Irradiation of chloranil and norbornadiene, however, gives rise to polarized signals of norbornadiene opposite in sign to those observed during sensitization with singlet 1-cyanonapthalene. Suggest a possible mechanism for the singlet and triplet sensitizations and a possible reason for the opposite polarizations for norbornadiene. Hint: triplet norbornadiene can isomerize to quadricyclane. (Roth, H. D., Schilling, M. L. M., and Jones, II, G. *J. Am. Chem. Soc.* **1981,** *103,* 1246.)

3. Photoinduced electron transfers between amines and ketones have been investigated by the CIDNP technique. For example, when 4-methylacetophenone is selectively excited in the presence of *N,N*-dimethyl-*p*-toluidine in an NMR cavity, the acetyl proton of the ketone is noted to display enhanced absorption in the absence of biphenyl but weak emission it its presence. Offer an explanation for this observation. Hint: biphenyl's triplet energy is E_T = 69.4 kcal mol^{-1}. (Hendriks, B. M. P., Walter, R. I., and Fischer, H. *J. Am. Chem. Soc.* **1979,** *101,* 2378.)
4. Confirm the values of k_{dis} for reactions 2.5 and 2.6 in acetonitrile and tetrahydrofuran, as given in thc tcxt. Now determine the effect of the center-to-center separation distance on k_{dis}. Calculate k_{dis} for d_{cc} = 10 Å and then for d_{cc} = 5 Å in acetonitrile and tetrahydrofuran. How do you explain these differences?
5. $\bar{\nu}_{EX}$ was measured and plotted for several aromatic hydrocarbon–amine exciplex pairs in a series of solvents. $2\mu^2/hc\rho^3$ was calculated for each pair from the slopes of these plots (ρ = 5 Å). Calculate the dipole moments of each pair from the tabulated values given below (DEA = N,N-diethylamine):

Reactants	$2\mu^2/hc\rho^3$
^{1}[pyrene]* + DEA	10.1×10^3
^{1}[DEA]* + biphenyl	15.1×10^3
^{1}[pyrene]* +pyrene	0

What conclusions can you reach concerning your calculated values? (Beens, H., Knibbe, H., and Weller, A. *J. Chem. Phys.* **1967,** *47,* 1183.)

6. Calculate ΔG_{c1} for the reactions in Table 1.6 in cyclohexane assuming $d_{cc} \sim 7$ Å. Refer to Problem 1 in Chapter 1.

CHAPTER

3

Electron-Transfer Photochemistry

The scope of photoinduced electron transfer includes an impressive array of photochemical transformations involving ion-pair states. To cite only a few examples, one may mention photoinduced isomerizations, decompositions, donor–acceptor additions and couplings, dimerizations, ring openings, eliminations, extrusions, and oxidations. It is now recognized that photoinduced electron transfer allows the organic chemist to achieve certain chemical transformations that by other routes may be difficult. For this reason, a familiarity with "electron-transfer photochemistry" is a desirable goal for the chemist who may want to understand, at least at a practical level, an important facet of the chemical effects of light on matter. In this chapter, we shall describe reactions that serve as examples of the major types of photoinduced electron-transfer reactions. Before delving into these examples, however, we shall introduce the reader to an overview of the most important pathways by which stable products can arise.

3.1. The Pathways of Photoinduced Electron-Transfer Reactions

These pathways can be classified as (1) sensitizations, (2) cosensitizations, (3) chain transfers, (4) additions, and (5) triplet recombinations. There is a repeating motif to these pathways. That is, in each case, the initial event is light absorption by one of the components of a reaction mixture. The generation of short-lived intermediates (usually an exciplex or radical ion pair when the reactants are neutral organic species) is followed by any one of several pathways. Here we encounter a rich tapestry of reaction mechanisms. The nature of these mechanisms is usually dictated by the structures of the reactants as well as the polarity of the solvent.

3.1.1. Photosensitization by Electron Transfer

In photosensitization by electron transfer, a chemical transformation is triggered by light absorption by a photoexcited electron donor or acceptor. We designate this species as a *primary* photosensitizer. Typically, the primary quenching step leads to an ion pair (although in some reactions involving charged reactants, electron transfer may lead to a neutral rather than charged ion pair). Following the quenching reaction, a series of steps follows that may result in additional charged ionic species and finally the formation of stable products. Eventually, the photosensitizer is regenerated and becomes available for another cycle. For a photosensitizer operating as an electron donor, we can write the pathways depicted in Scheme 3.1:

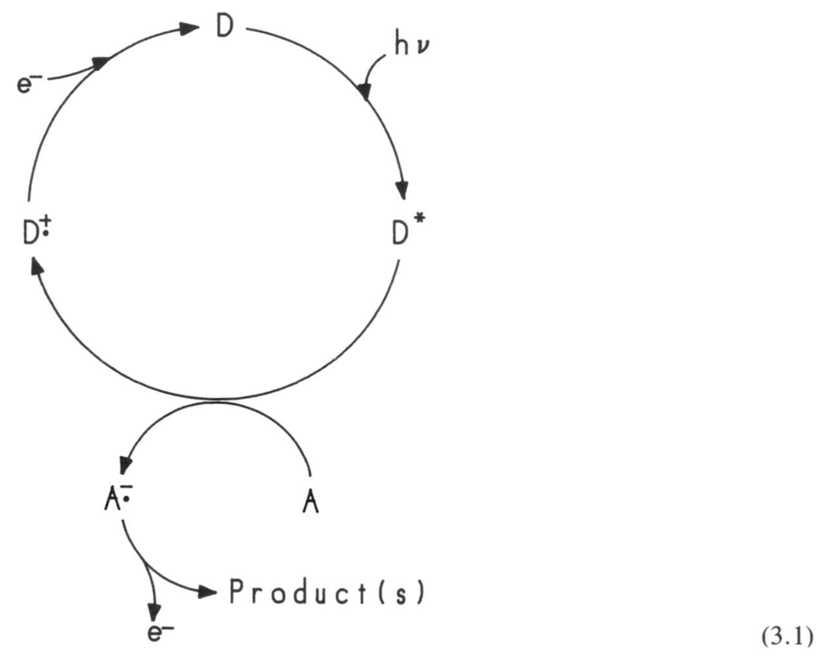

(3.1)

In 3.1, to keep our model simple, we have ignored intersystem crossing to triplet states and other pathways. Accordingly, we have designated the excited photosensitizer with an asterisk and omitted the spin configuration. In some reactions, we may have to specify the spin multiplicity of the sensitizer and, in fact, will do so when introducing certain examples where spin multiplicity plays a role.

One of our objectives in this chapter will be to examine the product transformations resulting from the reactions shown in Scheme 3.1

In the strictest sense, the photosensitizer does not experience a permanent chemical change after completion of the reaction. Although the sensitizer may temporarily be converted into a short-lived intermediate such as a radical cation or anion, it is eventually restored to its original state. The regenerated sensitizer can then be excited again to participate in another cycle. This process repeats itself up until a point where the quencher is totally consumed. It is desirable that side reactions such as coupling between the sensitizer and quencher be minimized, although under certain circumstances reaction between donor and acceptor may be a desired outcome (Section 3.1.5). In some photosensitizations, reaction between the sensitizer and quencher (such as a photoaddition) cannot be avoided. This may occur during long reaction times and may result in the consumption of the photosensitizer.

Deactivation pathways such as radiationless decay of the excited state or electron return can reduce the efficiency of product information. Conditions should be optimized to favor k_{dis} and inhibit k_{ret}. For example, the employment of highly polar media, salts, or organized assemblies to enhance ion dissociation or the use of a long-lived excited sensitizer, can enhance the efficiency.

3.1.2. Cosensitization

In some cases, two sensitizers working together can bring about a desired chemical transformation. One molecule, the *primary* sensitizer is needed to absorb or capture the incident light energy. It then can be quenched by a *secondary* sensitizer by an electron-transfer mechanism. In Scheme 3.2, there is shown a possible pathway by which a cosensitization may proceed (the species with the prime, D′, is designated as the cosensitizer):

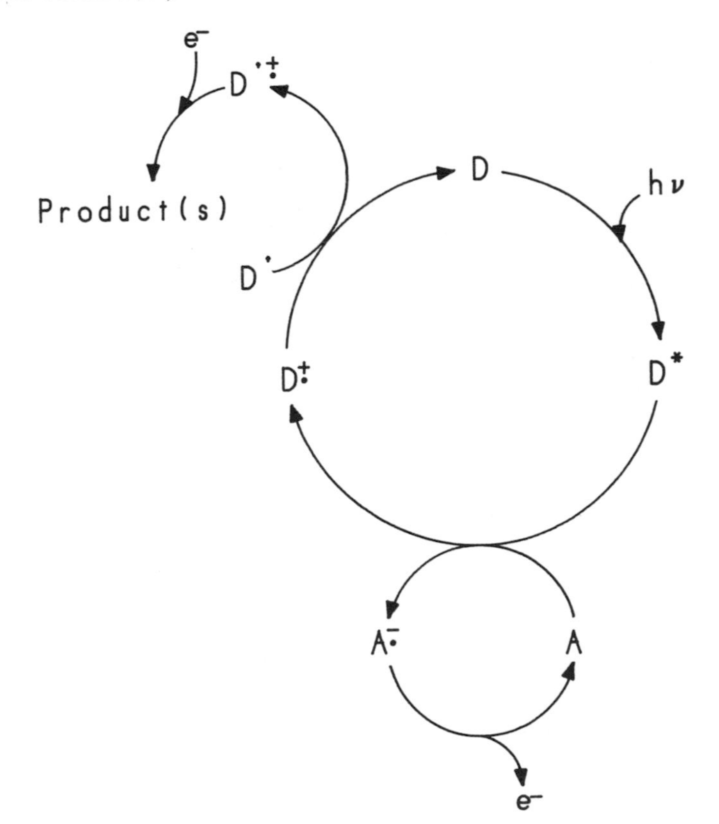

(3.2)

Note the requirement that the redox potential of D′ should be lower than that of D (i.e., D′ should be easier to oxidize). Reactions proceeding along the lines shown in Scheme 3.2 are fairly common.

3.1.3. Triplet Recombination

The reader will recall from Chapter 2 that triplet recombination may occur when the one of the reactant's triplet state lies lower than the energy of the ion-pair state. The reactant can be the sensitizer or the quencher. Following its formation, the triplet state can then decay to ground state or participate in chemical reaction.

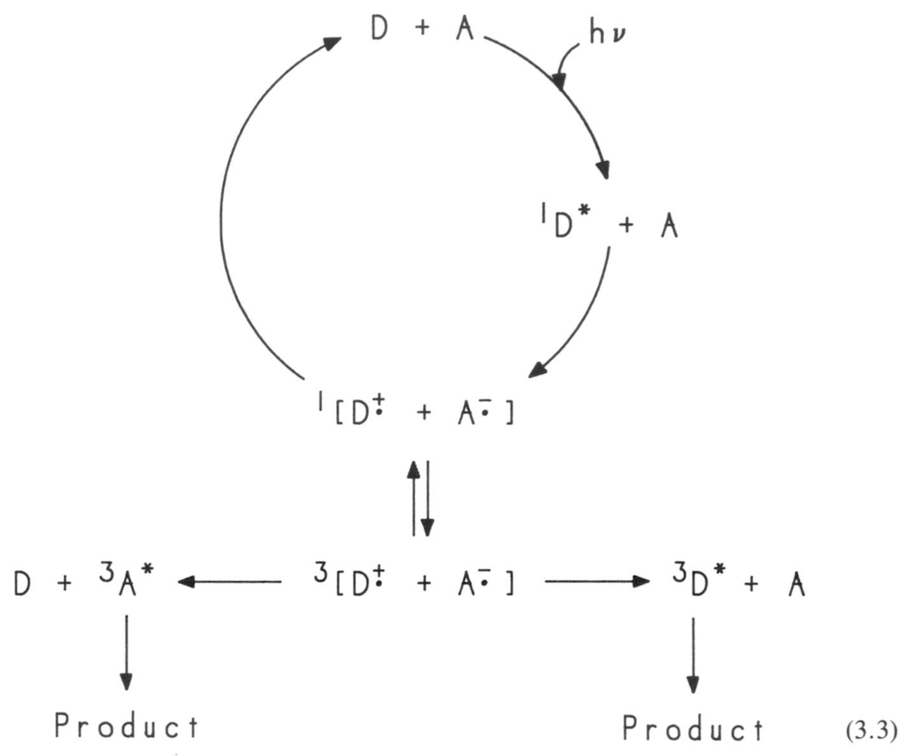

(3.3)

The feasibility of triplet recombination is determined by energetics (i.e., the energy of the triplet state must be lower than the energy stored in the ion-pair state). The newly formed triplet may then be converted into stable products. These products can be said to arise from an electron-transfer pathway since the initial step is quenching by electron transfer. CIDNP is one of the experimental probes used to deduce the presence of triplet recombination. As we learned in Chapter 2, the appearance of products having polarized NMR spectra during the quenching of singlet or triplet states is sometimes suggestive of triplet recombination. However, the energies of the various intermediates must be known accurately before a triplet recombination pathway can be postulated.

3.1.4. Chain Transfer

Chain transfer is analogous to a looping structure in a computer program. Consider the set of reactions in Scheme 3.4:

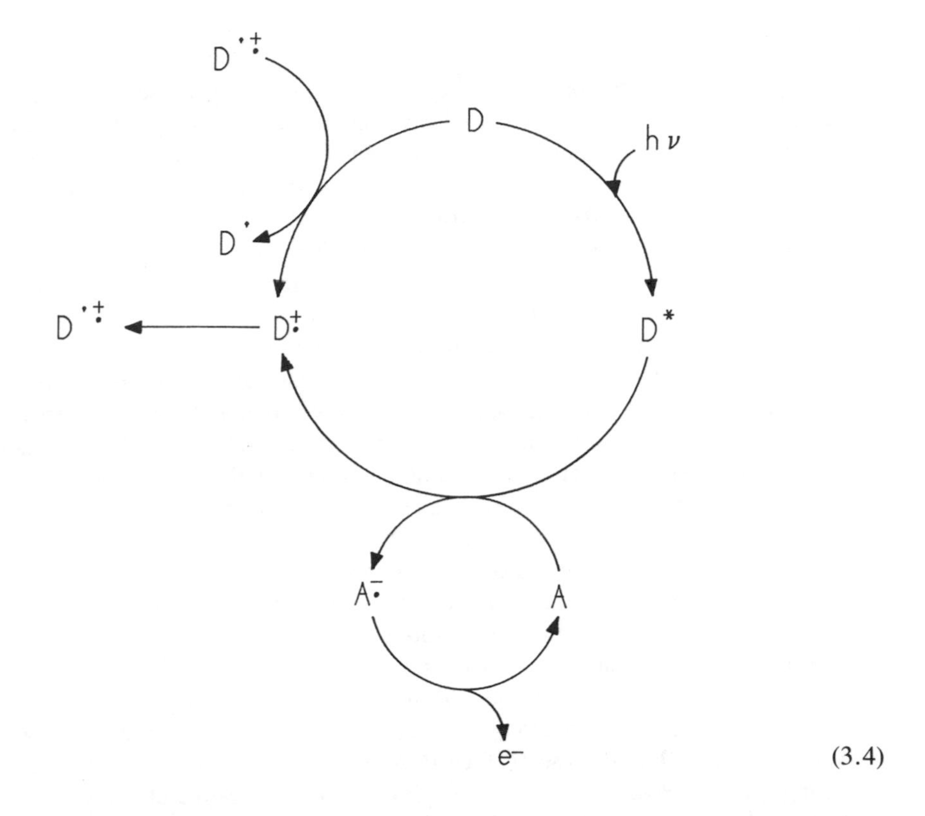

(3.4)

The efficiency of a chain mechanism depends on the thermodynamics of the repeating loop. The reaction terminates when most of the quencher is depleted. The quantum yields in chain transfers usually are much greater than unity since for each photon absorbed by the sensitizer, numerous molecules of D′ can be generated. In fact, a quantum yield much greater than 1 is a good diagnostic test of a chain process. A similar reaction scheme can be written for sensitization with an electron acceptor.

3.1.5. Photoinduced Electron-Transfer Additions

These are reactions involving an addition between the excited species and a ground-state compound. Strictly speaking, the excited species is not a photosensitizer since it becomes permanently consumed by the reaction pathway. As mentioned above, coupling reactions between donor and acceptor molecules may or may not be a desired outcome. The efficiency of these additions is especially sensitive to molecular structure and solvent polarity, as we shall show by examples later in the chapter.

3.2. Examples

With the pathways outlined in Section 3.1 as models, we can now discuss examples that exemplify the scope of photoinduced electron-transfer chemistry. The majority of these examples include the electron-transfer transformations of organic substrates. The sensitizers are organic, for the most part, but in some cases may include inorganic complexes. As the reader will observe, most of the examples in this chapter are concerned with the formation and reactions of radical cations. This is because much of the work appearing in the chemical literature has dealt with systems where radical cations can easily be generated. In contrast, there are relatively only a few examples involving the generation of radical anions.

The majority of the reactions covered in this chapter are *photosensitizations,* according to the definition presented earlier. In some examples, the photosensitization may be accompanied by changes in the photosensitizer, especially at longer reaction times. These cases are portrayed as photosensitizations, even though the irreversible destruction of the photosensitizer may be undesirable.

Another point to consider is that the majority of examples presented in this chapter involve reactions taking place in degassed (deoxygenated) solvents. This is because the presence of oxygen may interfere with a desired outcome, as oxygen is quite a reactive species. Photooxygenations (i.e., reactions in the presence of O_2) are, however, important in their own right and will be covered separately.

The discussion starts with photoinduced isomerizations occurring by an electron-transfer mechanism. The next group of reactions that is discussed involves bond cleavage reactions. The discussion then turns to donor–acceptor reactions. The remaining sections cover interesting examples that are suggestive of the scope of photoinduced electron transfer.

3.2.1. Photosensitized Electron-Transfer Isomerizations and Rearrangements

3.2.1.1. Geometric Isomerizations

Cis–trans isomerizations, also known as geometric isomerizations, take place in compounds containing double bonds. These reactions occupy an important place in photochemistry. When being introduced to photochemistry, the student is often taught how light can be effectively used to accomplish these isomerizations. It is shown that in ground-state unsaturated organic compounds, there are substantial energy barriers ($\sim$70 kcal mol^{-1}) required for the twisted motions to bring about *cis–trans* isomerization (Fig. 3.1). There is a way to get around this. For example, the unsaturated molecule can be directly photoexcited to create its excited state. In

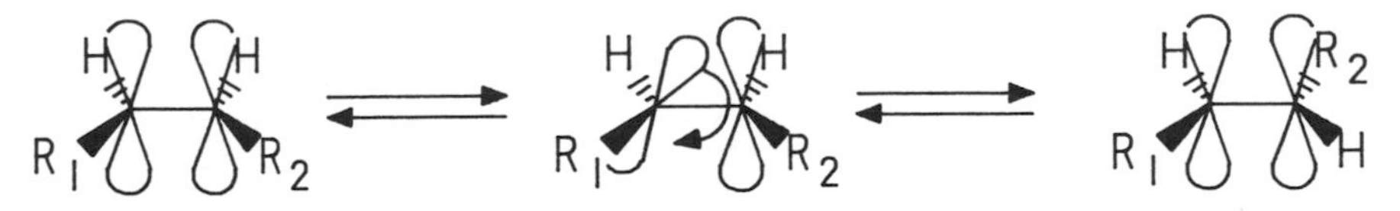

Figure 3.1 Representation of "twisting" motions accompanying isomerization of *cis*- and *trans*-olefins.

formation of the excited olefin, an electron in a highest occupied π-orbital is promoted to an antibonding π*-orbital. This "uncoupling" of the π-orbital leaves a σ-bond, which can undergo rotation about the bond axis much more easily than a π-bond. The rotation about the σ-bond in the excited state generates a more stable *twisted* geometry. The twisted state can relax to the planar ground state leading finally to the formation of *cis* and *trans* isomers.

The exact composition of the isomers formed during a photochemical isomerization depends in a complicated way on a number of factors, including the rates of formation of the *cis* and *trans* isomers from the twisted state as well as their energies.

Numerous geometric isomerizations take place by energy transfer from triplet sensitizers to the triplet states of the *cis* and *trans* isomers of certain olefins. Although sensitizers with triplet energies higher than the triplet state of either isomer can bring about the efficient formation of *cis* and *trans* isomer, the *cis/trans* ratio usually depends on the relationship of the excited-state energy of the sensitizer to that of the quenching olefin.

Cis–trans isomerizations can also be achieved via a pathway involving sensitized electron transfer. There are two pathways by which these isomerizations usually take place:

1. The excited sensitizer may remove or donate an electron to produce a radical cation or a radical anion, respectively. The isomerization may then be followed by bond rotation around the single bond in the radical cation or anion intermediate. Isomerization does not take place directly within the excited state.
2. A low-lying triplet state of an unsaturated compound may be populated from a high-lying ion-pair state by the triplet recombination pathway. The excited triplet state may then undergo bond rotation to a twisted state, which then converts into *cis* and *trans* isomers.

In the following sections, the discussion will involve photoisomerizations following sensitization by the above two mechanisms.

3.2.1.1.1. Geometric Isomerizations Involving Radical Ion Pairs

The mechanisms shown in 3.5 and 3.6 summarize *cis* to *trans* and *trans* to *cis* isomerizations in the presence of electron-donating

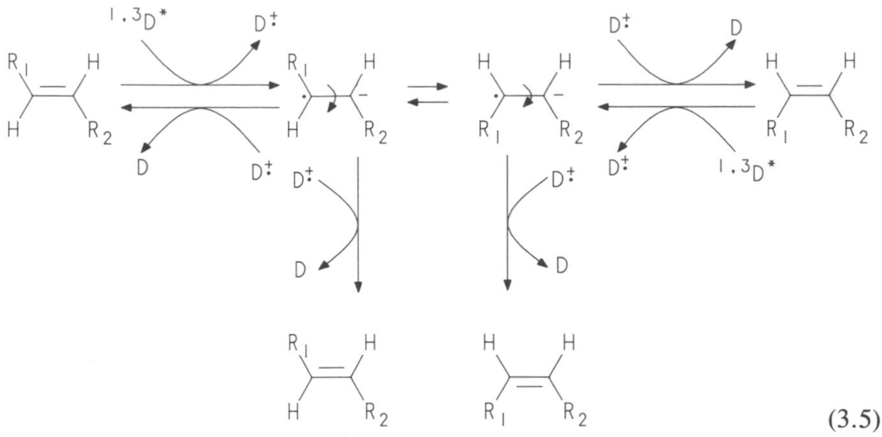

(3.5)

or -accepting

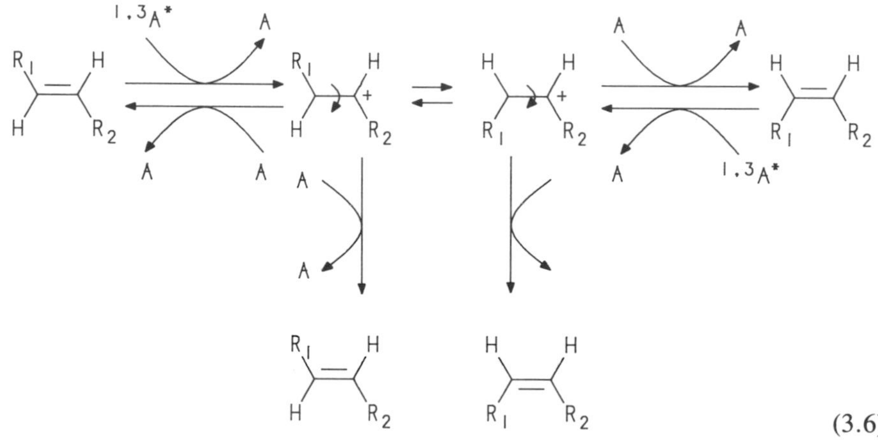

(3.6)

photosensitizers. The twisting motion in a radical cation or anion is driven by steric and electronic factors, which in turn determine the final ratios of *cis* and *trans* isomers. In general, the photostationary states involving cation or anion radical intermediates are enriched with the more stable *trans* isomer. The barrier to rotation about the single bond in the radical ion must not be so large as to compete with electron return, otherwise the isomerization may turn out to be inefficient.

An illustrative case of isomerization proceeding by photosensitized electron transfer is the reaction of *N*-methyl-4-(β-styryl)pyridinium (**1**) in the presence of zinc tetraphenylporphyrin (ZnTPP)[1]:

hν, ZnTPP

1 ⇌ 2 (3.7)

The isomerization of either *cis* or *trans* isomer leads predominantly to the *trans* isomer. An electron-transfer pathway is likely since quenching of excited ZnTPP by a series of compounds similar to **1** leads to curves where the rate of the quenching is correlated with the redox potentials of these molecules. Further support is provided by the fact that triplet energies of **1** and **2,** both equal to ~50 kcal mol^{-1}, are higher than the triplet energy of the porphyrin sensitizer (36.7 kcal mol^{-1}). This rules out triplet–triplet energy transfer. A pathway for the isomerization of **1** or **2** using ZnTPP may begin with electron abstraction by the sensitizer. The initial abstraction can be sensitized by either the singlet or triplet states of the porphyrin since in both cases the reaction is exothermic:

$^{1,3}ZnTPP^*$ → $ZnTPP^{+\cdot}$

1 ⇌ (radical)

ZnTPP ← $ZnTPP^{+\cdot}$ (3.8)

$^{1,3}ZnTPP^*$ → $ZnTPP^{+\cdot}$

2 → (radical) (3.9)

An equilibrium is established between the radical products of these quenching steps. As electron return completes the reaction sequence, the sensitizer is regenerated.

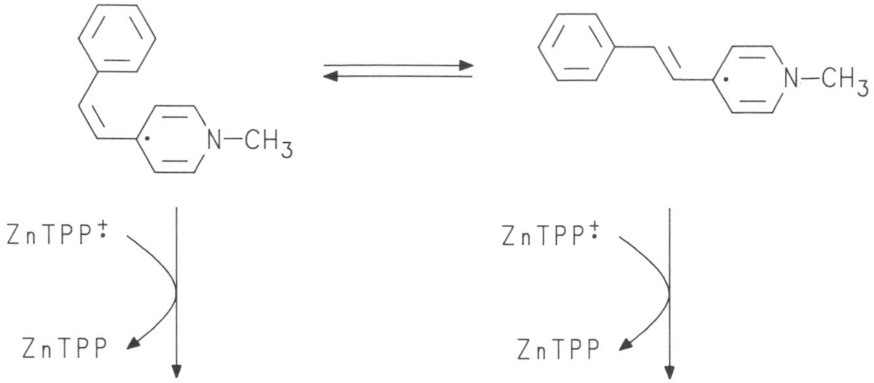
N—CH3
N—CH3
ZnTPP+·
ZnTPP
ZnTPP+·
ZnTPP

1 2 (3.10)

Another example is the case where 2,4,6-triphenylpyrylium tetrafluoroborate (TP^+) sensitizes the conversion of *trans*-stilbene to its *cis*-isomer[2]:

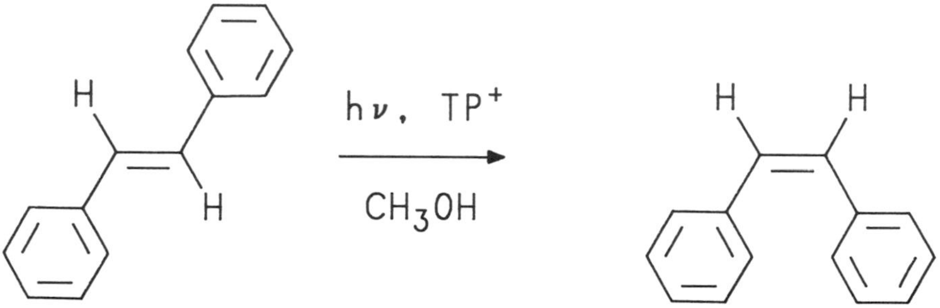
H
H
hν, TP+
CH3OH
H
H

3 4

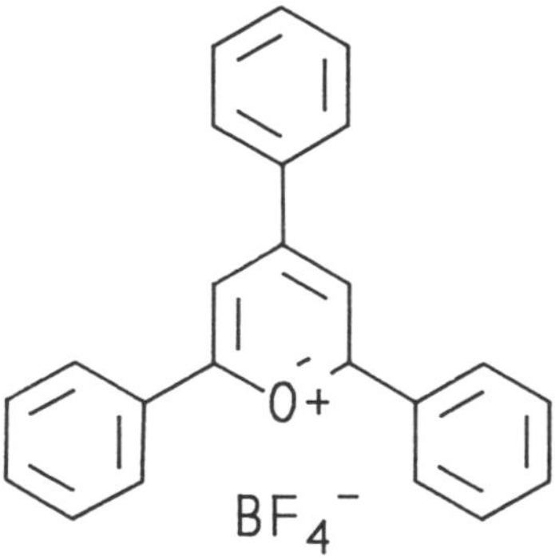
O+
BF4−

TP^+ (3.11)

The fluorescence of TP^+ is quenched efficiently by *trans*-stilbene (**3**) at near diffusion-controlled rates. As in the previous example, singlet and triplet states of the photosensitizer can abstract an electron from the olefin substrate. Quenching is thermodynamically favorable (i.e., $\Delta G_{el} = -23.7$ and -11.7 kcal mol^{-1} for singlet and triplet quenching, respectively). Furthermore, the transient spectrum of the sensitizer pyryl radical was identified by time-resolved spectroscopy. This observation coupled with the favorable energetics suggested an electron-transfer pathway leading to formation of the radical cation of *trans*-stilbene:

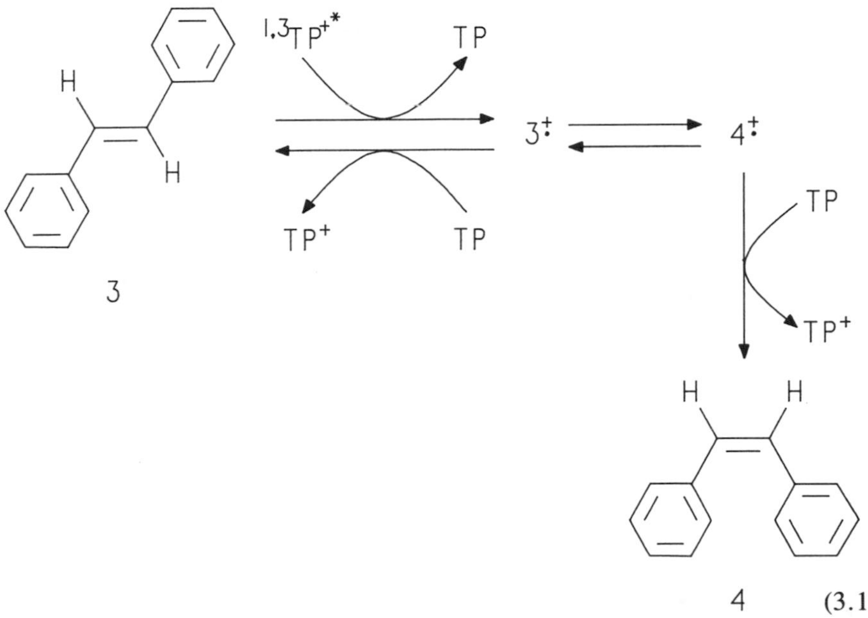

(3.12)

Several sensitized photoisomerizations have been documented where the energies of sensitizer and quencher lie so close together that energy and electron transfer may coexist. We bring this to the attention of the reader to emphasize that in many instances a clearcut distinction between energy and electron transfer cannot be made, and that one has to consider the possibility of both mechanisms occurring in the same system. The *cis/trans* ratio has been used as a diagnostic indicator of energy or electron transfer. Since *trans* radical ions are usually more stable than their *cis* counterparts, a mixture rich in the *trans* isomer may suggest an electron-transfer pathway. This is acceptable as long as it is recognized that in energy transfer, the *cis/trans* ratio can vary considerably depending on the energy of the excited sensitizer. For this reason, estimating *cis/trans* ratios may not yield definitive information. Other evidence such as direct spectroscopic observation of intermediates may be needed to characterize the details of the pathway.

One such example where electron and energy transfer may coexist is the photoisomerization of *cis*- and *trans*-dipyridylethylene salts (Fig. 3.2) in the presence of various sensitizers.[3] That energy transfer is a possible pathway with many sensitizers is suggested by the fact that the triplet energies of the dipyridylethylenes shown in Fig. 3.2 are only ~50 kcal mol^{-1}. With sensitizers having higher triplet energies, [e.g., benzophenone (E_T = ~69 kcal mol^{-1})], energy transfer is likely. Here the rates of quenching are diffusion-controlled. However, in the cases of olefins **7** and **9,** the isomerization efficiencies are not as high as one would expect if energy transfer were the exclusive pathway (Table 3.1). Competitive electron transfer, which can account for the loss in isomerization efficiency, may be an accompanying pathway. A similar conclusion applies to quenching of [Ru(bpy)$_3^{2+}$]* whose excited-state energy, ~49 kcal mol^{-1}, lies close to the triplet-state energies of the olefins. However, on sensitization by Ru(bpy)$_3^{2+}$, flash photolysis experi-

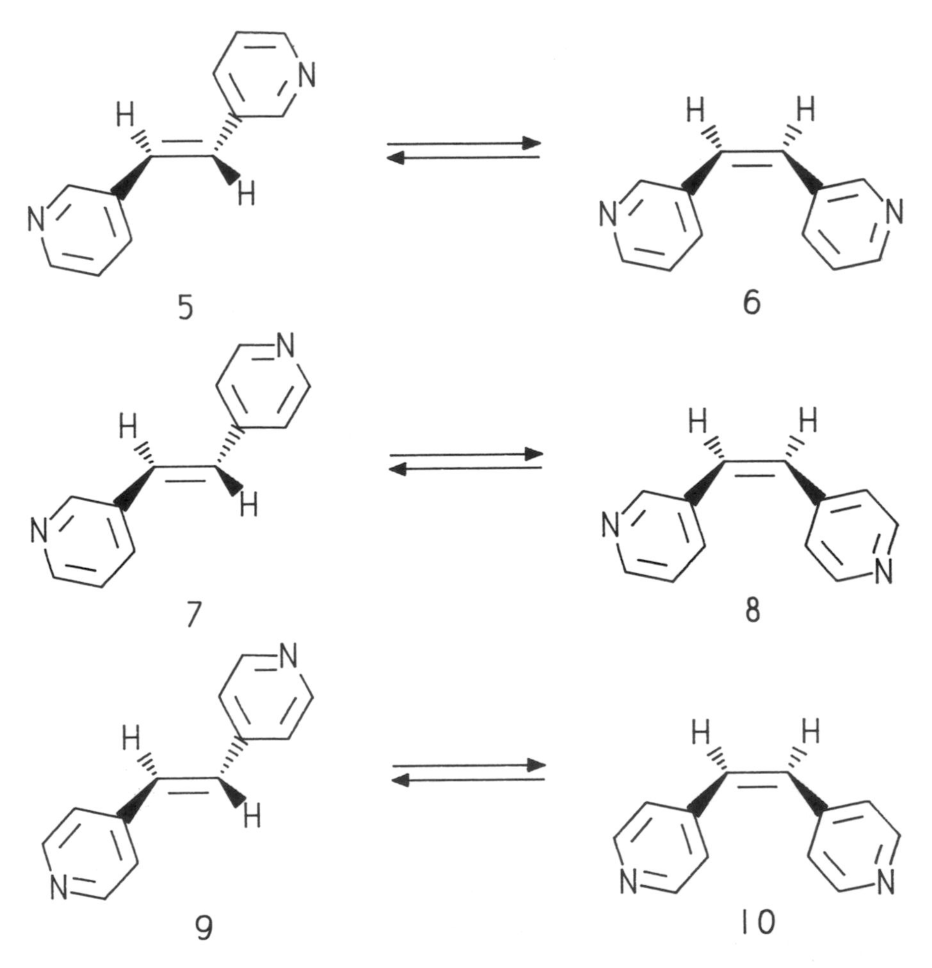

Figure 3.2 Structures of 1,2-bispyridylethylene quenchers used in the study of competitive energy and electron transfer quenching of benzophenone, Ru(bpy)$_3^{2+}$, and other sensitizers.

Table 3.1 QUENCHING RATE CONSTANTS AND ISOMERIZATION QUANTUM YIELDS FOR THE QUENCHING OF BENZOPHENONE AND $Ru(bpy)_3^{2+}$ BY BISPYRIDYLETHYLENE SALTS[a,b]

	Benzophenone		$Ru(bpy)_3^{2+}$	
	k_q (M^{-1} s^{-1})	Φ_{isom}	k_q (M^{-1} s^{-1})	Φ_{isom}
5 → **6**	~2 × 10^9	0.4	6 × 10^6	0.45
7 → **8**	—	0.3	9 × 10^8	0.02
9 → **10**	~1 × 10^{10}	0.2	2 × 10^9	0.007

[a] From reference 3.
[b] Refer to Fig. 3.1 for structures of quenchers.

ments provided spectroscopic evidence of electron-transfer products. The important point here is that both quenching mechanisms may be occurring and one of them may even play a dominant role (as probably in the quenching of $[Ru(bpy)_3^{2+}]^*$. This study underscores the importance of examining the thermodynamic properties of the reactants when specifying the nature of the quenching step.

3.2.1.1.2. Sensitized Geometric Isomerizations by Triplet Recombination

It was noted in Section 3.1.3. that recombination in the ion-pair states can generate reactive triplet states, and that some of these triplets may undergo isomerization. The *cis–trans* isomerization of 1,2-diphenylcyclopropane (**11** and **12**) in polar solvents is one example[4]:

$h\nu$, DCN
CH_3CN

11 12

DCN (3.13)

A CIDNP study of the *cis–trans* isomerization of **11** undertaken with singlet sensitizers provided an important clue about the state from which isomerization proceeds. The fluorescence of singlet sensitizers such as 1,4-dicyanonaphthalene was found to be quenched in polar solvents with accompanying isomerization. The free-energy change is -22 kcal mol^{-1}: the triplet recombination pathway is thermodynamically feasible. Polarized NMR signals were observed in the methine protons of **12.** These considerations are compatible with initial electron transfer to generate a singlet ion pair following by intersystem crossing to a triplet ion pair. Given that the triplet phosphorescent state of 1,2-diphenylcyclopropane is ~57 kcal mol^{-1} and that the triplet ion-pair state lies at 64 kcal mol^{-1}, a triplet-recombination pathway appears plausible:

$$11 + {}^{1}DCN^{*} \longrightarrow {}^{1}[11^{+\cdot} + DCN^{-\cdot}] \rightleftarrows {}^{3}[11^{+\cdot} + DCN^{-\cdot}] \longrightarrow 11^{+\cdot} + DCN^{-\cdot}$$

$$11^{+\cdot} + DCN^{-\cdot} \xrightarrow{-DCN} \text{"bond-cleavage" triplet}$$

$$12 \longleftarrow \text{(1,3-diphenylpropane-1,3-diyl)} \longleftarrow \text{"bond-cleavage" triplet} \longrightarrow 11 \qquad (3.14)$$

In some triplet recombinations, population of the triplet sensitizer may occur at the expense of triplet olefin production. This aspect of a triplet recombination pathway reduces the efficiency of the reaction. In fact, the inefficient isomerization of *trans*-stilbene (**3**) to the *cis*-isomer (**4**) in the presence of 9,10-dicyanoanthracene has been attributed to population of the sensitizer's triplet state.[5] The fact that the radical cation of **3** was detected by transient absorption spectroscopy was taken as evidence for an isomerization proceeding via a radical cation intermediate. However, the triplet state energy of 9,10-dicyanoanthracene is only $E_T \sim 42$ kcal mol^{-1} as compared to $E_T \sim 49$ kcal mol^{-1} for the stilbene substrate. Evidently, population of triplet 9,10-dicyanoanthracene can occur from a triplet radical ion pair. This may occur in competition with ion dissociation to the stilbene radical cation:

$$3 + {}^{1}DCA^{*} \longrightarrow {}^{1}[3^{\dagger} + DCA^{\bar{\cdot}}] \rightleftharpoons {}^{3}[3^{\dagger} + DCA^{\bar{\cdot}}] \longrightarrow 3 + {}^{3}DCA^{*}$$

$$^{3}[3^{\dagger} + DCA^{\bar{\cdot}}] \longrightarrow 3^{\dagger} + DCA^{\bar{\cdot}} \longrightarrow 4^{\dagger} + DCA^{\bar{\cdot}} \longrightarrow 4 + DCA \quad (3.15)$$

The important point to be emphasized here is that triplet recombination is a pathway that one should always consider when writing mechanisms for sensitized electron-transfer transformations.

3.2.1.1.3. Cosensitized and Chain-Transfer Geometric Isomerizations

We now turn to isomerizations that can proceed by cosensitization and chain transfer. The isomerization *cis*- or *trans*-1,2-diphenylcyclobutane (**13** and **14**), for example, can be sensitized in the presence of *m*-dicyanobenzene and phenanthrene (**15**)[6]:

$$13 \underset{CH_3CN}{\overset{h\nu,\ m\text{-DCB},\ 15}{\rightleftharpoons}} 14 \quad (3.16)$$

m-DCB 15

In this system, phenanthrene serves as the primary photosensitizer and *m*-dicyanobenzene as the secondary photosensitizer. Once the excited singlet of phenanthrene donates an electron to *m*-DCB, the radical cation of phenanthrene removes an electron from **13**$^{+\cdot}$. Thus, for isomerization of **13**, we have:

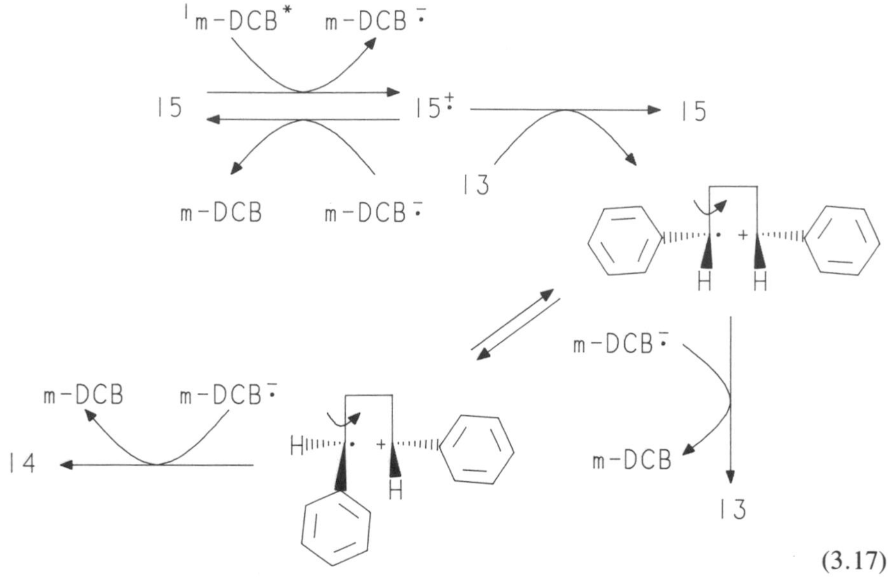

(3.17)

As noted earlier, isomerizations which proceed via a chain mechanism are associated with high quantum yields. The isomerization of 1-phenoxypropene (**16** and **17**) is one such example[7]:

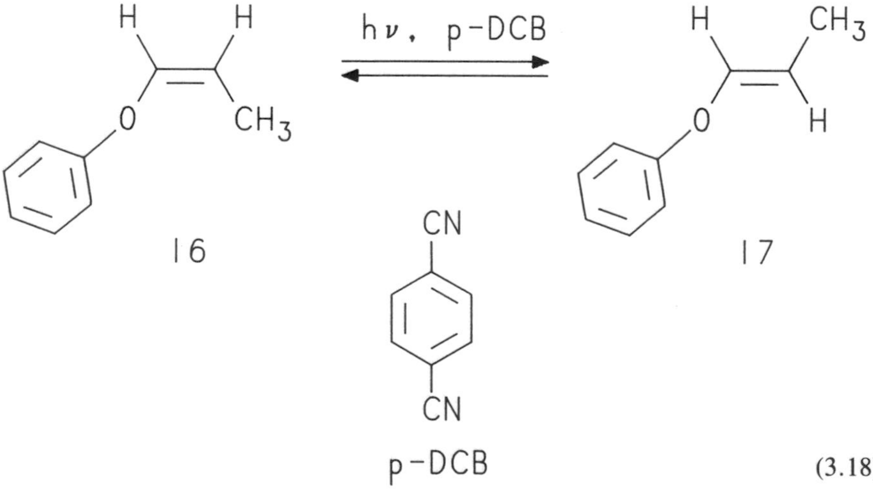

(3.18)

This sensitization can also be regarded as a cosensitization in the sense that it requires the presence of two sensitizers, *p*-dicyanobenzene and phenanthrene. After oxidation of phenanthrene, either olefin can donate an electron to the phenanthrene radical cation to form phenoxypropene radical cations:

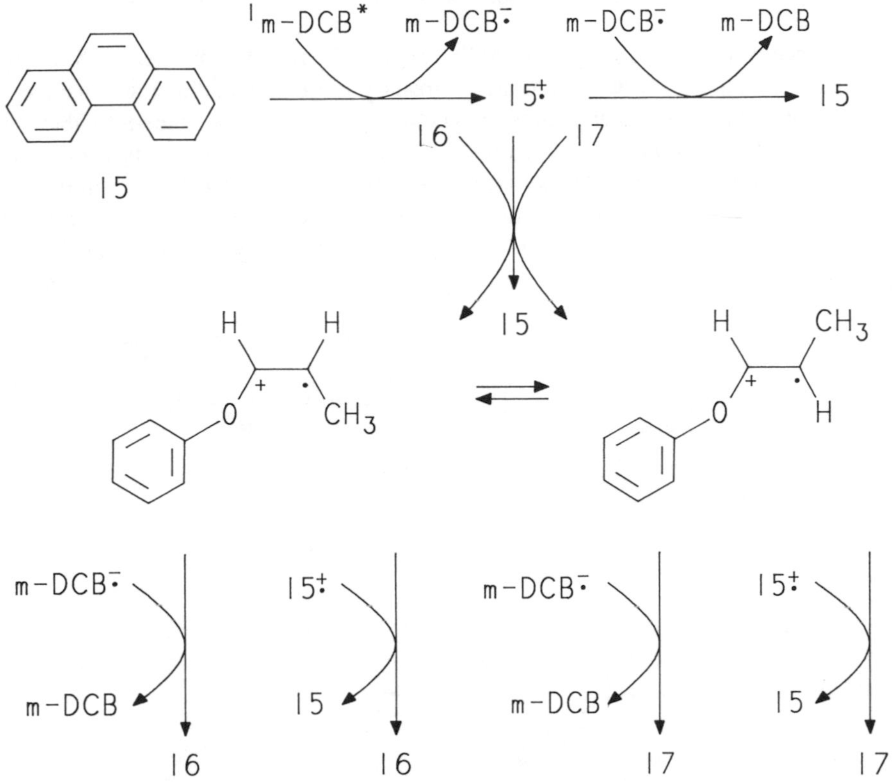

(3.19)

3.2.1.2. Valence Isomerizations and Rearrangements

Norbornadiene and quadricyclane are two well-known examples of valence isomers. Interconversions between these structural isomers are known as valence isomerizations.

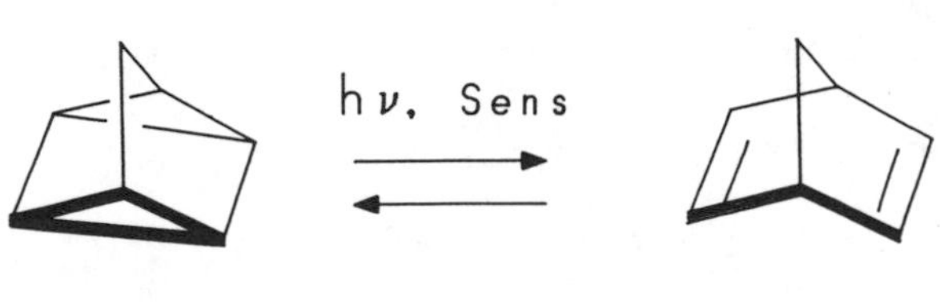

(3.20)

Quadricyclane lies at a higher energy than the norbornadiene. This feature has been one of the reasons for the interest in this reaction as a model for *solar energy collection and storage*. This approach might be quite practical if a way is found to effect the controlled release of energy from energy-rich molecules such as quadricyclane.[8]

Valence isomerizations may proceed thermally, by direct photoexcitation, or by sensitization with a suitable sensitizer. Strongly electron-withdrawing sensitizers such as quinones or cyanoaromatics molecules, for example, are known to abstract an electron from norbornadiene or quadricyclane to generate their radical cations. Isomerization subsequently takes place by mechanisms involving these intermediates. The isomerization of norbornadiene and the reverse process has been postulated to proceed via triplet recombination in the presence of high-lying singlet sensitizers (see Problem 2.2)[9,10]:

$$\mathbf{18} + {}^{1}\text{1-CN}^{*} \longrightarrow {}^{1}[\mathbf{18}^{+\cdot} + \text{1-CN}^{-\cdot}] \rightleftharpoons {}^{3}[\mathbf{18}^{+\cdot} + \text{1-CN}^{-\cdot}] \longrightarrow {}^{3}\mathbf{18}^{*} + \text{1-CN} \longrightarrow \mathbf{19}$$

1-CN (3.21)

The sensitized isomerization of Dewar benzene (**20**) to hexamethylbenzene (**21**) in the presence of naphthalene is another example[11]:

$$\mathbf{20} \xrightarrow[\text{CH}_3\text{CN}]{h\nu,\ \text{Naphthalene}} \mathbf{21}$$

Naphthalene (3.22)

20 quenches naphthalene fluorescence and subsequently isomerizes to **21.** In polar solvents, this sensitized isomerization proceeds with a quantum yield as high as $\Phi \sim 80$. This observation is consistent with the chain mechanism shown in 3.23:

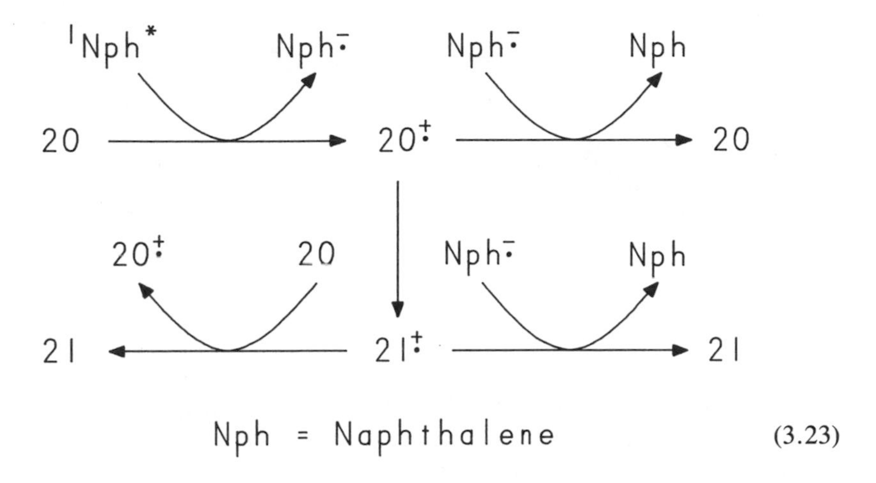

(3.23)

A CIDNP study involving sensitized isomerization with triplet chloranil led to the suggestion that two discrete radical cations of Dewar benzene are involved in the reaction[12]:

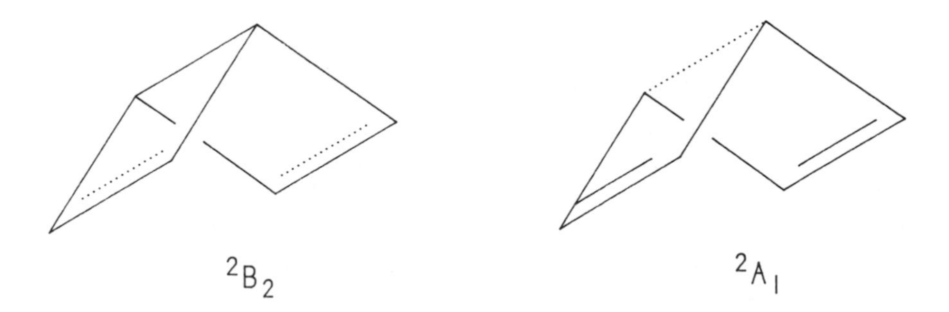

The observed polarization patterns suggested the presence of positive spin density on olefinic and bridgehead carbons. Moreover, since the 1,4 bond (1.583 Å) in the $^2B_2$ radical cation is not as long as the same bond (2.085 Å) in the $^2A_1$ radical cation, the former is predicted to be more stable. A proposal involving two different cation intermediates was also made for the norbornadiene $\rightleftharpoons$ quadricyclane isomerization.[9]

The valence isomerizations of certain phenylated molecules are thought also to involve a chain mechanism because of unusually high quantum yields. For example, **22** isomerizes to **23** in the presence of a pyrrylium salt sensitizer, TP^+ [13]:

CH3 CH3 hν, TP+ CH3 CH3

22 23 (3.24)

Evidence supporting electron transfer is as follows: (1) $\Delta G_{el} = -26$ kcal mol^{-1} for quenching of singlet TP^+ and -14 kcal mol^{-1} for quenching of triplet TP^+; and (2) cleavage does not take place in the presence of a quencher with a lower redox potential [e.g., $E^0(D^+/D) = 0.81$ V for tetramethoxybenzene vs. 1.40 V for **22**]. These observations are consistent with the following pathway:

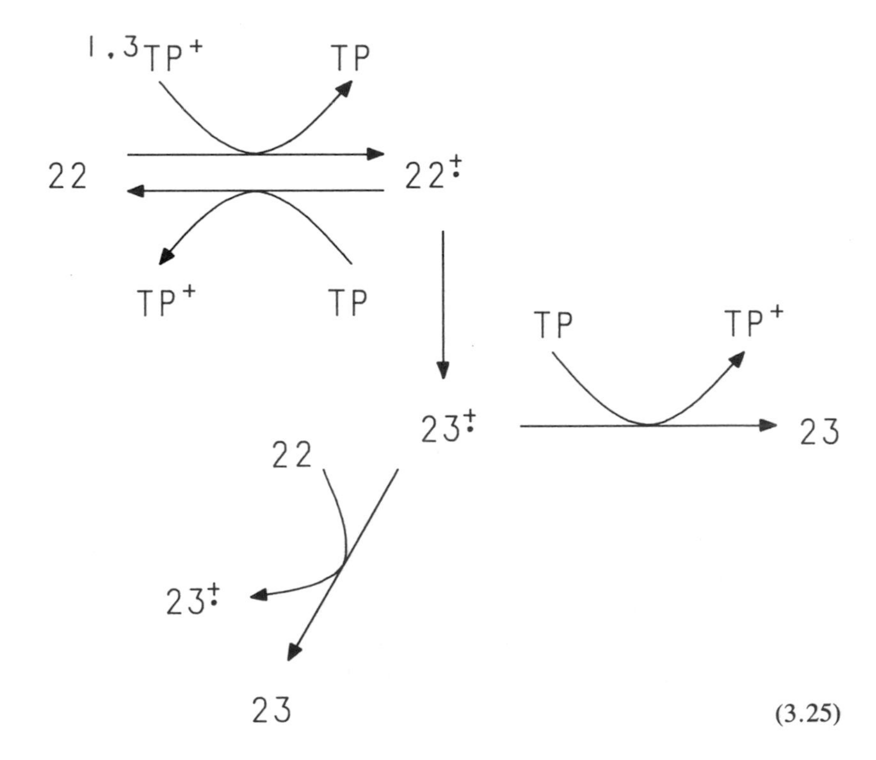

(3.25)

A novel photoinduced electron-transfer rearrangement has been reported for spirofluorenebicyclo[6.1.0]nonatriene (**24**)[14]:

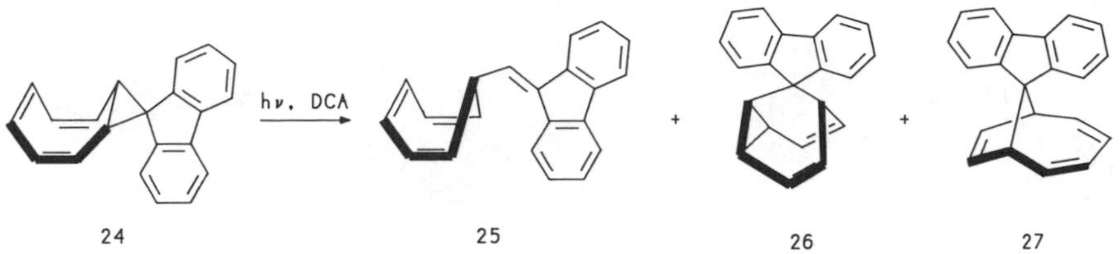

(3.26)

In polar solvents, the product distribution is **25** (47%), **26** (16%), and **27** (6%). The rearrangement is thought to involve a full electron transfer followed by bond cleavage in the cyclopropane group:

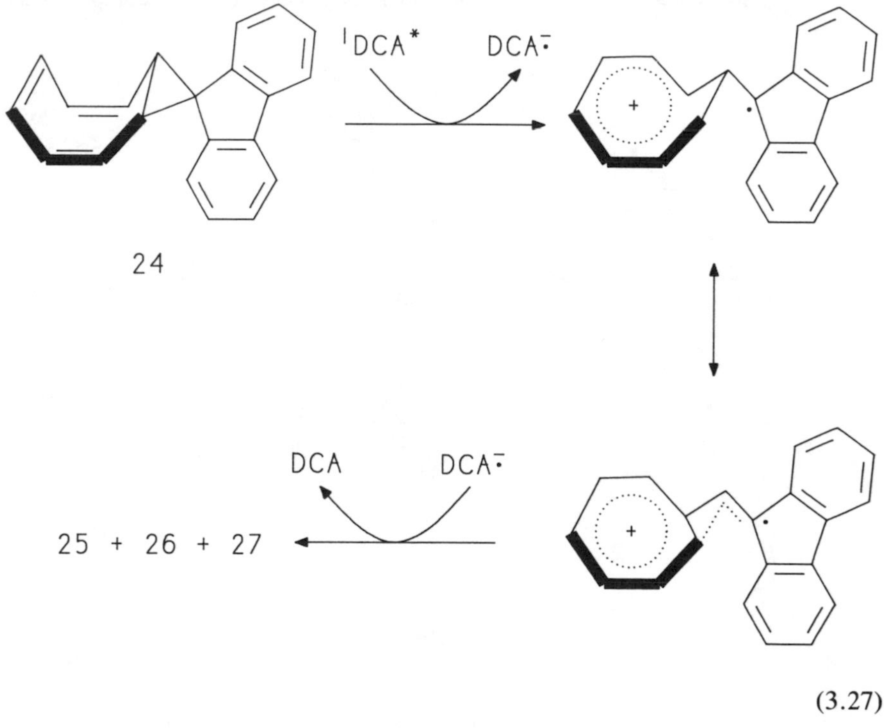

(3.27)

The 9,10-dicyanoanthracene-sensitized rearrangement of phenylated tricyclo [4.2.0.0^{2,5}]octane (**28**) yields 1,5-cyclooctadiene (**29**) and tricyclo[3.3.0.0^{2,6}] octane (**30**)[15]:

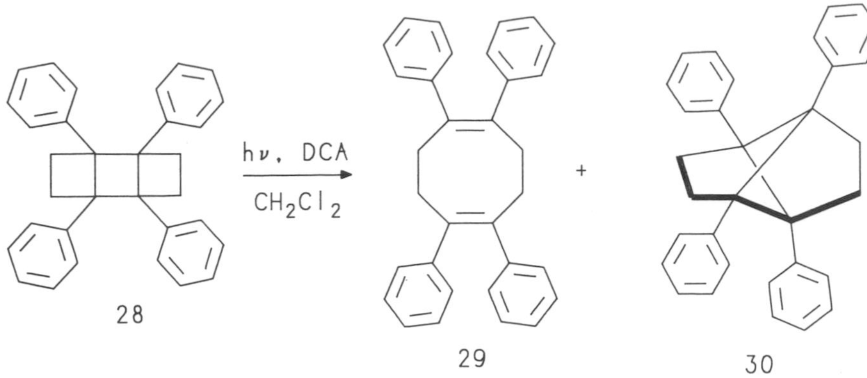

28 $\xrightarrow[CH_2Cl_2]{h\nu,\ DCA}$ 29 + 30

(3.28)

Electron transfer from **28** to singlet 9,10-dicyanoanthracene is estimated to be exothermic by ~ -3 kcal mol^{-1}. The salient feature in this rearrangement is cleavage of one of the cyclobutane carbon–carbon bonds. The fragmentation of this bond to a 1,4-cation radical apparently drives the reaction. A subsequent isomerization followed by transannular bond formation and finally electron return completes the reaction:

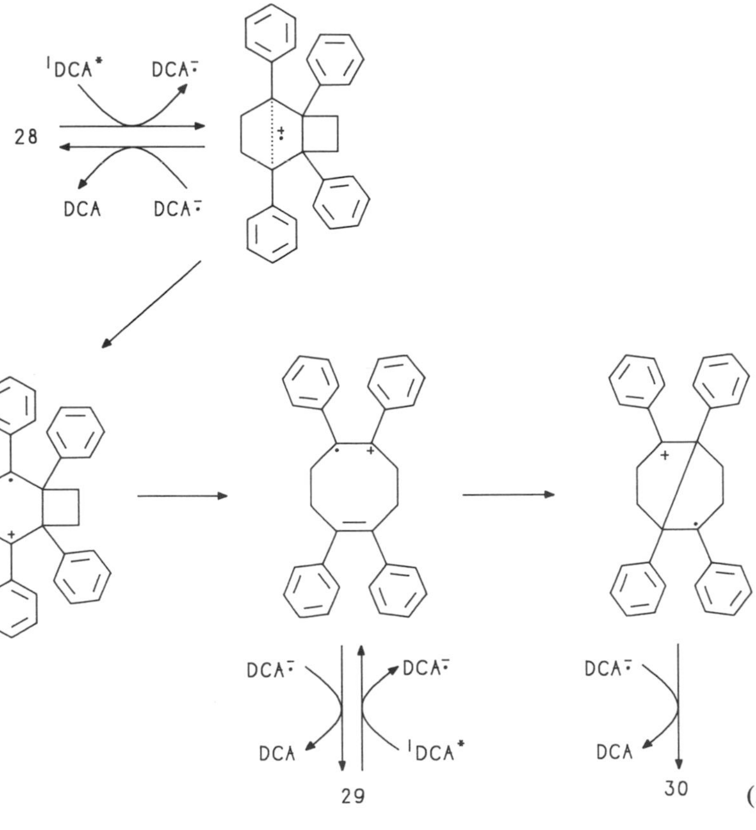

28 ⇄ (^{1}DCA*, DCA$^{\cdot-}$; DCA$^{\cdot-}$, DCA)

DCA$^{\cdot-}$ DCA ⇄ ^{1}DCA* DCA$^{\cdot-}$ 29

DCA$^{\cdot-}$ DCA 30 (3.29)

As shown in 3.29, there is also the possibility of sensitized rearrangement from **29** to **30.**

In concluding this section, mention must be given to valence isomerizations involving radical anion intermediates. Although not as numerous as those with cation radicals, a few examples have been documented. One such reaction is the valence isomerization of dimethylbicyclo[2.2.1]hepta-2,5-diene-2,3-dicarboxylate (**31**) and **32** in the presence of singlet aromatic sensitizers.[16,17]:

CO_2CH_3 CO_2CH_3 $\xrightleftharpoons{h\nu,\ Sens}$ CO_2CH_3 CO_2CH_3

31 32 (3.30)

Triplet recombination has been implicated as the major pathway prior to isomerization with several singlet sensitizers. Presumably, isomerization of the valence isomers proceeds through their triplet states.

3.3.2. Photosensitized C—C Bond Cleavages

The cleavage, or breaking, of saturated carbon–carbon (C—C) bonds is an important facet of photochemistry. These reactions can be achieved by direct photoexcitation or by sensitization. In the following section, we direct our attention to C—C cleavages proceeding by sensitized electron transfer.

3.2.2.1. Cleavage of Saturated Carbon–Carbon Bonds via Radical Cations

Numerous photoexcited electron acceptors have the ability to remove an electron from carbon–carbon σ-bonds in certain organic substrates. These electron abstractions are often followed by cleavage of these bonds. Thus, strong electron acceptors can oxidize saturated hydrocarbons to their radical cation intermediates, which in turn may undergo bond cleavage to form more stable compounds. Even strong covalent C—C bonds perhaps with dissociation energies as large as 100 kcal mol^{-1} can suffer bond cleavage. Typically, the radical cation is formed as the excited electron acceptor removes an electron from the highest occupied molecular orbital of the carbon–carbon bond:

σ^* π^* σ π $\longrightarrow$ σ^* π^* σ π

A^* D $A^{\overline{\cdot}}$ $D^{\stackrel{+}{\cdot}}$

(3.31)

The radical cation thus formed can be stabilized by substituted electron-donating groups such as aromatic molecules and alkoxy groups.

In guaranteeing the success of a sensitized bond cleavage, thermodynamic criteria, which by now the reader is well familiar with, must be satisfied. In the case of saturated organic quenchers, energy transfer is not important since the singlet and triplet states of these molecules usually lie well above the singlet and triplet states of the sensitizers.

For an example of a sensitized cleavage, consider the ring opening of tetraphenylcyclopropane (**33**) in the presence of 9,10-dicyanoanthracene in acetonitrile[18]:

$h\nu$, DCA; CH_3CN

33 34 (3.32)

The mechanism that seems to explain the formation of **34** involves an initial removal of an electron from the carbon–carbon bond in **33** by the excited singlet state of 9,10-dicyanoanthracene. This is followed either by electron return to reactants or by a hydrogen migration. The final product **34** is formed as the radical anion of the sensitizer returns an electron to the intermediate radical cation:

$^1DCA^*$ $DCA^{\cdot -}$

33

DCA $DCA^{\cdot -}$

H

DCA $DCA^{\cdot -}$

34

H

(3.33)

Another example involves the singlet 9,10-dicyanoanthracene-sensitized bond cleavage of biphenyl-fused cyclobutanes. The sensitized cleavage of compounds possessing *cis*-fused biphenyl groups leads to anthracene and stilbene isomers[19]:

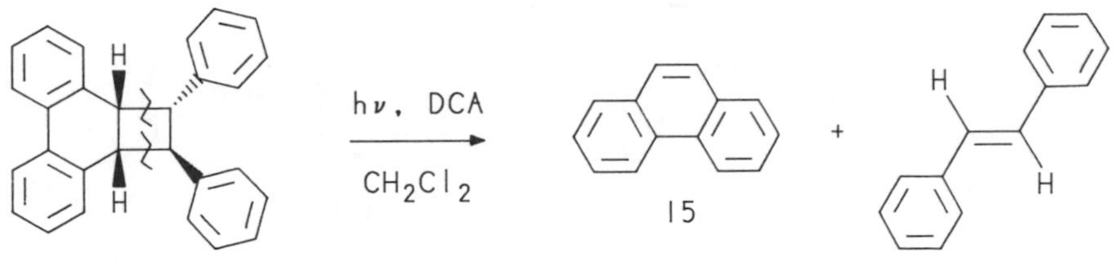

35 $\xrightarrow[CH_2Cl_2]{h\nu,\ DCA}$ 15 + 3

(3.34)

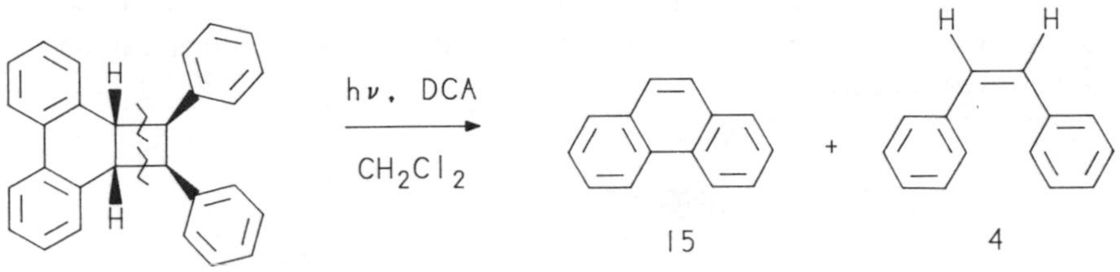

36 $\xrightarrow[CH_2Cl_2]{h\nu,\ DCA}$ 15 + 4

(3.35)

That these reactions proceed by electron transfer is supported by negative free energies and the observation that the quenching of DCA fluorescence occurs at near diffusion-controlled rates. After electron removal from the substrates, cleavage occurs across the bonds marked in 3.34 and 3.35. In contrast, the cleavage of **37**, a *trans*-fused isomer, leads to distyrylbiphenyl:

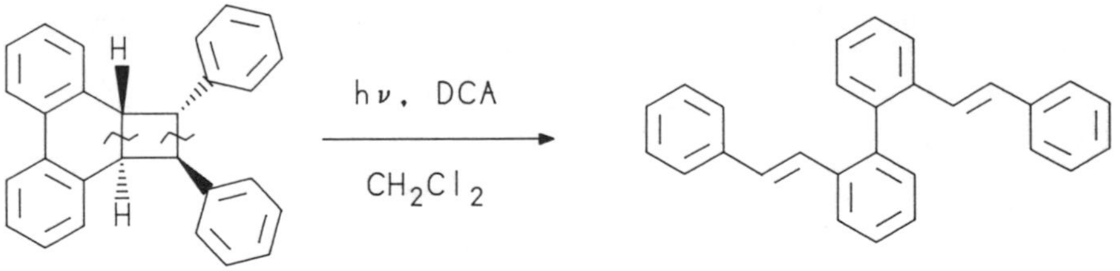

37 $\xrightarrow[CH_2Cl_2]{h\nu,\ DCA}$ 38

(3.36)

Sensitized bond cleavage of 1,1,2,2-tetraphenylethane (**39**) in acetonitrile/ methanol solutions leads to the formation of diphenylmethane (**40**) and methyl diphenylmethyl ether (**41**)[20]:

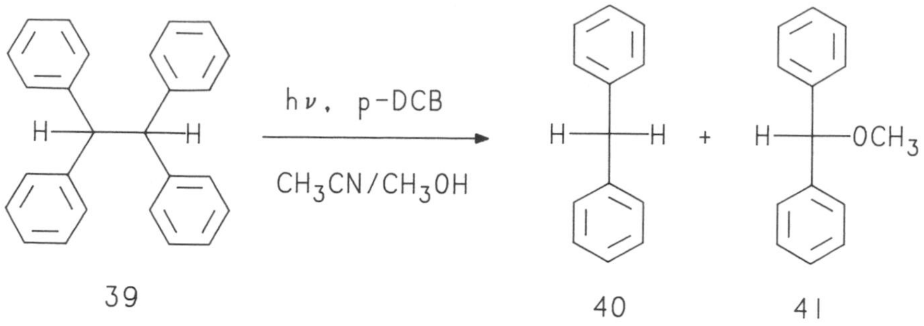

(3.37)

The initial step is formation of the radical cation of the ethane derivative followed by cleavage of this radical cation to diphenylmethyl radical and a carbocation. Subsequent electron-transfer, proton-transfer, and an alcohol addition culminate in the observed products:

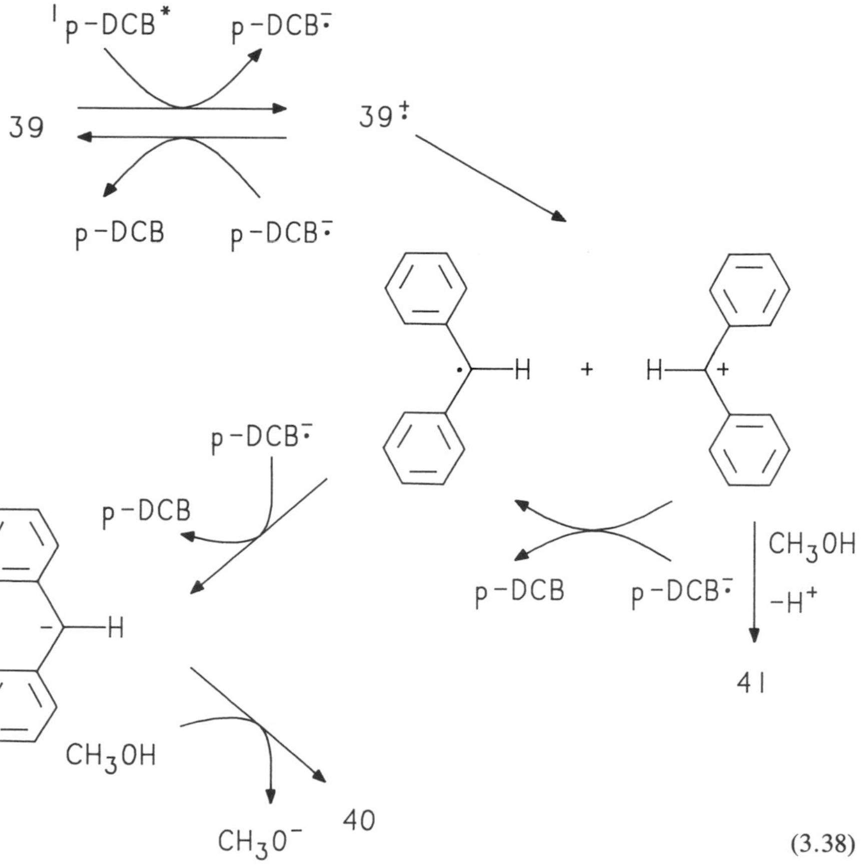

(3.38)

Cleavages of saturated bonds in aromatic pinacols and pinacol ethers have also been shown to proceed via sensitized electron-transfer pathways. In the case of these substrates, the hydroxy and methoxy groups exert a stabilizing effect on the intermediate radical cations. These reactions usually give rise to the formation of a complex product mixture. An example of this type of cleavage is shown below[21]:

$$\mathbf{42} \xrightarrow[CH_3CN]{h\nu,\ DCN} \mathbf{43} + \mathbf{44} + \mathbf{45} \qquad (3.39)$$

Irradiation of acetonitrile mixtures of **42** and 1,4-dicyanonaphthalene for 3 h yields benzophenone and reduced sensitizer. The initial step involves excitation of 1,4-dicyanonaphthalene followed by oxidation of the pinacol. Ketyl radicals can be generated by direct cleavage of the benzpinacol radical cation and electron return from the sensitizer's anion[22]:

$$42 + {}^1DCN^* \longrightarrow [42^{+\bullet} + DCN^{-\bullet}] \longrightarrow 2\ Ph_2\dot{C}OH + DCN \qquad (3.40)$$

Alternatively, ketyl radicals may be generated as shown in 3.41:

$$42 + {}^1DCN^* \longrightarrow [42^{+\bullet} + DCN^{-\bullet}] \longrightarrow \bullet O\text{-}CPh_2\text{-}CPh_2\text{-}OH + DCNH^{\bullet}$$

$$\bullet O\text{-}CPh_2\text{-}CPh_2\text{-}OH \longrightarrow Ph_2\dot{C}OH + 43; \quad Ph_2\dot{C}OH \xrightarrow{-H\bullet} 43; \quad DCNH^{\bullet} \xrightarrow{H\bullet} 44 + 45 \qquad (3.41)$$

With either 3.40 or 3.41, the formation of ketyl radicals is supported by flash photolysis studies. The majority of ketyl radicals disappear by coupling to form the starting benzpinacol substrate:

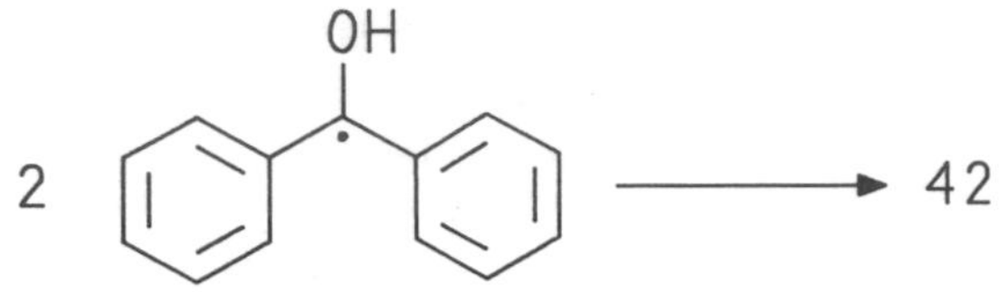

(3.42)

Several examples of sensitized bond cleavages which proceed by cosensitization have been documented. The stereospecific ring cleavage of **46** in the presence of *p*-dicyanobenzene and phenanthrene (**15**) is a typical case[23]:

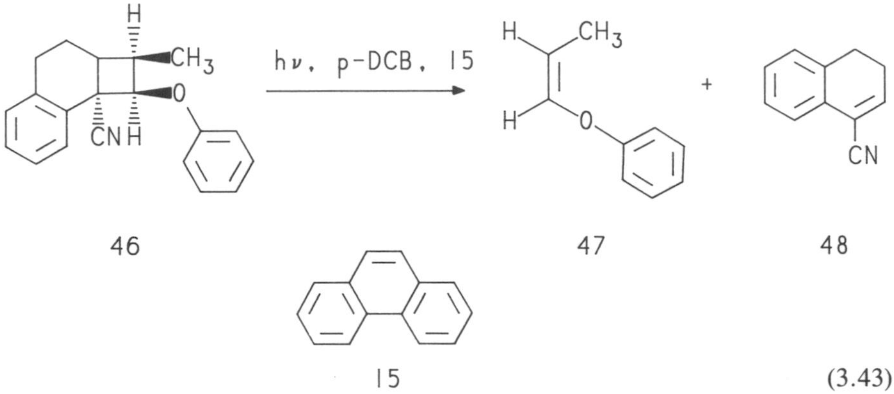

(3.43)

In 3.43 phenanthranene is the light-absorbing species. After excitation, its singlet state donates an electron to the cyanobenzene sensitizer (the cyclobutane substrate **46** does not quench phenanthrene fluorescence):

$$15 \xrightarrow{h\nu} {}^{1}15^{*} \quad (3.44)$$

$$^{1}15^{*} + \mathrm{DCB} \xrightarrow{k_{el}} [15^{\ddagger} + \mathrm{DCB}^{\bar{\cdot}}] \xrightarrow{k_{ret}} \mathbf{15} + \mathrm{DCB} \quad (3.45)$$

Given the large positive redox potential of **46,** oxidation of the cyclobutane derivative by $\mathbf{15}^{\ddagger}$ is expected to proceed with difficulty. Instead of a full electron-transfer, a charge-transfer interaction such as that shown in 3.46 was proposed:

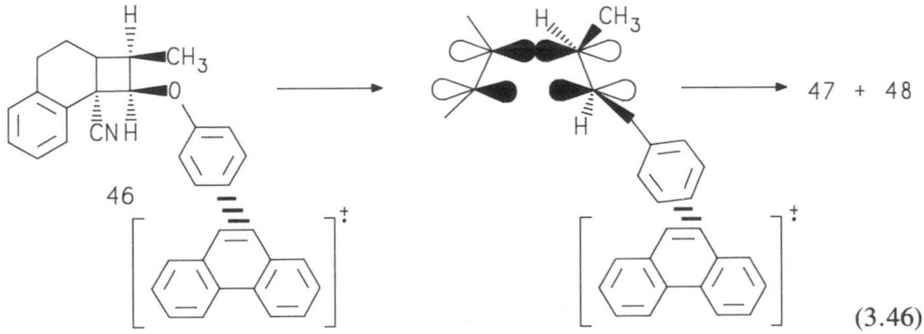

(3.46)

Stereoretention in olefin **47** is high, especially in the initial stages of the reaction. Consequently, the formation of a discrete radical cation capable of undergoing bond rotations is not likely. If a discrete radical cation were involved, $\mathbf{46}^{+\cdot}$ would likely isomerize to its more stable *trans* isomer. It appears, however, that the reaction may be assisted by formation of a π-complex between **46** and the phenanthrene radical cation. As bonds in the π-complex are weakened by the positive charge in phenanthrene radical cation, bond cleavage is favored.

3.2.2.2. Cleavage of Saturated Carbon–Carbon Bonds via Radical Anions

Although examples of sensitized bond cleavages of organic radical anions are not as numerous as those involving organic radical cations, certain amine donors have been shown to photosensitize the decomposition of some organic substrates (Table 3.2). These decompositions apparently proceed via bond-cleavage reactions of radical anion intermediates. For example, *N,N,N′,N′*-tetramethylbenzidine sensitizes the decomposition of the cyclic aryl pinacol sulfate **49** in acetonitrile[24]:

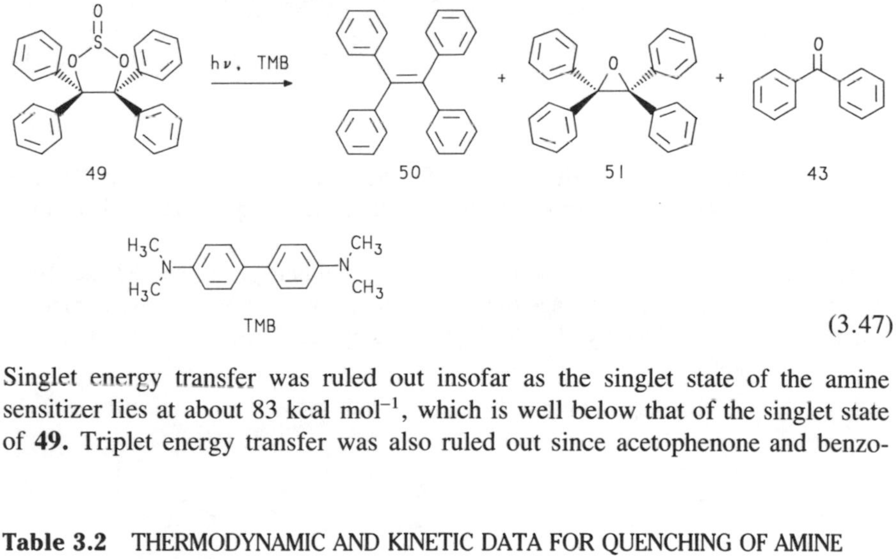

(3.47)

Singlet energy transfer was ruled out insofar as the singlet state of the amine sensitizer lies at about 83 kcal mol^{-1}, which is well below that of the singlet state of **49.** Triplet energy transfer was also ruled out since acetophenone and benzo-

Table 3.2 THERMODYNAMIC AND KINETIC DATA FOR QUENCHING OF AMINE SENSITIZERS[a]

Donor	Acceptor	ΔG_{el} (kcal mol^{-1})	k_q (M^{-1} s^{-1})
N,N,N′N′-Tetramethylbenzidine (S)	Dibenzylsulfone	−71.6	1×10^{10}
N,N,N′N′-Tetramethylbenzidine (T)	Dibenzylsulfone	−51.6	2×10^{7}
N,N-Dimethylaniline (S)	Acetophenone	−18.6	—
N,N-Dimethylaniline (T)	Acetophenone	−13.7	—

[a] Values from tables in Chapter 1 and references therein.

phenone, whose triplet energies are higher than the triplet of *N,N,N′,N′*-tetramethylbenzidine do not sensitize the reaction. With these considerations, it is likely that electron transfer is the dominant mechanism:

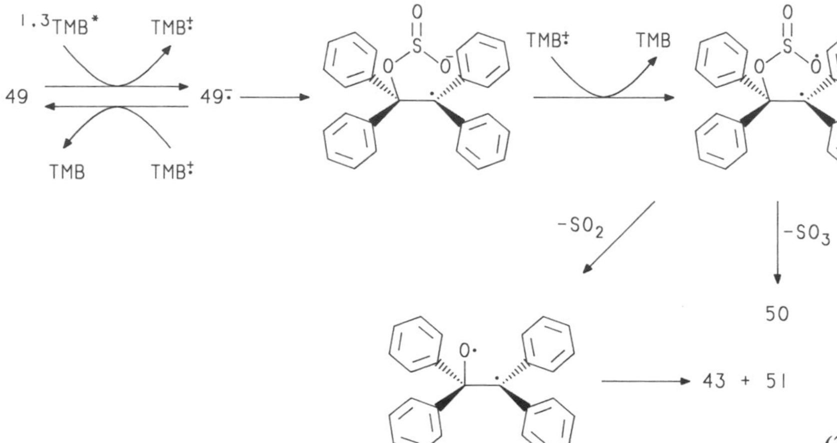

1,3TMB* TMB+· 49 49-· TMB TMB+· TMB+· TMB -SO2 -SO3 50 43 + 51

(3.48)

The reaction proceeds with electron transfer from the excited amine to **49.** The radical anion of **49** then undergoes cleavage to generate a ring-opened radical anion intermediate.

The sensitization of sulfur dioxide extrusion from dibenzyl sulfone (**52**) has been shown to occur via a ring-cleavage of a radical anion charge-transfer intermediate.[25] The reaction is sensitized by *N,N,N′,N′*-tetramethylbenzidine:

hν, TMB + SO_2

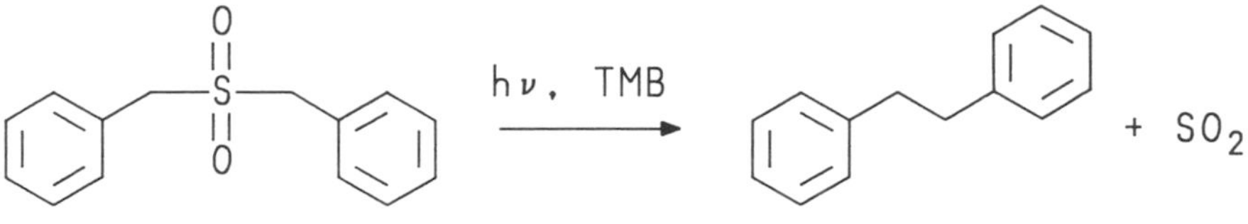

52 53 (3.49)

The free energies of quenching by electron transfer were estimated to be -67 kcal mol^{-1} (singlet quenching) and -47 kcal mol^{-1} (triplet quenching). A mechanism involving a charge-transfer complex between TMB and the sulfone rather than full electron transfer was proposed:

52 + ^{1,3}TMB ⟶ [$52^{\delta-}$ + $TMB^{\delta+}$] $\xrightarrow{-TMB}$ $C_6H_5\dot{C}H_2$ + $\cdot SO_2CH_2C_6H_5$

[$52^{\delta-}$ + $TMB^{\delta+}$] ⟶ 52 + TMB

$\cdot SO_2CH_2C_6H_5$ ⟶ $H_2\dot{C}C_6H_5$ + SO_2

$C_6H_5\dot{C}H_2$ + $H_2\dot{C}C_6H_5$ ⟶ 53

(3.50)

Product formation involving quenching of the triplet state was predicted to be more efficient than quenching of the singlet state where spin-allowed electron return to reactants should be more favorable.

3.2.3. Photosensitized Additions via Electron Transfer

The addition reactions of radical cations derived from alkene and other electron-rich molecules in the presence of photoexcited electron acceptors constitute one of the most important examples of electron-transfer photosensitizations. These photoadditions include dimerizations, cross-cycloadditions, and nucleophilic substitutions. In general, these conversions require highly polar media such as acetonitrile to aid in ion dissociation. In many instances, the addition of a salt can also aid in ion dissociation. We first consider the dimerization and cross-cycloadditions involving radical cations. The subsequent discussion will include nucleophilic additions. Our strategy will be to examine certain novel aspects of donor–acceptor additions paying special attention to the role of solvent polarity.

3.2.3.1. Sensitized [2π + 2π] and [4π + 2π] Dimerizations and Cycloadditions via Radical Cations

When solutions of a strong electron acceptor and an electron-rich olefin are irradiated in strongly polar solvents, the olefin may transfer an electron to the sensitizer. The newly created olefin cation radical may then combine with another electron-rich olefin. Frequently, the starting olefin is the only electron-rich substrate available so that it adds to itself. Ring closure of the newly formed radical cation dimer is followed by an electron return from the acceptor's radical anion, ultimately leading to a stable cyclobutane-like adduct:

$^1A^*$ → $A^{\cdot-}$; $A^{\cdot-}$ → A

(3.51)

Table 3.3 THERMODYNAMIC AND KINETIC DATA FOR QUENCHING OF 1-CYANONAPHTHALENE[a]

Donor	Acceptor	ΔG_{el} (kcal mol^{-1})	k_q (M^{-1} s^{-1})
Indene	1-Cyanonaphthalene (S)	−16.7	1×10^{10}
1,1-Diphenylethylene	1-Cyanonaphthalene (S)	−10	1.3×10^{10}
Indene	1-Cyanonaphthalene (T)	14.6	—
1,1-Diphenylethylene	1-Cyanonaphthalene (T)	21.6	—

[a] Values from tables in Chapter 1 and references therein.

This reaction is classified as a sensitized [2π + 2π] addition. It represents the simplest type of ring-closure where two ethylenic-like molecules combine to form two σ-bonds. When the olefin adds to itself, the cycloaddition is referred to as a dimerization. Photocycloadditions sensitized by electron transfer are stepwise reactions involving polar radical ion intermediates. Similar comments apply to [4π + 2π] cycloadditions.

The involvement of radical ion pair intermediates in sensitized [2π + 2π] and [4π + 2π] cycloadditions is consistent with an electron-transfer mechanism. This is confirmed by an examination of the redox potentials of the sensitizers and olefins. Table 3.3 summarizes the free energies accompanying several reactions where a singlet electron acceptor is quenched by an electron-rich olefin to give the latter's radical cation.

The dimerization of *N*-vinylcarbazole (**54**) in the presence of triplet chloranil was one of the early examples to be investigated[26]:

2 (54) —hν, CA→ (55)

CA

(3.52)

A chain mechanism is most likely involved here because quantum yields exceeding unity were measured:

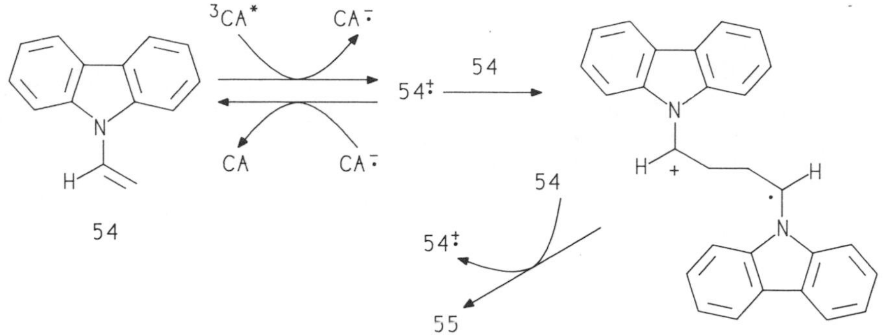

(3.53)

Because carbazole has a high triplet energy (E_T = 70 kcal mol^{-1}), triplet–triplet energy transfer involving sensitization with chloranil (E_T = 62.3 kcal mol^{-1}) is virtually ruled out. In fact, electron transfer is exothermic since $\Delta G_{el} = -36.3$ kcal mol^{-1} using the values from Tables 1.7 and 1.9. Reaction 3.52 can also be sensitized with benzophenone, methylene blue, and anthraquinone.

The dimerization of dimethylindene (**56**) in the presence of singlet tripyridyl-tetrafluoroborate (TP^+) is shown in 3.54[27]:

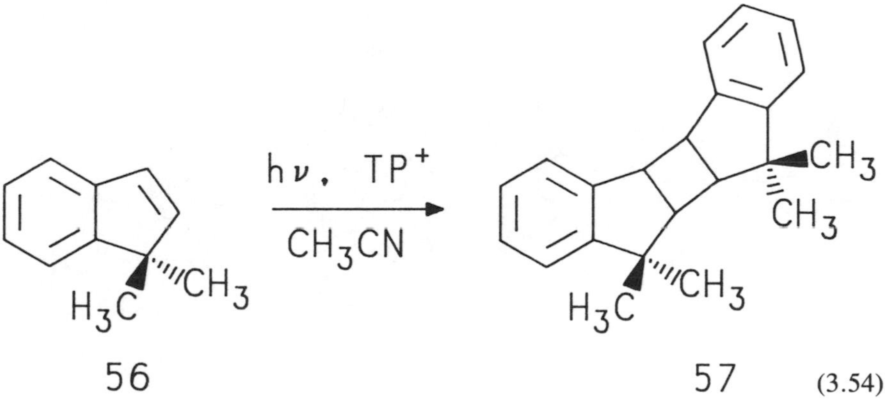

(3.54)

The predominant product formed here is the *anti*-head-to-head dimer (**57**).

When 1-cyanonaphthalene is irradiated in the presence of indene (**58**) and ethyl vinyl ether (**59**), cross-cycloaddition products and [$2\pi + 2\pi$] dimers are formed[28]:

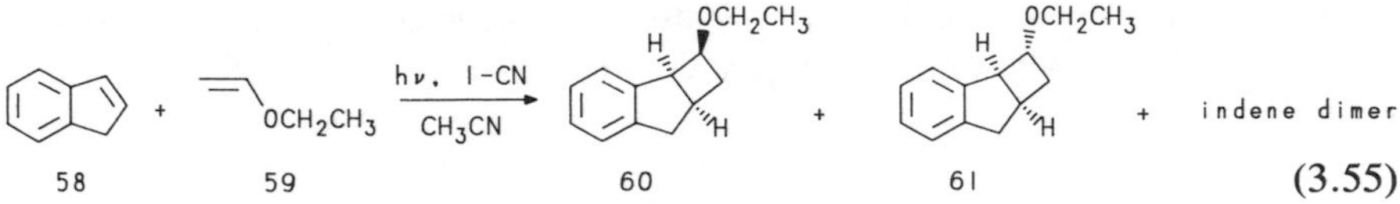

(3.55)

Oxidation of the indene, which has the lower ionization potential, is favored over oxidation of **59:**

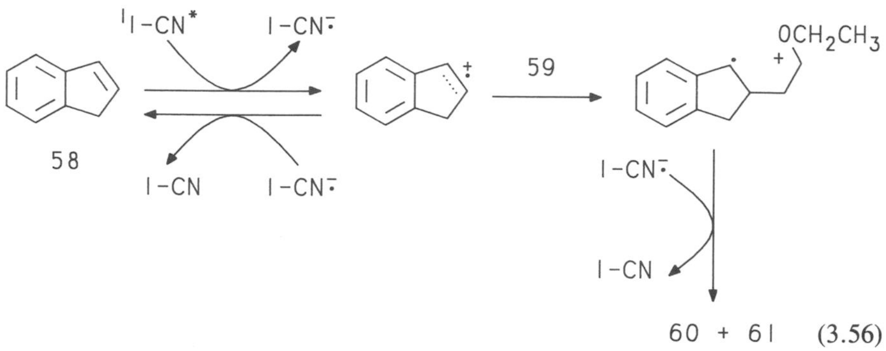

^{1}I-CN* I-CN$^{\cdot -}$ I-CN I-CN$^{\cdot -}$ 58 59 OCH_2CH_3 I-CN$^{\cdot -}$ I-CN

60 + 61 (3.56)

The proposed mechanism involving electron transfer is supported by the observations that reaction does not take place in nonpolar solvents, in the absence of the sensitizer, or in the presence of triethylamine. The amine, of course, is a relatively good electron donor and therefore can suppress product formation.

1,1-Diphenylethylene (**62**) dimerizes in acetonitrile via a radical cation intermediate in the presence of electron acceptors[29,30]:

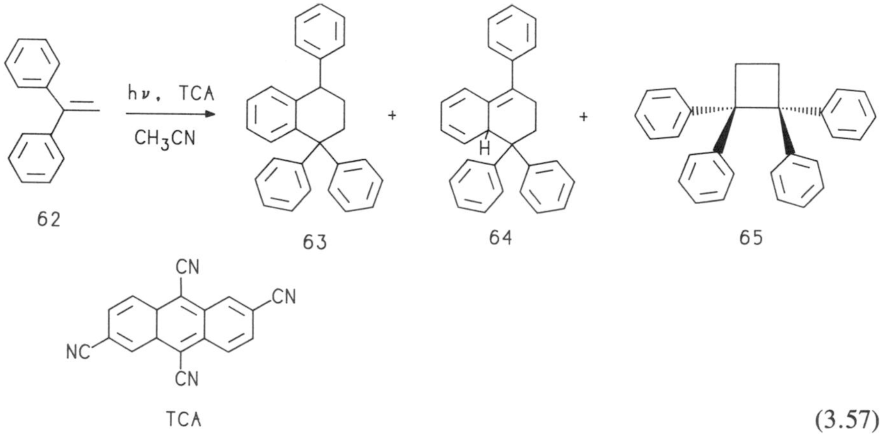

hν, TCA; CH_3CN; 62; 63; 64; 65; CN; CN; NC; CN; TCA

(3.57)

In this case, the product distribution depends on the olefin concentration. Farid has interpreted this dependence as due to partitioning between *in-cage* and *out-of-cage* products[30]:

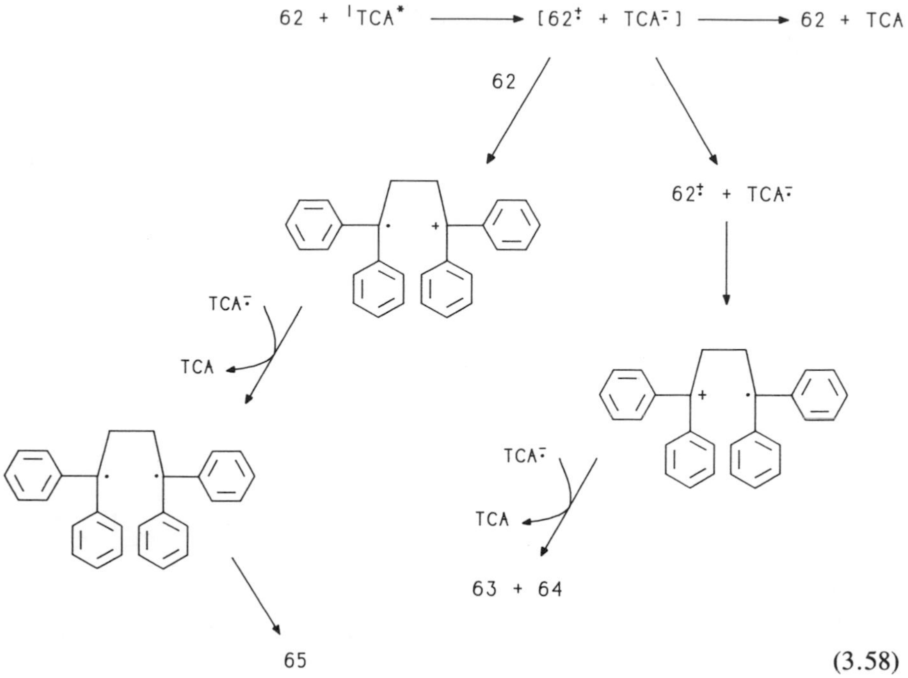

$$62 + {}^{1}TCA^{*} \longrightarrow [62^{\dagger} + TCA^{\overline{\cdot}}] \longrightarrow 62 + TCA$$

62

$62^{\dagger} + TCA^{\overline{\cdot}}$

$TCA^{\overline{\cdot}}$

TCA

$TCA^{\overline{\cdot}}$

TCA

63 + 64

65 (3.58)

With high olefin concentrations, increasing levels of the cyclobutane product are formed. This concentration dependence can be understood if the existence of two different types of radical ion-pair intermediates is assumed as shown in 3.58. Quenching by electron transfer initially takes place within a solvent cage generating a contact ion pair. The ion pair can subsequently undergo ionic dissociation to give a dissociated radical cation derivative of the olefin. Cyclization in the dissociated radical cation then leads to **63** and **64.** On the other hand, at higher olefin concentrations the contact ion pair can be intercepted by a neutral olefin yielding a biradical intermediate.

Proceeding to other examples of [2π + 2π] cycloadditions, we note that the 9,10-dicyanoanthracene-sensitized cyclization of benzyl vinyl ethers leads to [2π + 2π] cycloadducts[31]:

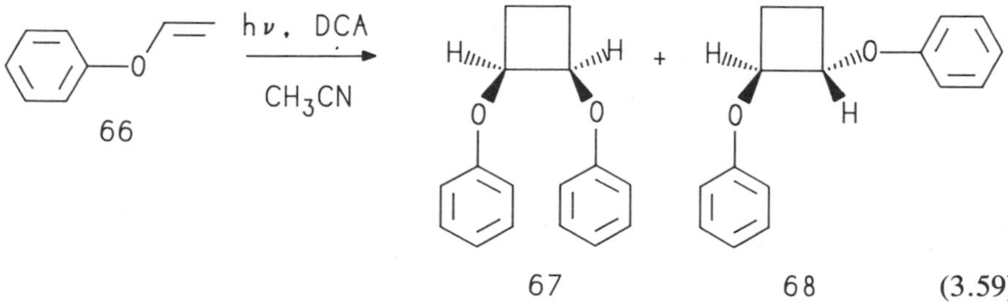

$$\xrightarrow[CH_3CN]{h\nu,\ DCA}$$

66 67 68 (3.59)

The reaction proceeds efficiently in acetonitrile. The formation of cyclodimers of trimethylsilyl vinyl ethers is shown in 3.60:

$(H_3C)_3SiO$—O (69) $\xrightarrow[CH_3CN]{h\nu,\ DCN}$ 70 + 71

(3.60)

The following pathway has been proposed:

$$69 + {}^{1}DCN^{*} \longrightarrow [69^{\dot{+}} + DCN^{\dot{-}}] \longrightarrow 69 + DCN$$

$$[69^{\dot{+}} + DCN^{\dot{-}}] \xrightarrow{69} 70 + 71 + DCN \qquad (3.61)$$

Electron-transfer-sensitized [4π + 2π] cycloadditions invite comparison with the well-known thermal Diels–Alder [4π + 2π] cycloaddition between ethylene and butadiene:

$$72 + 73 \xrightarrow{\Delta} 74 \qquad (3.62)$$

Diels–Alder reactions are usually concerted. Under ideal conditions, they proceed with high reaction rates. These reactions are favored by strong orbital overlap between the HOMOs and LUMOs of the reacting olefins. It is known that conversion of one of the reactants to its radical cation by *nonphotochemical* means can in fact enhance reaction rates. For example, conversion of 1,3-cyclohexadiene (**75**) to its radical cation in the presence of a cation-radical salt does indeed enhance the formation of Diels–Alder type dimer over the uncatalyzed reaction[32]:

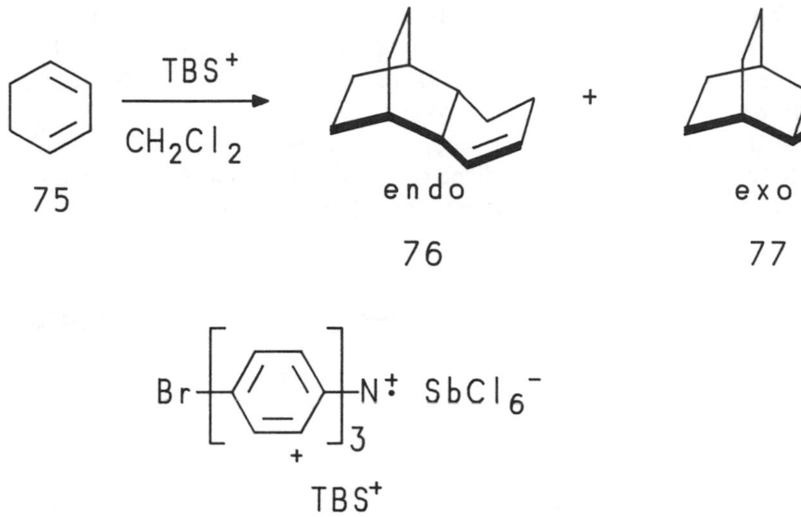

TBS⁺ / CH_2Cl_2

75 → endo 76 + exo 77

$Br\text{-}[C_6H_4]_3N^{+\cdot}\ SbCl_6^-$

TBS⁺

(3.63)

As compared to the thermal dimerization shown in 3.62, which results in only 30% yield in 20 h, reaction 3.63 proceeds with 70% yield in only 15 min! The *endo/exo* ratio is about 5. Similarly, Diels–Alder cycloadditions can occur in the presence of photoexcited electron acceptors, which convert one of the olefins to its radical cation. Thus, in the presence of singlet 9,10-dicyanoanthracene, 1,3-cyclohexadiene (**75**) forms the [$4\pi + 2\pi$] adducts **76** and **77**.[33]

The formation of [$4\pi + 2\pi$] adducts may also be accompanied by the formation of [$2\pi + 2\pi$] products. The dimerization of 1-acetoxy-1,3-cyclohexadiene (**78**), for example, can be sensitized by 1,4-dicyanonaphthalene[33]:

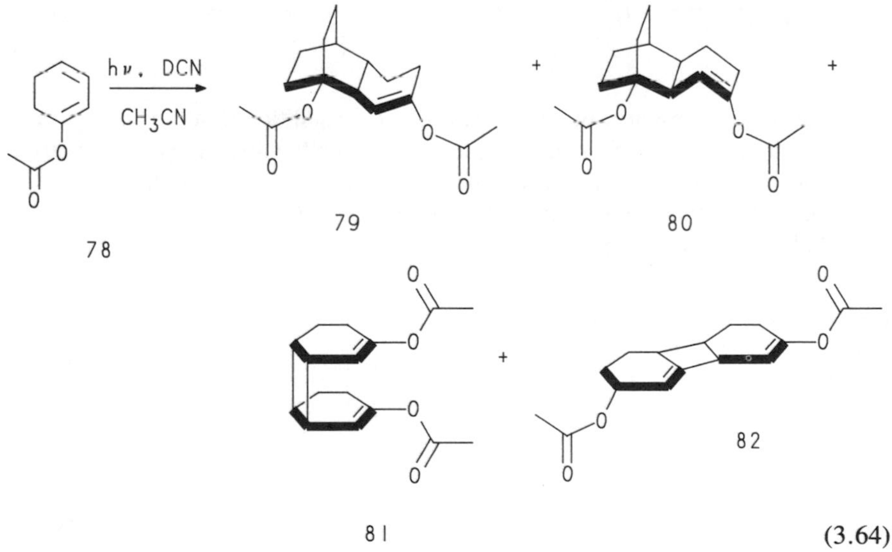

hν, DCN / CH_3CN

78 → 79 + 80 + 81 + 82

(3.64)

The product distribution depends on the concentration of olefin substrate. The *endo* dimer (**80**) is favored at low concentrations of **78** and the *exo* dimer at higher olefin concentrations. The ratio of [$2\pi + 2\pi$] (**81** and **82**) to [$4\pi + 2\pi$] (**79** and **80**) products increases with increasing olefin concentration.

Substituted electron-rich cyclohexadienes can cross-react with other olefins. As an example, 2-acetoxy-1,3-cyclohexadiene (**83**) and 1,4-dioxene (**84**) combine in the presence of certain electron acceptors to give [$4\pi + 2\pi$] products[33]:

83 + 84 —($h\nu$, TP^+ / CH_3CN)→ 85 + 86

(3.65)

In 3.65, oxidation of 1,4-dioxene is favored because of its lower ionization potential.

3.2.3.2. Sensitized Nucleophilic Additions via Radical Cations

It is well known that radical cations are easy targets for attack by nucleophiles. Thus, alcohols, cyanides, or other electron-rich agents can function as powerful nucleophiles in electron-transfer photosensitizations. The general mechanism is

$^1A^*$ → $A^{\cdot-}$; Nu:; $A^{\cdot-}$ → A; Products

(3.66)

This pathway is an example of an *anti-Markovnikov* addition (Markovnikov's rule states that an unsymmetrical reagent adds to a double bond such that the positive portion of the reagent attaches to the end of the olefin with the larger number of hydrogen atoms). We can write the following pathway for a sensitized addition of an alcohol to an olefin such as 1-propene (**87**):

87 —($^1A^*$ → $A^{\cdot-}$)→ radical cation + $H\ddot{O}CH_3$ → H—$\overset{+}{O}CH_3$ adduct —($A^{\cdot-}$ → A, $-H^+$)→ anion —($+H^+$)→ $CH_3CH_2CH_2OCH_3$

88 (3.67)

Note that the orientation in a photosensitized addition reflects the stability of the "*anti*-Markovnikov" intermediate. Irradiation of acetonitrile/methanol solutions containing methyl *p*-cyanobenzoate and 1-phenylcyclohexene (**89**) results in two *anti*-Markovnikov addition products (**90** and **91**)[34]:

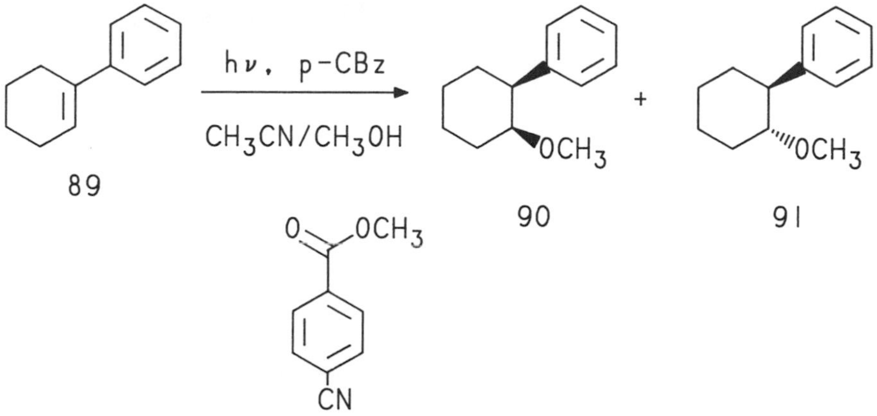

$$89 \xrightarrow[CH_3CN/CH_3OH]{h\nu,\ p\text{-CBz}} 90 + 91 \quad (3.68)$$

89 90 91

p-CBz

Similarly, 1,1-diphenylethylene (**62**) can be converted to an ether (**92**)[29]:

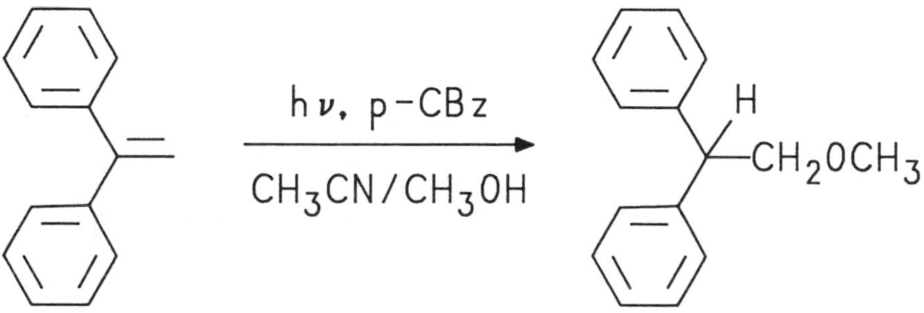

$$62 \xrightarrow[CH_3CN/CH_3OH]{h\nu,\ p\text{-CBz}} 92 \quad (3.69)$$

62 92

Cyanide ion can also attack certain olefins, converting them to their nitrile derivatives. Since aromatic nitriles are important precursors in many organic synthetic schemes, the preparation of these compounds is a subject of great interest to chemists. Cyanation using cyanide salts can be carried out by electron-transfer photosensitization[35]:

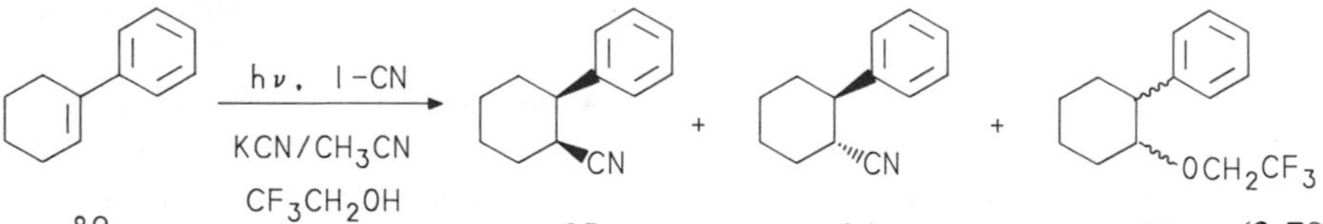

$$89 \xrightarrow[KCN/CH_3CN,\ CF_3CH_2OH]{h\nu,\ 1\text{-CN}} 93 + 94 + 95 \quad (3.70)$$

89 93 94 95

As in the case of the sensitized addition of alcohols, the mechanism proceeds via a nucleophilic attack:

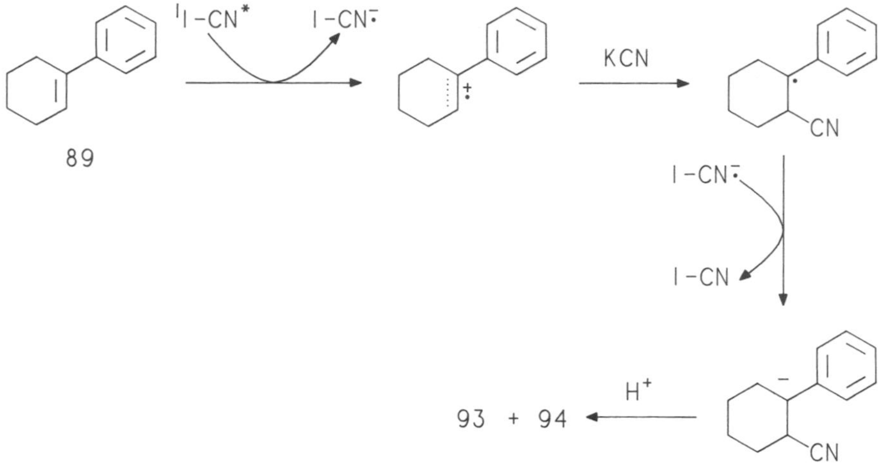

¹I-CN*
I-CN⁻·
89
KCN
CN
I-CN⁻·
I-CN
H⁺
93 + 94
CN

(3.71)

Nucleophilic attack can also proceed via cosensitized pathways. For example, irradiation of acetonitrile/methanol solutions of phenanthrene, *p*-dicyanobenzene, and indene results in nucleophilic attack by methanol[36]:

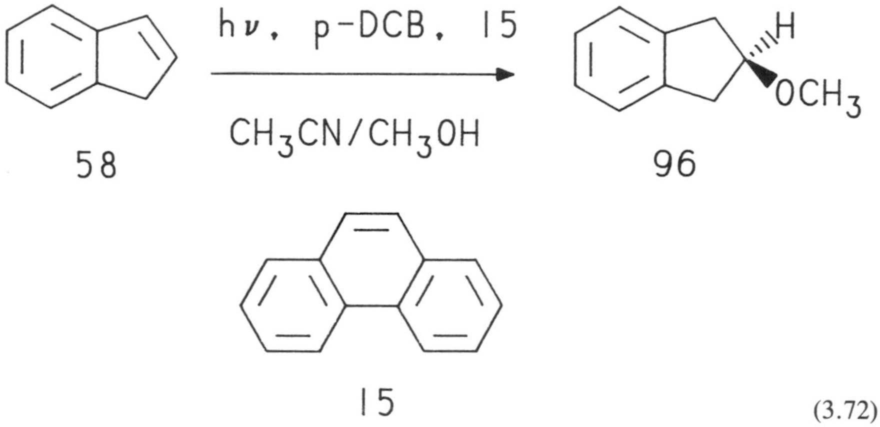

hν, p-DCB, 15
CH₃CN/CH₃OH
58
H
OCH₃
96
15

(3.72)

The mechanism is similar to the cosensitized pathways discussed earlier in this chapter. The primary electron step is electron donation from singlet phenanthrene to the *p*-dicyanobenzene.

There is an interesting class of related reactions where a bond cleavage of a σ-bond is accompanied by nucleophilic addition to a radical cation intermediate: the photosensitized reactions of strained cyclic organic molecules in alcoholic solutions. These highly strained polycyclic molecules are known to quench the

Table 3.4 THERMODYNAMIC DATA FOR THE QUENCHING OF SINGLET 1-CYANONAPHTHALENE BY HIGHLY STRAINED CYCLIC HYDROCARBONS[a]

Quencher	$E^0(D^{+\cdot}/D)$ (eV)	ΔG_{el} (kcal mol^{-1})
Tetracyclo[3.2.0.0^{2,7}.0^{4,6}]-heptane	0.91	−24.1
1,2,2-Trimethylbicyclo[1.1.0]-butane	1.23	−16.7
Tricyclo[4.1.0.0^{2,7}]heptane	1.50	−10.5

[a] From reference 38.

fluorescence of certain aromatic hydrocarbon sensitizers quite efficiently. For example, the quenching of 1-cyanonaphthalene fluorescence by highly strained hydrocarbons usually proceeds by exothermic electron transfer (Table 3.4).[37] Once oxidized, these strained molecules tend to undergo rearrangements. When these rearrangements take place in alcoholic solutions, nucleophilic addition products are often isolated. For example, in methanolic solutions, 1-cyanonaphthalene sensitizes the addition of methanol to tricyclo[4.1.0.0]heptane (**97**)[38]:

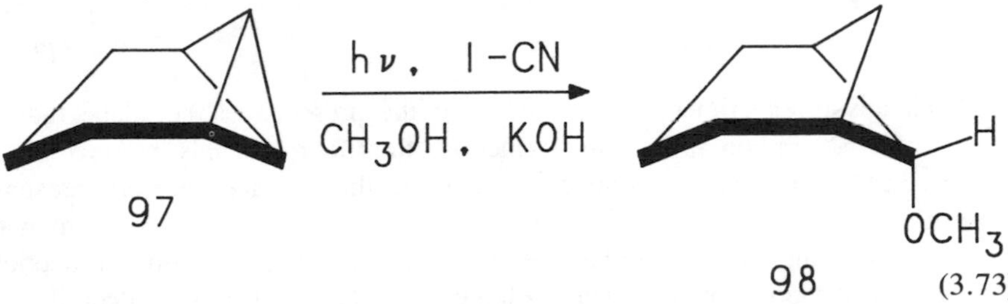

hν, 1-CN
CH_3OH, KOH
97
98
OCH_3
H
(3.73)

The reaction starts with formation of the radical cation of the strained tricyclo[4.1.0.0]heptane followed by nucleophilic addition of the alcohol:

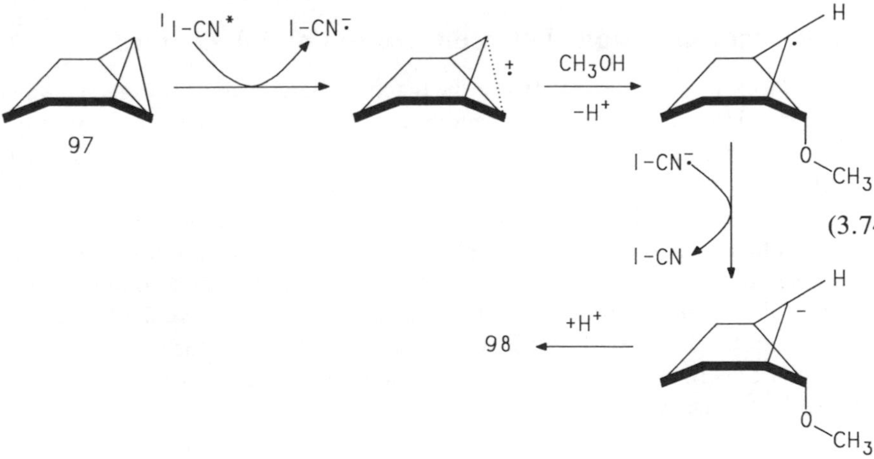

^{1}1-CN*
1-CN$^{-\cdot}$
97
CH_3OH
$-H^+$
1-CN$^{-\cdot}$
1-CN
H
O
CH_3
(3.74)
H
98
$+H^+$
O
CH_3

The scope of sensitized cleavages accompanied by nucleophilic addition has been extended to silicon-containing substrates. It has been shown, for example, that silane cation radicals undergo cleavage in the presence of certain nucleophiles. This has been demonstrated for the 9,10-dicyanoanthracene-sensitized oxidation of a series of substituted (*p*-methoxybenzyl) trialkylsilanes. The singlet state of the electron acceptor oxidizes the silane molecule to produce a silane cation radical, which is susceptible to attack by a nucleophilic solvent[39]:

$$\left[H_3CO{-}C_6H_4{-}CH_2Si(CH_3)_3 \right]^{\ddagger} \ (\mathbf{99}) \xrightarrow{Nu:} \left[H_3CO{-}C_6H_4{-}CH_2Si(CH_3)_3\cdots Nu \right]^{\ddagger} \longrightarrow H_3CO{-}C_6H_4{-}\dot{C}H_2 + Si(CH_3)_3\cdots Nu^+ \quad (3.75)$$

Flash photolysis experiments have confirmed the presence of the silane radical cation by its absorption at 500 nm. Evidently, the lifetime of this intermediate is increased as larger alkyl groups are substituted on silicon, since, for steric reasons, these groups can cause nucleophilic attack to be less favorable. Consistent with this proposal is the observation that the rate constant with the addition of alcohols decreases in the order of increasing bulkiness: methanol, isopropyl alcohol, and *tert*-butyl alcohol.

3.2.4. Photoadditions between Donors and Acceptors

With several exceptions, the majority of the reactions described in previous sections fall with the classic definition of an electron-transfer photosensitization: after oxidizing or reducing a substrate, the sensitizer is eventually restored to its original state. However, to give a complete account of the scope of photoinduced electron transfer, we must include reactions where an excited state and quencher undergo an addition to give stable products. In photoadditions of donor and acceptors, the excited electron donor or acceptor does not operate as a sensitizer according to the conventional definition of sensitizer. Photoadditions usually involve the formation of charge-transfer complexes between the excited state and the quencher. The nature of the product distribution arising from a photoaddition is often subject to the nature of the solvent polarity. This dependence reflects the control that solvent polarity

Table 3.5 SUMMARY OF WOODWARD–HOFFMANN RULES[a]

$[m + n]$	Allowed	Forbidden
$4q$	$m_s + n_a$ (GS)	$m_s + n_a$ (ES)
	$m_a + n_s$ (GS)	$m_a + n_s$ (ES)
	$m_s + n_s$ (ES)	$m_s + n_s$ (GS)
	$m_a + n_a$ (ES)	$m_a + n_a$ (GS)
$4q + 2$	$m_s + n_s$ (GS)	$m_s + n_s$ (ES)
	$m_a + n_a$ (GS)	$m_a + n_a$ (ES)
	$m_s + n_a$ (ES)	$m_s + n_a$ (GS)
	$m_a + n_s$ (ES)	$m_a + n_s$ (GS)

[a] GS, ground state; ES, excited state; s, suprafacial, a, antarafacial; $q, = 1, 2, 3 \ldots$.

exerts on the delicate competition between formation of polar exciplexes and radical ions. Frequently, these pathways lead to conspicuously different product distributions. In photoadditions proceeding via exciplexes, the exciplex pathway is usually concerted and accompanied by the well-known structureless, broad band-like emission characteristic of exciplexes. These exciplexes can involve singlet or triplet excited states. In either case, the efficiency of product formation competes with emission and nonradiative decay pathways.

Concerted photocycloadditions involving the formation of exciplexes are, in many cases, governed by the Woodward–Hoffmann selection rules. These rules, which are summarized in Table 3.5, allow one to predict the "allowedness" or "forbiddenness" of $[m\pi + n\pi]$ concerted cycloadditions where m and n are the number of π-electrons participating in the reaction (the reader who is not familiar with the Woodward–Hoffmann rules may want to consult the references listed at the end of this chapter for further discussion). In Table 3.5, the subscripts "s" and "a" refer to *suprafacial,* where the bonds being formed are on the same face of the reactants, and *antarafacial,* where the bonds being formed are on opposite faces. The rules in Table 3.5 apply to a cycloaddition of an olefin having m π electrons to one having n π electrons. For example, for bond formation taking place on the same faces of the reactants (one of which is a photoexcited state), reaction is allowed when $m + n = 4q$ where $q = 1, 2, 3. \ldots$ A thermal cycloaddition between these reactants is forbidden.

In cycloadditions involving singlet excited states, the stereochemistry of the reactants is often preserved during the course of the reaction. These reactions are concerted and can give sterically hindered, thermodynamically unstable products. In contrast, sensitized photoadditions involving polar radical ion pair intermediates, such as those encountered in the previous section, are step-wise, nonconcerted processes. The nature of the photoaddition products often reflects the structure of the donor and acceptor ions within the solvent cage. These products are usually stable and not sterically crowded, as in the case of concerted cycloadditions. In stepwise photoadditions, the Woodward–Hoffmann rules usually do not apply.

3.2.4.1. Photoadditions between Aromatic Compounds and Olefins

3.2.4.1.1. Photoadditions via Exciplexes

These reactions are nonionic processes. The efficiencies of the formation of donor–acceptor additions generally decrease with an increase in solvent polarity. The cycloadducts are usually sterically congested and suggestive of face-to-face contact within an exciplex. In this section, we consider examples of [$2\pi + 2\pi$] photocycloadditions of excited aromatic hydrocarbons with olefins. One example is the reaction between singlet phenanthrene and *cis*- and *trans*-methyl cinnamate (**100** and **102**) in nonpolar solvents[40]:

OCH3 O 100 hν, 15 → 101 (O, OCH3) 15

(3.76)

O OCH3 102 hν, 15 → 103 (O, OCH3)

(3.77)

The reaction is accompanied by quenching of phenanthrene fluorescence. This [$2\pi + 2\pi$] Woodward–Hoffmann allowed cycloaddition turns out to be stereospecific. Since the stereochemistry of the olefin is preserved, the reaction apparently involves a face-to-face exciplex between the π-orbitals of the olefin and singlet phenanthrene. Although no exciplex emission could be detected, the structure of the addition products strongly suggests the intervention of a sandwich-like exciplex. This reaction is an example of a cycloaddition leading to the formation of a sterically crowded product.

Another well-known example is the [$2\pi + 2\pi$] photocycloaddition reaction of 9-cyanophenanthrene and *trans*- and *cis*-stilbene (**3**) and (**4**).[41] This reaction is accompanied by geometric isomerization of the stilbene isomers:

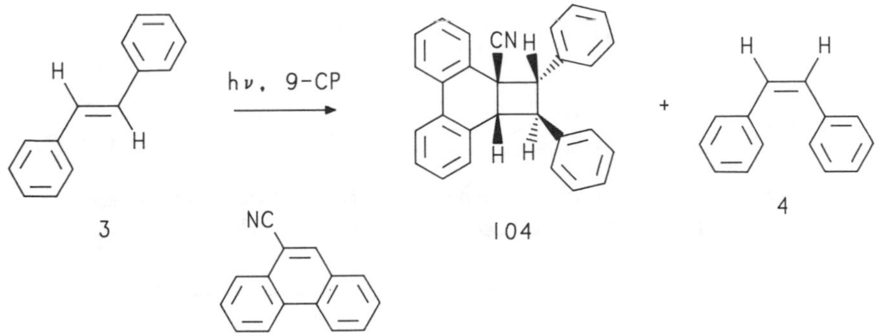

(3.78)

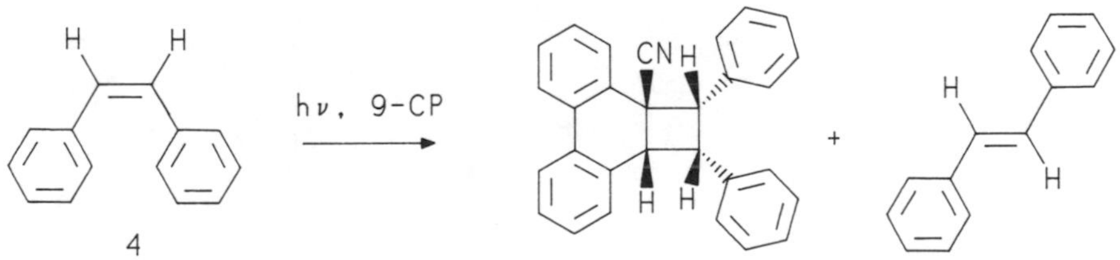

(3.79)

The product of the cycloaddition with *cis*-stilbene is the sterically crowded *cis, endo* isomer.

3.2.4.1.2. Photocycloadditions via Radical Ions

In contrast to the concerted nature of photocycloadditions via exciplexes, cycloadditions proceeding via radical ions are stepwise processes involving radical ion intermediates. The products of these ionic cycloadditions are generally in the more thermodynamically *anti* configuration. The rates of these reactions are enhanced in the presence of polar solvents.

The reactions of singlet aromatic cyanoaromatic compounds with olefins are typical of this class of cycloadditions. As noted previously, these reactions involve quenching by exothermic electron transfer and usually proceed by photosensitization pathways. Occasionally, these sensitizations are accompanied by additions between sensitizer and quencher. For example, addition products between *p*-dicyanobenzene and 2,3-dimethyl-2-butene (**106**) in acetonitrile have been identified[42]:

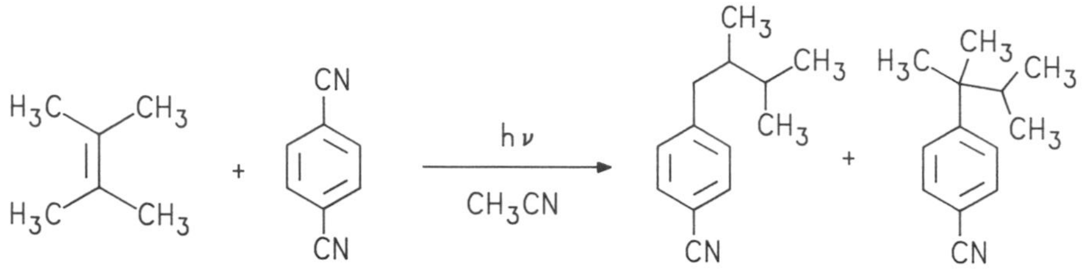

106 + p-DCB $\xrightarrow[CH_3CN]{h\nu}$ 107 + 108 (3.80)

The quenching of singlet *p*-dicyanobenzene by **106** is exothermic in acetonitrile (ΔG_{el} = -25.4 kcal mol^{-1}). The reaction sequence is shown below:

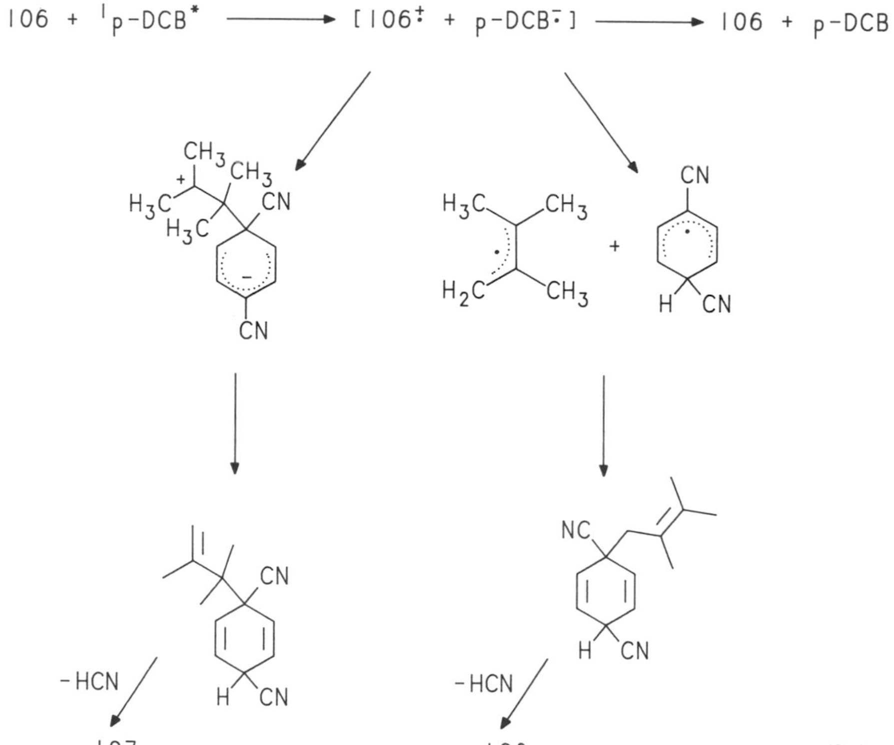

$$106 + {}^{1}p\text{-DCB}^{*} \longrightarrow [106^{+\cdot} + p\text{-DCB}^{-\cdot}] \longrightarrow 106 + p\text{-DCB}$$

$-$HCN → 107 $-$HCN → 108 (3.81)

Photoadditions similar to 3.80 appear to be quite general. That radical cations are formed is consistent with the strongly polar conditions in these systems, although these charge-transfer intermediates are probably similar to contact ion pairs. The strongly ionic character of these intermediates is supported by the observation that in the presence of nucleophilic alcohols, substituted products often are formed, e.g.[43]:

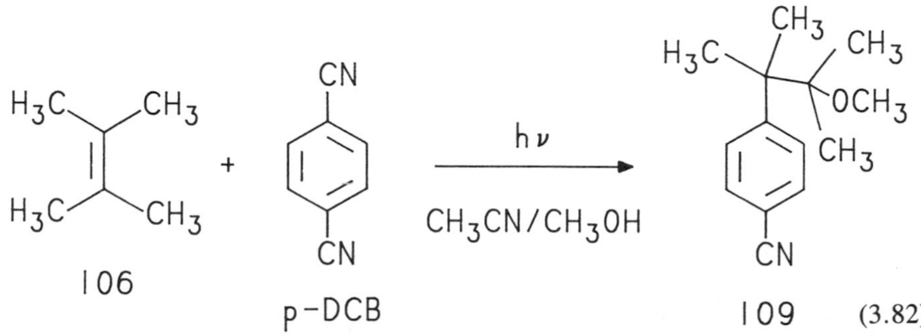

(3.82)

Reaction 3.82 implies the existence of radical cations, which can subsequently be trapped by nucleophiles. But completely dissociated ion pairs are probably not involved. Given the nature of the product, the ion-pair intermediate probably remains within the cage as a contact ion pair. Donor–acceptor photoadditions tend to be in-cage processes where the donor and acceptor intermediates remain in intimate contact.

Photoadditions between excited cyanoaromatic molecules and organic substrates such as the reaction of cumene (**110**) with 1,4-dicyanonaphthalene are also typical of this class of cycloadditions[44]:

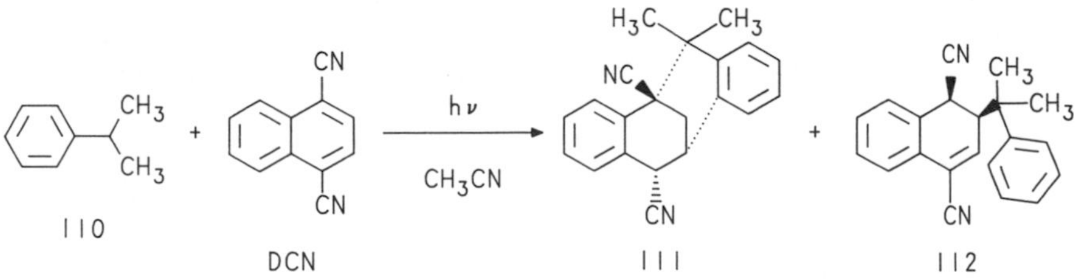

(3.83)

A key step in 3.83 is thought to be a proton transfer from **110** to the radical anion of 1,4-dicyanonaphthalene. This step is followed by coupling of the radicals. The authors of this study suggested that the radical cation and anion can be positioned parallel to one another in such a way as to favor proton transfer:

111 (3.84)

Evidence for a cage reaction was suggested by the high stereoselectivity in the products.

3.2.4.1.3. Directing the Course of Photoadditions

There has been intense interest concerning the effect of solvent in photocycloadditions. For example, Mattay observed that by using the modified Weller equation (Eq. 2.12), one can predict stereoselectivity.[45] The model he chose to investigate was the reaction of α,α,α-trifluorotoluene (**113**) with unsaturated molecules. Photoexcitation of **113** in the presence of olefins with high redox potentials leads to several cycloaddition products, e.g.,

113 + 114 → 115 + 116 (3.85)

The formation of the cycloadduct can be explained as a concerted [2π + 2π] cycloaddition. However, photoexcitation of **113** in the presence of an olefin with a lower redox potential such as **117** favors the formation of a substitution product:

113 + 117 → 118 (3.86)

Substitution via electron transfer is the preferred route when $\Delta G_{el} < 0$. The substitution product is formed through the formation of a contact ion pair that ultimately dissociates:

$$[C_6H_5CF_3]^* + 117 \longrightarrow [113^{\cdot-} + 117^{\cdot+}] \longrightarrow C_6H_5\dot{C}F_2 + \text{(dioxolanyl radical)} \longrightarrow 118 \quad (3.87)$$

$$[113^{\cdot-} + 117^{\cdot+}] \longrightarrow 113 + 117 \qquad [113^{\cdot-} + 117^{\cdot+}] \longrightarrow 113^{\cdot-} + 117^{\cdot+}$$

This example is particularly instructive in the sense that it highlights the differences between the concerted, *exciplex* pathway and the stepwise, *radical-cation* pathway. Although these pathways are influenced by reactant structure, solvent polarity also plays an important role by influencing the thermodynamics of ion-pair formation.

3.2.4.2. Photoadditions between Aromatic Compounds and Amines

The quenching of electron acceptors by amines may also be accompanied by formation of stable photoaddition products. The structures of these products depend, to a large degree, on the nature of the amine as well as the polarity of the solvent. The reaction of *trans*-stilbene (**3**) with tertiary amines illustrates this point.[46] *trans*-Stilbene forms exciplexes with tertiary amines in nonpolar solvents, as evidenced by exciplex emission. These exciplexes are unreactive in nonpolar solvents and normally form adducts. With increasing polarity, however, there is enhanced formation of radical ions (i.e., increasing polarity favors dissociation of the exciplex):

$$^1[3]^* + R_3N: \longrightarrow {}^1[3^{\cdot-}\,R_3N^{\cdot+}]^* \longrightarrow {}^1[3^{\cdot-} + R_3N^{\cdot+}] \longrightarrow 3 + R_3N$$

$$^1[3^{\cdot-}\,R_3N^{\cdot+}]^* \xrightarrow[\text{Nonpolar solvents}]{\text{Proton transfer}} \xrightarrow{\text{Coupling}} \text{Adduct}$$

$$^1[3^{\cdot-} + R_3N^{\cdot+}] \rightleftharpoons {}^3[3^{\cdot-} + R_3N^{\cdot+}] \xrightarrow{\text{Polar solvents}} {}^33^{\cdot-} + R_3N^{\cdot+} \longrightarrow 3 + 4 \quad (3.88)$$

Time-resolved resonance Raman spectroscopy (Section 2.3.3) has provided direct evidence that an ion pair is formed after dissociation of the exciplex. Once the exciplex dissociates into an ion pair, proton transfer is unlikely; rather, electron return accompanied by isomerization is more favorable for the ion pair.

In the reaction of unsymmetrical α-substituted trimethylamines with singlet *trans*-stilbene, the formation of two stilbene–amine adducts occurs.[47,48] With diisopropylmethyl amine, addition to the less substituted α—C—H bond predominates:

1 [stilbene radical cation] + 119 ⟶ 4⁻ + [amine radical cation] ⟶ [benzyl radical] + [α-aminoalkyl radical] ⟶ 120

H₃C, CH₃, N, H, H₂C, N⁺, H₂Ċ

119

120

(3.89)

It was postulated that deprotonation is favored by overlap between the half-vacant *p*-orbital of nitrogen and the developing orbital on the methyl carbon. Since there is less steric congestion for deprotonation of the methyl group, the hydrogen on the methyl group is favored to leave. Thus, it appears that steric factors rather than radical stability direct the course of this reaction. However, in other cases, resonance stabilization of the α-aminoalkyl radical is important. For example, the reaction of *trans*-stilbene with dimethylpropargyl amine (**121**) proceeds as shown below:

1 [stilbene radical cation] + 121 ⟶ 3⁻ + [amine radical cation] ⟶ [benzyl radical] + [α-aminoalkyl radical] ⟶ 122

H₃C, CH₃, N, CH₂, C, CH, H—CH, ·CH, H

3

121

122

(3.90)

The appearance of **122** as the major product suggests that the acetylene group enhances the acidity of the α—C—H proton.

The photoaddition reactions of excited electron acceptors with primary and secondary amines involve bond breaking of the N—H bond to form aminyl radicals as well as the α—C—H bond to form α-aminoalkyl radicals[49]:

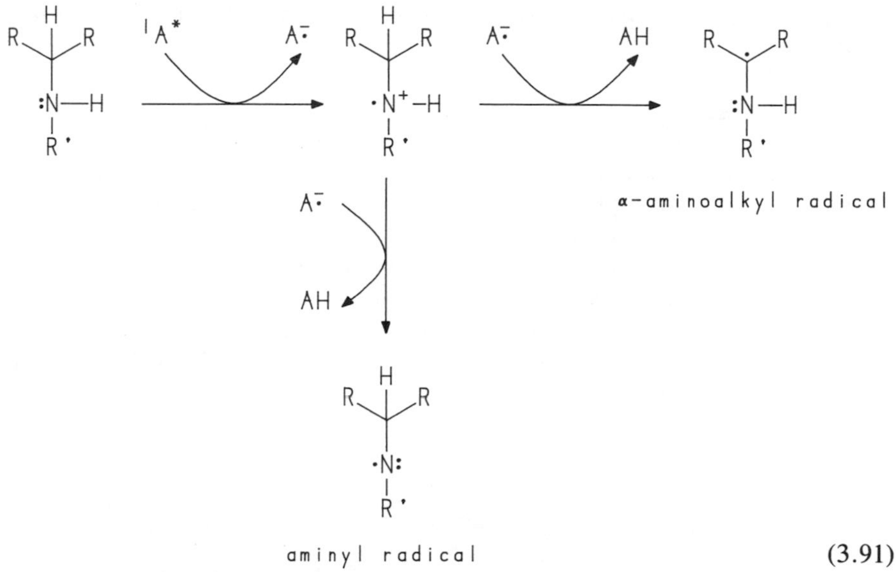

α-aminoalkyl radical

aminyl radical (3.91)

The formation of aminyl radicals is favored in nonpolar solvents. Probably this is influenced by hydrogen bonding between the N—H hydrogen and the singlet acceptor within an exciplex intermediate. Formation of α-aminoalkyl radicals is favored by resonance stabilization.

Thus, in nonpolar media, quenching of singlet 9-cyanophenanthrene by diethylamine (**123**) leads mainly to deprotonation of the N—H bond:

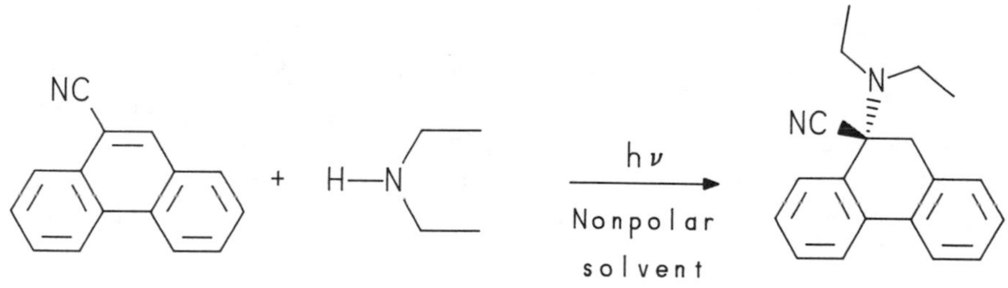

9-CP 123 124 (3.92)

As the strength of the α—C—H bond is reduced, the possibility of α-aminoalkyl radical formation is increased, as evidenced by the appearance of **127** in 3.93:

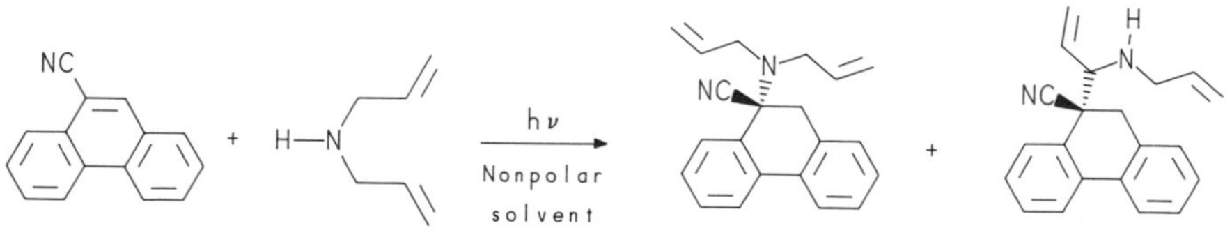

9-CP 125 126 127 (3.93)

α-Deprotonation in diallylamine is expected to be favorable because of resonance stabilization.

3.2.4.3. Photoadditions of Carbonyl Compounds

Carbonyl compounds such as ketones, aldehydes, and quinones possess the property to act as effective electron acceptors in the excited state. Quenching by electron transfer involving the singlet and triplet states of these sensitizers is often exothermic (Table 3.6). These quenching processes may be accompanied by the formation of stable coupling products. Many carbonyl compounds form long-lived triplets. n,π^* triplet ketones can react with olefinic molecules to form $[2\pi + 2\pi]$ cycloadducts known as oxetanes. If the excited ketone is a sufficiently strong oxidizing agent, it can remove an electron from the olefin to form a charge-transfer complex:

Polar solvent

Nonpolar solvent

(3.94)

In nonpolar solvents, the intermediate in brackets is an exciplex. The pathway may be regarded as an *exciplex–biradical* mechanism. In highly polar solvents, charge transfer is more complete resulting initially in the formation of a radical ion pair.

There are numerous examples where electron rich olefins are known to quench ketone emission. These quenching steps often lead to the formation of stable prod-

Table 3.6 THERMODYNAMIC AND KINETIC DATA FOR THE QUENCHING OF CARBONYL SENSITIZERS[a]

Donor	Acceptor	ΔG_{el} (kcal mol^{-1})	k_q (M^{-1} s^{-1})
N,N-Dimethylaniline	Benzophenone (T)	−7.3	—
N,N-Dimethylaniline	Acetophenone (T)	−13.7	—
Norbornadiene	Chloranil (T)	−28.5	—
Quadricyclane	Chloranil (T)	−43.1	—

[a] Values from tables in Chapter 1 and references therein.

ucts. Biacetyl (**128**), for example, reacts with electron-rich 1,3-dioxoles such as **129**[50]:

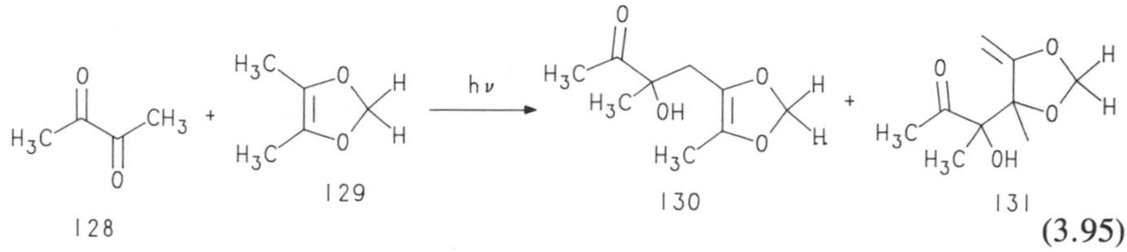

O, H_3C, CH_3, O (128) + H_3C, H_3C, O, O, H, H (129) —hν→ O, H_3C, H_3C, OH, H_3C, O, O, H, H (130) + O, H_3C, H_3C, OH, O, O, H, H (131)

(3.95)

The allylic hydrogen in **129** is labile so that deprotonation readily occurs within the radical pair. This is immediately followed by coupling:

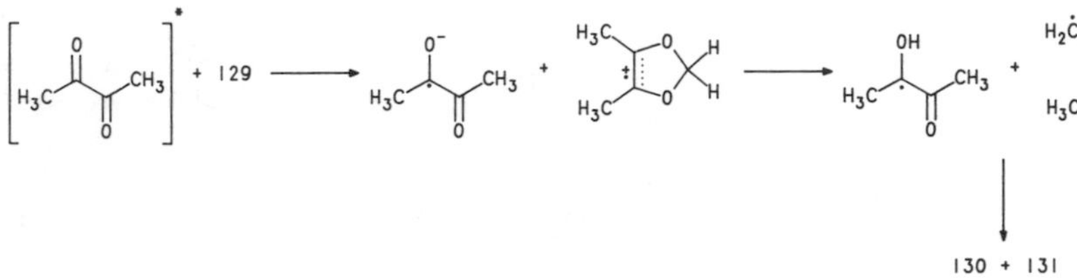

[H_3C, O, CH_3, O]$^{\bullet}$ + 129 ⟶ O^-, H_3C, CH_3, O + H_3C, H_3C, O, O, H, H ⟶ OH, H_3C, CH_3, O + $H_2\dot{C}$, H_3C, O, O, H, H

↓

130 + 131 (3.96)

Similarly, the reaction of triplet benzophenone (**43**) and **132** gives rise to an oxetane product:

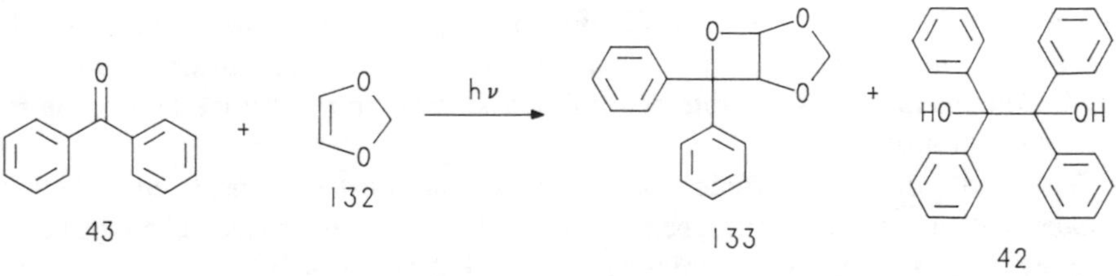

O (43) + O, O (132) —hν→ O, O, O (133) + HO, OH (42)

(3.97)

Increasing the solvent polarity results in decreasing yields of oxetane, probably due to an increase in the dissociation of an intermediate exciplex to radical ions. In fact, evidence for the benzophenone radical anion has been provided by nanosecond laser flash spectroscopy. The addition of a salt such as $LiClO_4$ was also shown to increase the yield of the radical ion pair.

Coupling products such as bibenzyls, pinacols, and benzpinacols are formed during the photoreduction of certain phenyl ketones by alkyl benzenes. In these reactions, the formation of charge-transfer intermediates competes directly with direct hydrogen atom abstraction. Quenching of the triplet states of benzophenones (n,π^*), acetophenones (n,π^* and π,π^*), and α,α,α-trifluoroacetophenones (π,π^*)

leads to the formation of modest yields of coupling products in the presence of toluene.[51] For instance, in the case of benzophenone, we have

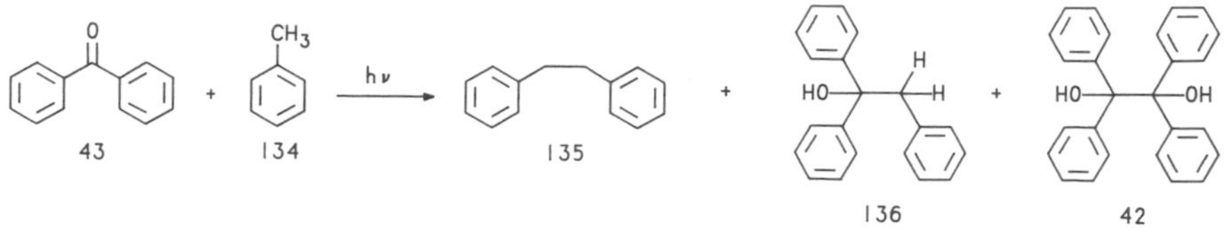

43 + 134 —hν→ 135 + 136 + 42

(3.98)

The following scheme summarizes the initial steps involved in the formation of **42, 135,** and **136:**

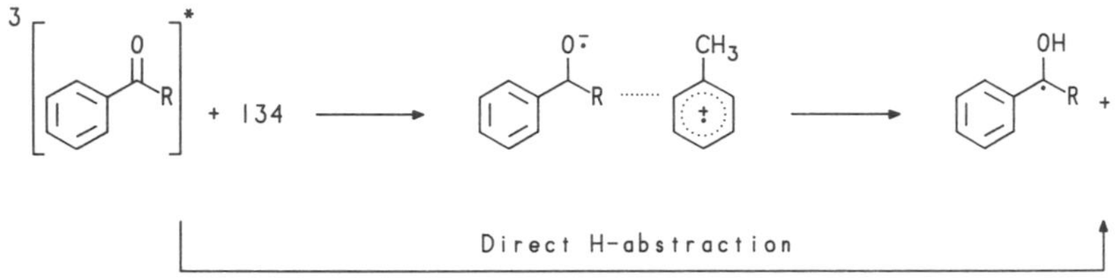

^{3}[PhCOR]* + 134 → [PhC(O$^{\cdot-}$)R ······ toluene$^{\cdot+}$] → PhC$^{\cdot}$(OH)R + PhCH$_2^{\cdot}$

Direct H-abstraction

(3.99)

As shown here, a one-step direct hydrogen abstraction competes with a two-step electron-transfer pathway. Wagner has used a similar scheme to analyze some of the features of ketone photoreduction reactions. For ketones with large redox potentials (e.g., those easily reduced), the rate constants of these reactions are usually much faster than can be expected if hydrogen abstraction were the dominant mechanism. For this reason, electron transfer is postulated to play an important role. Hydrogen abstraction is the rate-determining step when the ketone undergoes reduction with difficulty.

The nature of the electronic state of the ketone must be considered in ketone photoreductions. For example, the quenching of α,α,α-trifluoroacetophenone (**137**) by *p*-cymene (**138**) leads to a mixture of coupling products[51]:

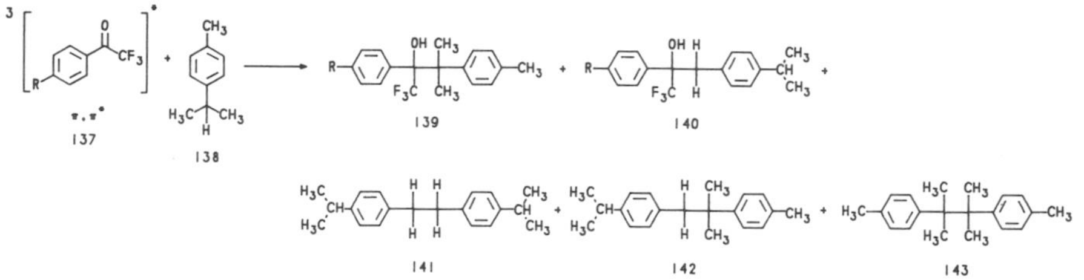

137 (π,π*) + 138 → 139 + 140 + 141 + 142 + 143

(3.100)

Under these conditions, *p*-cymene can fragment into both primary and tertiary radicals:

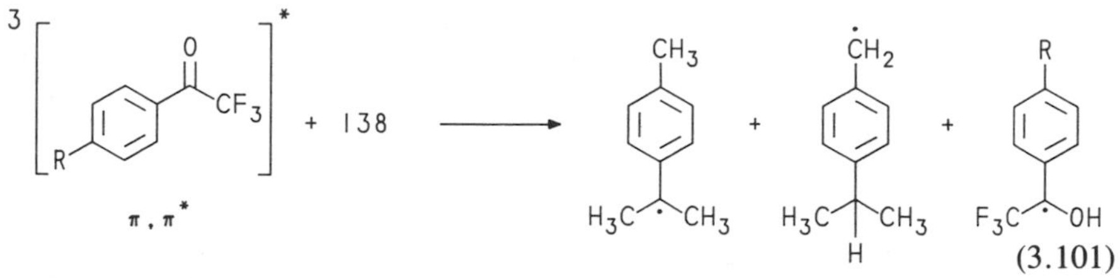

(3.101)

In comparing the product selectivity resulting from the quenching of n,π^* and π,π^* triplets, Wagner observed that quenching of π,π^* triplet ketones favors formation of primary radical products. He proposed that the high primary/tertiary ratios observed in the quenching of π,π^* triplets may be due to the preferred alignment of the methyl group over the acyl group[51]:

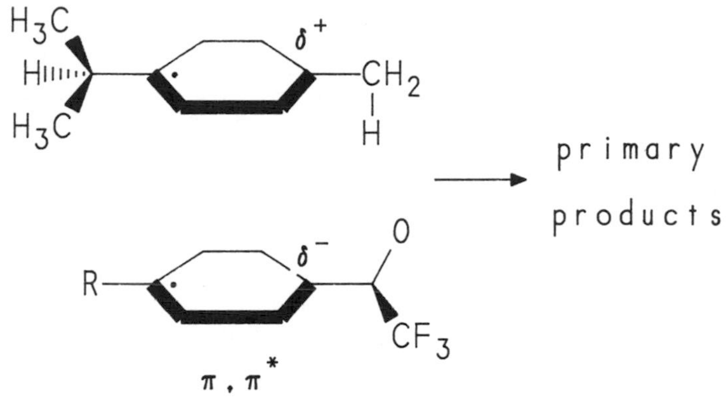

(3.101a)

In appreciating the difference in reactivity displayed by n,π^* and π,π^* triplet states, it is useful to recall that the aromatic group is electron deficient in aromatic carbonyl π,π^* states. This suggests that a face-to-face geometry between the π,π^* sensitizer and the aromatic quencher is favored. The preferred alignment is that shown above where the methyl group can better stabilize the partial positive charge on the alkyl benzene by hyperconjugation than can the isopropyl group. Since the preferred alignment brings the methyl hydrogen closer to the carbonyl group, formation of primary radicals and the resulting formation of primary products is favored.

Unlike the stereoelectronic reaction of π,π^* triplet ketones, which is governed by the positioning of the carbonyl group and the quencher, the reactions of n,π^* favor formation of tertiary radicals. In n,π^* triplet states, the carbonyl oxygen is more electron deficient than the aromatic group. Therefore, a face-to-face geometry is not as favored as in the case of π,π^* triplets. With n,π^* triplet states, the complex between the sensitizer and quencher assumes a less rigid orientation. Accordingly, radical stability rather than the relative orientation of the two reactants governs the course of the reaction (recall that the stability of free radicals goes as follows: $3^0 > 2^0 > 1^0$).

Excited carbonyls can also be quenched by amines. These reactions are normally exothermic (Table 3.6) and give rise to coupling products. The rate constants usually approach diffusion-controlled values. The evidence suggests that the quenching of excited carbonyls by amines is an electron-transfer process. As an example, the quenching of triplet acetophenone (**144**) by α-methylbenzylamine (**145**) was shown to be accompanied by formation of coupling products[52]:

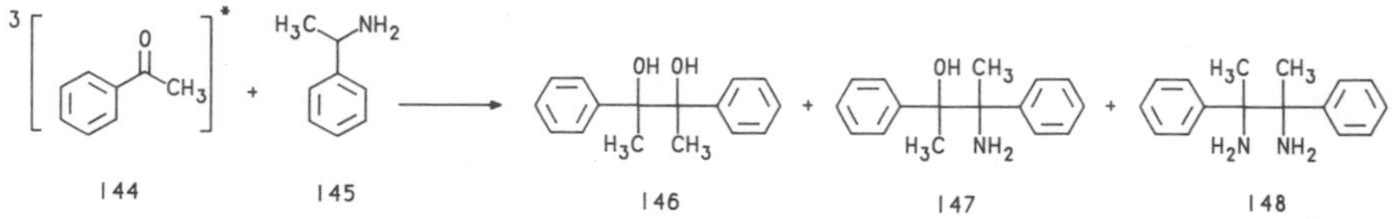

3 O CH3 * + H3C NH2 → OH OH H3C CH3 + OH CH3 H3C NH2 + H3C CH3 H2N NH2

144 145 146 147 148

(3.102)

N-Methylphthalimides are also known to undergo photoinduced electron-transfer additions. With electron-rich olefins, electron transfer from the olefin to the excited phthalimide leads to the formation of a radical ion pair followed by the formation of photoreduction products[53,54]:

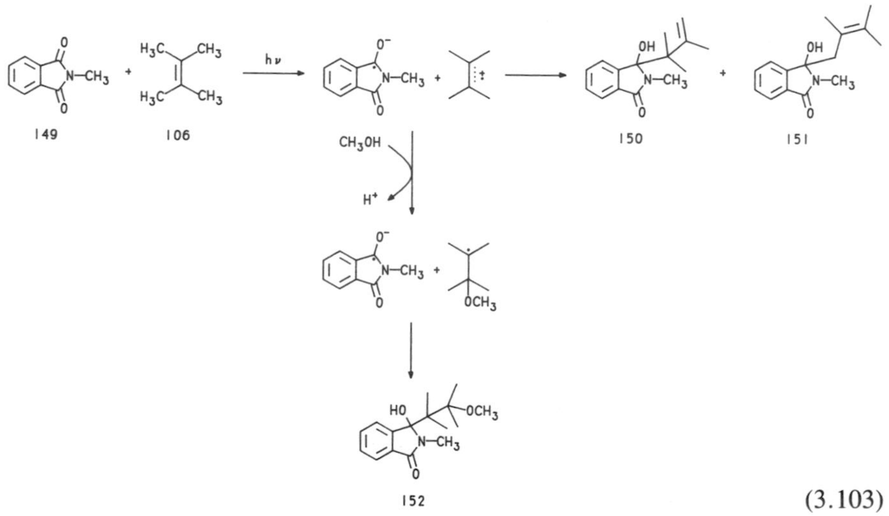

O N—CH3 O + H3C CH3 H3C CH3 —hν→ O⁻ N—CH3 O + (radical cation) → OH N—CH3 O + OH N—CH3 O

149 106 150 151

CH3OH H⁺

O⁻ N—CH3 O + OCH3

HO OCH3 N—CH3 O

152

(3.103)

The addition of a nucleophile such as methanol can trap the radical cation.

Similarly, irradiation of *N*-(2-methyl-2-propenyl)phthalimide in the presence of methanol leads to ring-closure products **154** and **155**[55]:

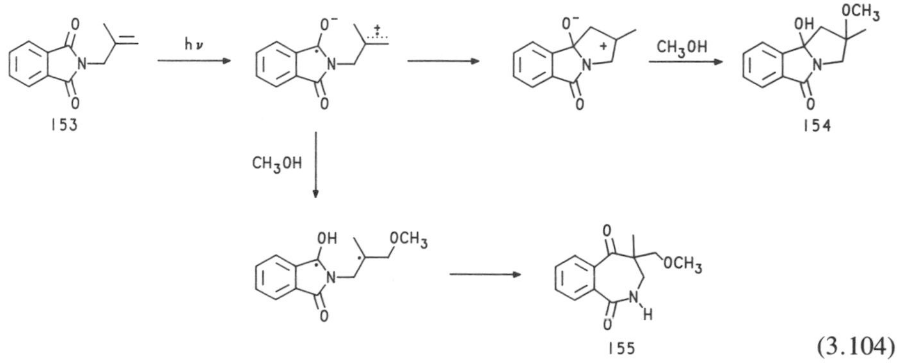

O N O —hν→ O⁻ N O → O⁻ + N O —CH3OH→ OH OCH3 N O

153 154

CH3OH

OH OCH3 N O → O OCH3 N H O

155

(3.104)

The formation of **154** and **155** lends support to the participation of an ion-pair intermediate.

3.2.4.4. Photoadditions to Iminium Salts

Mariano has undertaken an intensive investigation of the photoadditions of various substrates to the excited states of iminium salts.[56,57] These reactions suggest new approaches for the synthesis of *N*-heterocyclic molecules.

The LU orbitals of iminium salts are lower in energy than the orbitals of isoelectronic olefins. The enhanced electron affinity allows these compounds to perform as effective electron acceptors in their excited states[56]:

D D^{+}

(3.105)

From 3.105, we can appreciate that iminium salts are particularly reactive in the presence of electron donors such as olefins and arenes, and in the presence of certain nucleophilic agents. The photoaddition of isobutylene (**157**) to 2-phenyl-1-pyrrolinium (**156**) exemplifies the reaction[56]:

$h\nu$, CH_3OH

156 157 158 (3.106)

Similar reactions have been described for the addition of numerous olefins to excited pyrrolinium salts. These reactions are generally accompanied by the quenching of the fluorescence of the pyrrolinium salt with nearly diffusion-controlled rate constants. Support for an electron-transfer pathway has been provided by ESR and flash photolysis.

The electron-transfer photoadditions of certain allylsilanes to singlet 2-phenylpyrrolinium salts have also been documented. These reactions may involve the loss of an electrofugal group on the β position to the charged center. For an illustration, we note that irradiation of solutions of **159** and allylsilane (**160**) leads to **161.**[58,59] The pathway is shown in 3.107:

159 160 161 (3.107)

The formation of product can be rationalized if we consider σ- and π-orbital interactions between the silicon and olefin bonds. It is thought that the effect of these interactions is to lower the ionization potential of allylsilane so that it functions as a stronger electron donor. This notion receives some support from the observation that allylsilane is known to effectively quench the fluorescence of singlet pyrrolinium salts. Desilylation of the allylsilane radical cation is promoted by the presence of weak nucleophiles such as acetonitrile in the solvent. Reaction 3.107 represents a highly specific route for adding allylic groups to iminium salts. The efficiency of the reaction is limited only by the rate of electron return to starting reactants.

The photoaddition of allylstannane to iminium salts has also been investigated, e.g.,[60]

$$\mathbf{159} + CH_2{=}CH{-}CH_2Sn(CH_3)_3\ (\mathbf{162}) \xrightarrow[CH_3OH]{h\nu} \mathbf{163} \qquad (3.108)$$

When compared to the reactions of allylsilane derivatives to iminium salts, the efficiency of the allylstannane addition is about 10 times greater. This enhancement is a result of greater orbital interaction between the σ-metal–carbon bond and the π-orbital. This should lead to low ionization potentials and therefore to an enhancement in the efficiency of product formation.

Some pyrrolinium salts may experience intramolecular cyclization on photoexcitation. For example, irradiation of *N*-xylylpyrrolinium salt (**164**) in acetonitrile leads to near quantitative yields of a dimethylbenzopyrrolizidine (**165**), which further converts to **166**[61]:

$$\mathbf{164} \xrightarrow[CH_3CN]{h\nu} \mathbf{165} \xrightarrow{-H^+} \mathbf{166} \qquad (3.109)$$

Iminium salts also add to arenes. In the presence of toluene, for example, coupling products have been observed[62]:

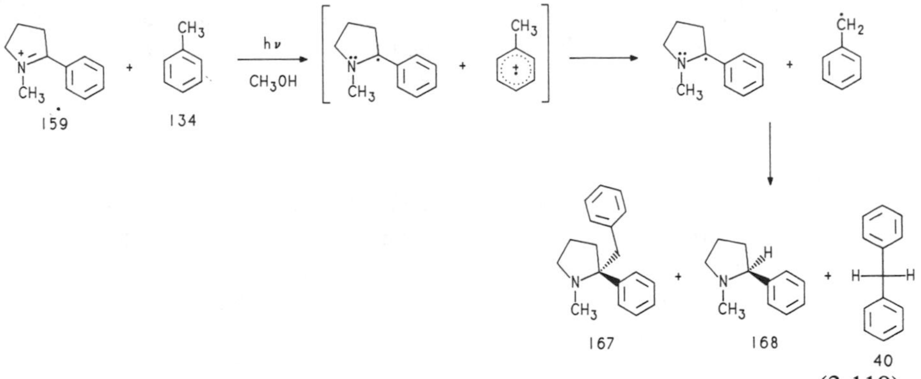

159 134 167 168 40

(3.110)

The reaction of **159** with benzyltrimethylsilane or benzyltrimethylstannane (**169**) proceeds as shown in 3.111:

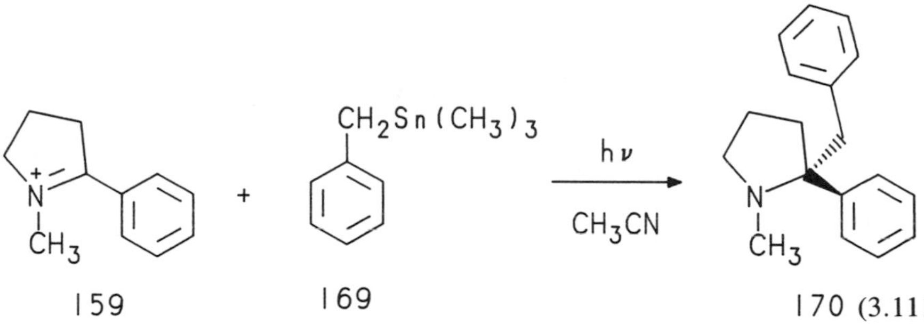

159 169 170 (3.111)

3.2.5. Photooxygenations

When aerated solutions of certain photosensitizers and organic substrates are exposed to light, a complex mixture of oxygenated products may be formed. Oxygen participates in the photosensitization either by directly reacting with the photosensitizer or by reacting with the substrate. Photooxygenations can involve energy transfer between a triplet sensitizer and molecular oxygen. Since molecular ground-state oxygen itself is paramagnetic and thus a triplet-state species, the quenching step can be regarded as an spin-allowed triplet–triplet energy transfer. The triplet sensitizer converts ground-state oxygen into a short-lived and highly reactive species called singlet oxygen. This species is designated by $^1O_2$. Singlet oxygen is capable of reacting with many electron-rich olefins and other organic substrates:

$$O_2 \xrightarrow{^3S^*} {}^1O_2 \qquad (3.112)$$

$$\text{Organic substrate} \xrightarrow{^1O_2} \text{Oxidized products} \qquad (3.113)$$

The reactions involving singlet oxygen are extensive and will not be covered here.[63]

Photooxygenations can also proceed via the photoinduced generation of an oxygen species called superoxide. In forming superoxide, an excited sensitizer interacts directly with molecular oxygen by functioning as an electron donor[64]:

$$^{1,3}D^* + O_2 \rightarrow D^{\dot{+}} + O_2^{\dot{-}} \quad (3.114)$$

$O_2^{\dot{-}}$ is the designation for superoxide. Certain porphyrins are known to sensitize the direct formation of superoxide in this manner.[65] Our attention, however, will be directed to photooxygenations involving the formation of $O_2^{\dot{-}}$ following the quenching of an excited electron acceptor. In this pathway, a singlet sensitizer initially abstracts an electron from an electron-rich substrate. The anion of the sensitizer subsequently donates an electron to oxygen transforming the latter into $O_2^{\dot{-}}$. Superoxide can then combine with the substrate's radical cation to form oxygenated products:

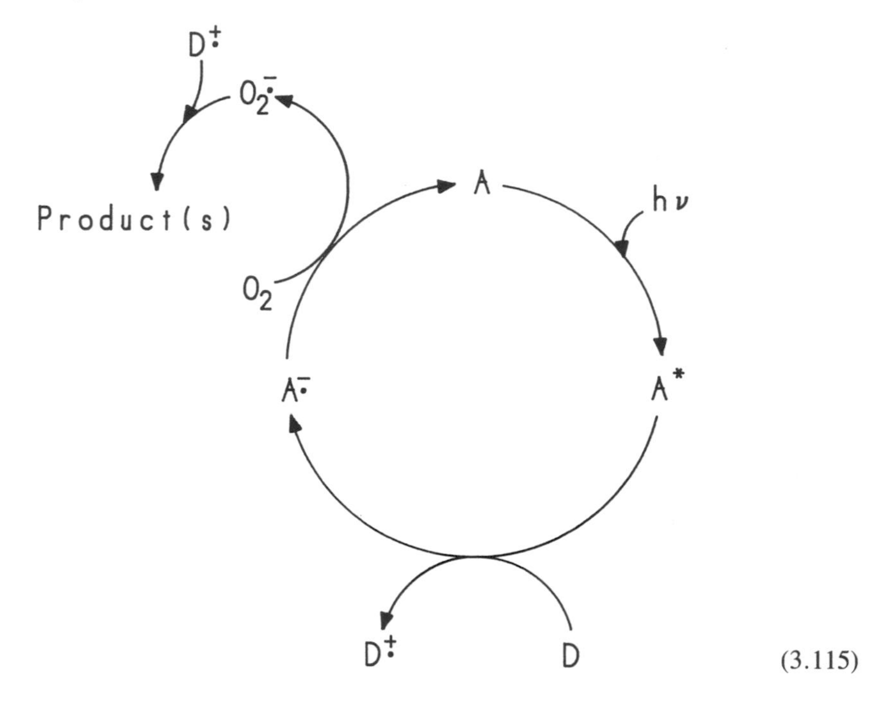

(3.115)

The participation of *singlet oxygen* and *superoxide* is dictated by the energy changes involved in the formation of these intermediates. Singlet oxygen can be generated with sensitizers having triplet levels exceeding 22.5 kcal mol^{-1} (the triplet energy of oxygen). On the other hand, the generation of superoxide requires that the reaction of the aromatic hydrocarbon radical anion with oxygen should be exothermic. From the redox potential of oxygen, $E^0(A/A^{\dot{-}}) = -0.33$ V vs. SCE, the free energy changes can be estimated. In analyzing reactions involving superoxide, the reader should be aware of the possible intervention of singlet oxygen. Often singlet oxygen and superoxide pathways coexist.

To gain better insight into the nature of the short-lived oxygen intermediates that play a role in photooxygenations, the Lewis structures of singlet oxygen and superoxide are shown below:

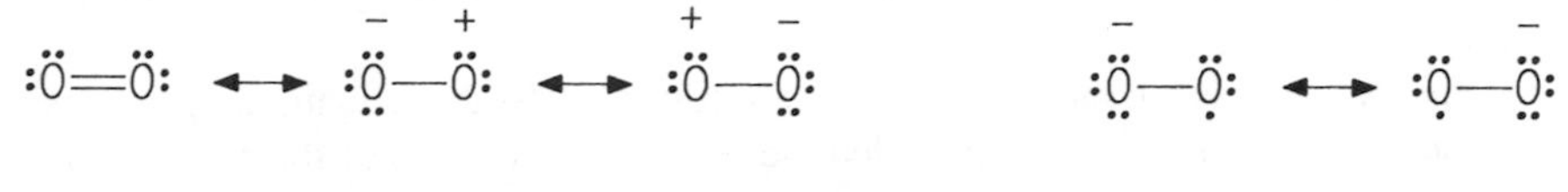

singlet oxygen superoxide

It appears that the best Lewis acid structure for singlet oxygen is a zwitterion (all electrons are paired). Superoxide is a more electron-rich species and therefore more reactive with positive atomic centers.

There are numerous examples of photooxygenations involving the participation of superoxide. A well-known example is the formation of tetraphenyloxirane (**51**) and benzophenone (**43**) when oxygen-rich solutions of a light-absorbing electron acceptor such as 9,10-dicyanoanthracene and tetraphenylethylene (**50**) are exposed to light[66]:

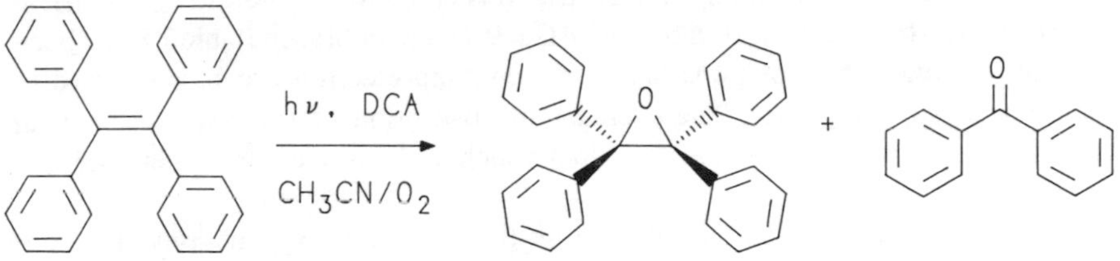

$$50 \xrightarrow[CH_3CN/O_2]{h\nu,\ DCA} 51 + 43 \qquad (3.116)$$

During the initial quenching step, the sensitizer oxidizes the olefin. Once superoxide is formed by the reaction of the sensitizer's anion with oxygen, it subsequently reacts with the radical cation:

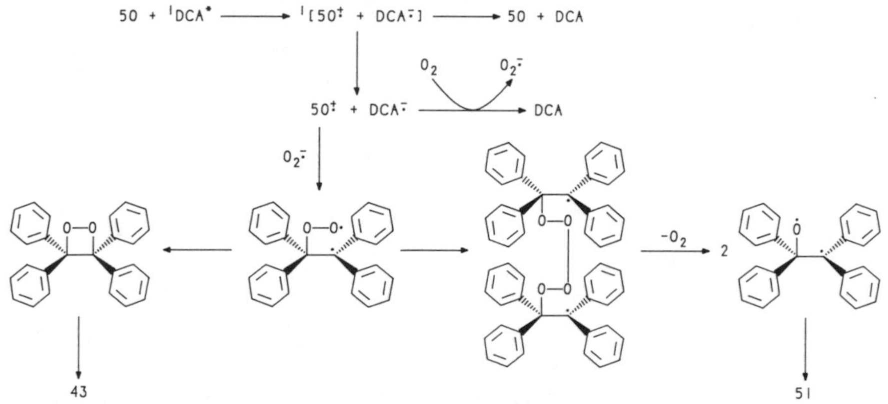

$$50 + {}^1DCA^* \longrightarrow {}^1[50^{\ddagger} + DCA^{\bar{\cdot}}] \longrightarrow 50 + DCA$$

$$50^{\ddagger} + DCA^{\bar{\cdot}} \xrightarrow{O_2 \rightarrow O_2^{\bar{\cdot}}} DCA$$

(3.117)

The fact that the initial quenching step is via electron transfer is supported by the observation that the reaction is efficient only in polar solvents. In addition, the reaction can be quenched by molecules having low redox potentials such as *p*-dimethoxybenzene.

In the presence of aerated methanolic solutions, 9,10-dicyanoanthracene sensitizes the oxygenation of 1,1-diphenylethylene (**62**).[66] Analysis of the product mixture suggests the formation of radical cation intermediates with subsequent reaction with superoxide and methanol:

hν, DCN; CH_3OH/O_2 — 62 → 43 (83%) + 171 (7%) + 172 (9%) (CH_2OCH_3, OH) + 173 (CH_2OCH_3, H)

(3.118)

In the photosensitized oxygenation of *trans*-stilbene (**3**), evidence marking the role of superoxide has been inferred from laser flash photolysis studies.[67,68] Transient spectra of the stilbene radical cation and the 9-cyanoanthracene anion have been recorded in oxygenated solutions of the stilbene and electron acceptor exposed to light from a Nd-YAG laser. The decrease in absorption of *trans*-stilbene radical cation was noted when a good electron donor such as 1,2,4-trimethoxybenzene was added.

Here are a few more examples of electron-transfer photooxygenations. The sensitized photolysis of 5,5-diphenylpent-4-en-1-ol (**174**) in the presence of oxygen and methanol involves nucleophilic trapping of a radical cation intermediate[69]:

hν, DCA; CH_3CN/O_2 — 174 (CH_2OH) → 43 + 175 (OH) + 176 (H)

(3.119)

Various cyclopropane derivatives can also be photooxygenated[68]:

hν, DCA; CH_3CN/O_2 — 177 (OCH_3, H_3CO) → 178 + 179 (CH_3O, OCH_3, O—O)

(3.120)

Electron transfer has been implicated in the photooxygenation of 3,5-cycloheptadienone derivatives in the presence of 9,10-dicyanoanthracene[70]:

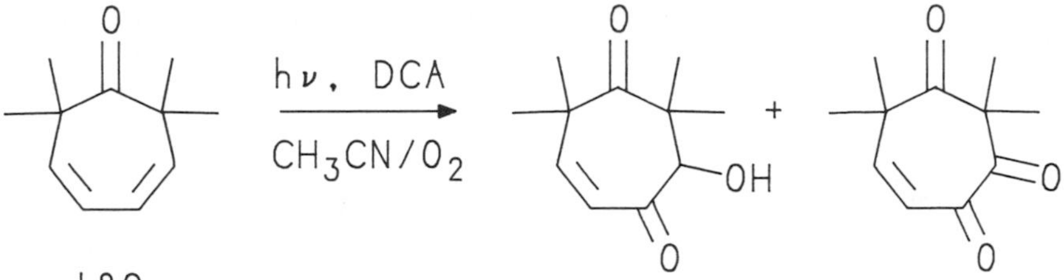

hν, DCA
CH_3CN/O_2

180 181 182 (3.121)

Interesting heterocyclic derivatives can be transformed during electron-transfer photooxygenations, e.g.,[71]

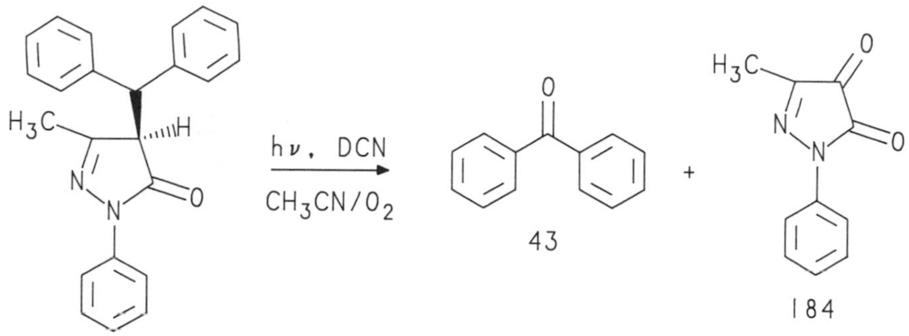

hν, DCN
CH_3CN/O_2

183 43 184 (3.122)

Consider the the oxygenation of tetraphenyloxirane (**51**) in the presence of 9,10-dicyanoanthracene and biphenyl (BP).[72] This reaction yields an ozonide (**185**) and benzophenone (**43**)[73]:

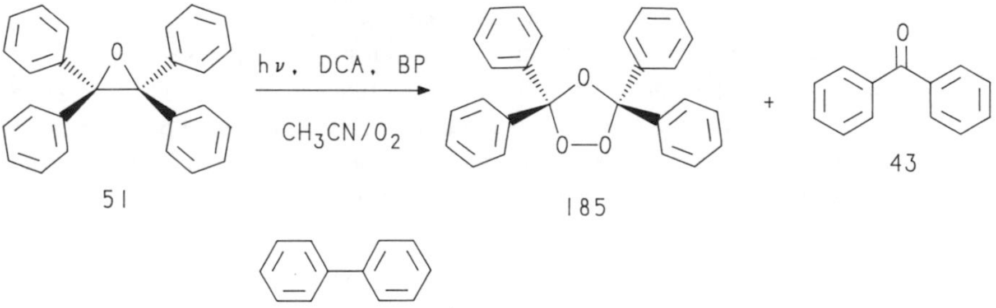

hν, DCA, BP
CH_3CN/O_2

51 185 43

BP (3.123)

The reaction proceeds sluggishly without biphenyl. However, in the presence of biphenyl, there is a significant enhancement in the rate of photooxygenation. Schaap proposes that biphenyl acts as a cosensitizer. According to Schaap, biphenyl can quench singlet 9,10-dicyanoanthracene. This step is followed by electron transfer between the biphenyl radical cation and the oxirane. The oxirane radical cation experiences bond breaking. Superoxide then adds to the radical cation to form the ozonide (**185**):

$$51 + {}^1DCA^* \longrightarrow {}^1[51^{+\cdot} + DCA^{-\cdot}] \longrightarrow 51 + DCA$$

$$\downarrow$$

$$43 + 185 \xleftarrow{O_2^{-\cdot}} 51^{+\cdot} + DCA^{-\cdot} \xrightarrow{O_2 \to O_2^{-\cdot}} DCA$$

$$\uparrow \; -BP^{+\cdot}, \; +51$$

$$BP^{+\cdot} + DCA^{-\cdot} \xrightarrow{O_2 \to O_2^{-\cdot}} DCA$$

$$\uparrow$$

$$BP + {}^1DCA^* \longrightarrow {}^1[BP^{+\cdot} + DCA^{-\cdot}] \longrightarrow BP + DCA \qquad (3.124)$$

This pathway is reasonable given that **51** by itself is known to be an inefficient quencher of 9,10-dicyanoanthracene fluorescence.

Some workers have suggested that singlet oxygen may be involved in 3.123.[74] This conclusion rests in part on spectroscopic studies that have revealed the presence of triplet states of biphenyl and 9,10-dicyanoanthracene. It has been proposed that singlet oxygen can be formed by triplet sensitization and can participate in the photooxygenation—that the rate enhancement in the presence of biphenyl is due to a singlet oxygen pathway.

In fact, distinguishing between electron-transfer (superoxide) and dye-sensitized (singlet oxygen) photooxygenations has been a major challenge to chemists working in this field. This has spurred on several efforts to develop procedures for identifying these pathways. A novel approach has been proposed by Ando and Akasaka, who showed that the stereospecific oxygenation of 3-adamantylidenetricyclo[3.2.1.$0^{2,4}$]octane (**186**) in solutions of a cosensitizer (*trans*-stilbene) leads to formation of a product that differs stereochemically from a product *known* to result from singlet oxygenation.[75] Thus, in the presence of *trans*-stilbene and 9,10-dicyanoanthracene, **186** is converted to an *exo* derivative **187**:

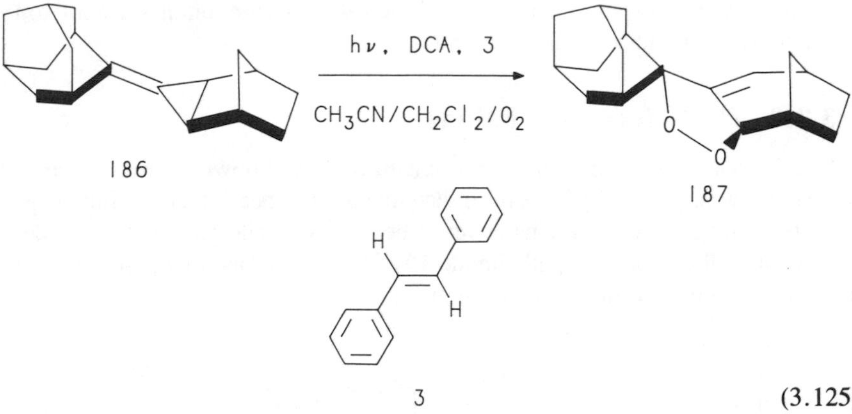

186 $\xrightarrow[CH_3CN/CH_2Cl_2/O_2]{h\nu,\ DCA,\ 3}$ 187

3 (3.125)

Ando and Akasaka compared the product resulting in 3.125 with the product generated from a singlet oxygen reaction. Thus, when **186** reacts with 1,4-dimethylnaphthalene endoperoxide (**188**), which is known to produce singlet oxygen, *endo*-bicyclo[3.2.1]octene (**189**) is formed:

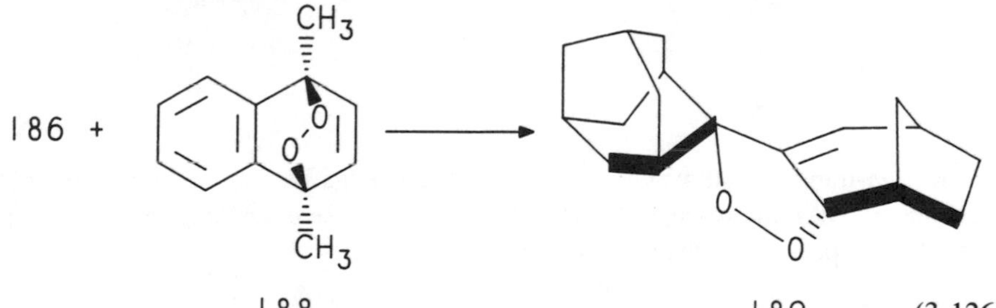

186 + 188 ⟶ 189 (3.126)

It is concluded that 3.125 proceeds by an electron-transfer pathway. This reaction can be regarded as a *mechanistic* probe that "tests" for a singlet oxygen or superoxide pathway. Were **189** formed in 3.125, we would suspect a singlet oxygen pathway.

Before closing this section, we draw attention to the salt effect in photooxygenations. Addition of $Mg(ClO_4)_2$, $LiClO_4$, $NaClO_4$, or tetra-*n*-butylammonium perchlorate (TBAP) has been shown to enhance the yield of products in the cosensitized photooxygenation of 1,1-diphenyl-2-vinylcyclopropane (**190**)[76,77]:

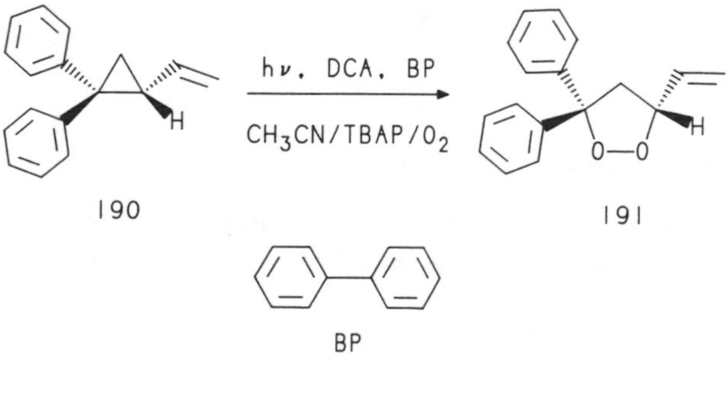

190 $\xrightarrow[CH_3CN/TBAP/O_2]{h\nu,\ DCA,\ BP}$ 191

BP (3.127)

As pointed out previously, the presence of the salt assists in ion dissociation to free ion pairs, thereby blocking electron return.

3.2.6. Photodecarboxylations

Although decarboxylations using photosensitizers have known for many years,[78,79] relatively few applications have been developed. It appears that the full scope of this interesting class of reactions is yet to be realized. The photosensitized decarboxylation of the *N*-acyloxyphthalimide **192** in the presence of the sensitizer 1,6-bis(dimethylamino)pyrene is one example[80]:

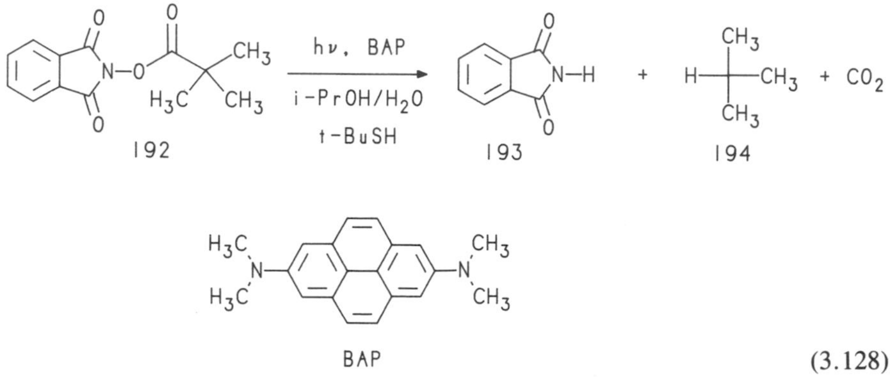

hν, BAP; i-PrOH/H_2O; t-BuSH

192 → 193 + 194 + CO_2

BAP

(3.128)

An electron-transfer pathway is supported by the negative free energies for the quenching of BAP fluorescence by **192.** The rates of these quenchings are diffusion controlled. A possible pathway is shown below:

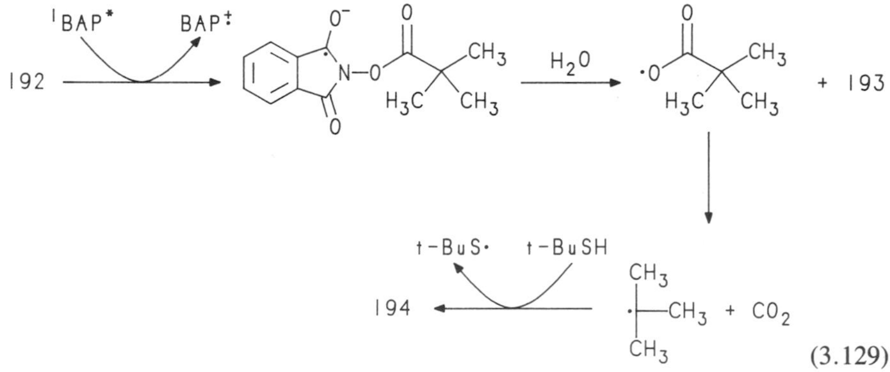

^{1}BAP* BAP$^{+\cdot}$

192 → ; H_2O → + 193

t-BuS· t-BuSH

194 ← + CO_2

(3.129)

3.2.7. Photosensitized Aminations

Another reaction demonstrating the synthetic utility of photosensitized electron transfer is the direct amination of arenes. It has been established that irradiation of acetonitrile/water solutions containing certain arenes, ammonia, and *m*-dicyanobenzene leads to the formation of dihydroaminoarenes, e.g.,[81]

$$15 + NH_3 \xrightarrow[CH_3CN/H_2O]{h\nu,\ m\text{-}DCB} 195 \quad (3.130)$$

Because phenanthrene effectively quenches the fluorescence of *m*-dicyanobenzene (ammonia and the amines do not), the primary electron-transfer step is formation of the arene radical cation followed by nucleophilic attack by amines or ammonia:

$$15 \underset{m\text{-}DCB \leftarrow m\text{-}DCB^{\bar{\cdot}}}{\overset{^1m\text{-}DCB^* \rightarrow m\text{-}DCB^{\bar{\cdot}}}{\rightleftarrows}} 15^{+\cdot} \xrightarrow{+NH_3} [\text{radical}] \xrightarrow{m\text{-}DCB^{\bar{\cdot}} \rightarrow m\text{-}DCB} 195 \quad (3.131)$$

Photoaminations of arenes by primary organic amines such as methyl and ethyl amine as well as allyl amine also generate respectable yields of amine products.

3.2.8. Photoinduced Polymerizations

There are several pathways by which photoinduced electron transfer can be used to bring about the polymerization of certain organic substrates. For example, on photoexcitation, a donor (or acceptor) in combination with an acceptor (or donor), may form radical products capable of initiating a polymerization. The quenching of excited ketones by amines usually leads to radical intermediates that can initiate the polymerization of monomers such as acrylonitrile (**197**)[82]:

$$^3[43]^* + (CH_3CH_2)_3N\ (196) \longrightarrow [43^{\bar{\cdot}} + 196^{+\cdot}] \longrightarrow Ph_2\dot{C}OH + (CH_3CH_2)_2N\dot{C}HCH_3\ (I\cdot) \quad (3.132)$$

In 3.132, the amine radicals formed by quenching of the ketone are the initiators (we use the designation I˙ to represent the initiator). The propagation proceeds as follows:

$$n\ CH_2{=}CH\text{-}CN\ (197) \xrightarrow[\text{propagation}]{I\cdot} I\text{-}[CH_2\text{-}CH(CN)]_n\cdot \xrightarrow[\text{termination}]{T\cdot} I\text{-}[CH_2\text{-}CH(CN)]_n\text{-}T \quad (3.133)$$

(Here T˙ "terminates" the reaction.)

A donor and acceptor may be excited to produce a pair of ions, one of which directly converts a monomer to an ionic state. This technique has been applied to the synthesis of polypyrrole from its monomer, pyrrole (**198**). (Polypyrrole has attracted interest because it belongs to a family of polymers called *conducting polymers.*) It has shown that on photoexcitation, $Ru(bpy)_3^{2+}$ can donate an electron to $Co(NH_3)_5Cl^{2+}$ to generate $Ru(bpy)_3^{3+}$.[83] The latter can remove an electron from pyrrole, which subsequently polymerizes:

$$[Ru(bpy)_3^{2+}]^* + Co(NH_3)_5Cl^{2+} \longrightarrow Ru(bpy)_3^{3+} + Co(NH_3)_5Cl^{+}$$

198 → (pyrrole radical cation, $N^{+\cdot}$–H) + $Ru(bpy)_3^{2+}$

polypyrrole ← propagation

(3.134)

Similar work has already been done to exploit this reaction in the synthesis of other polymer systems. The examples shown in 3.132 and 3.134 are only suggestive of the enormous scope of photoinduced polymerization by electron transfer.

3.2.9. Sensitized Electron-Transfer Reactions of Bioorganic Substrates

In many biochemical transformations, electron-transfer reactions are a common theme. The breadth of these reactions has fundamental implications. For example, electron transfer is one of the major pathways in photosynthesis, and it plays a significant role in metabolic processes of living organisms. Electron-transfer pathways are also of considerable importance when evaluating the effects of light on human beings.

3.2.9.1. The Photosensitized Reactions of NAD(P)H and NAD(P)$^+$

Of primary importance to living organisms are the metabolic pathways required to transform chemical energy into useful work. These reactions also include the biosynthesis of necessary cellular constituents. Many of these processes are oxidation and reduction reactions involving electron transfer and electron-transport. A complicated series of oxidation reactions of carbohydrates, for example, releases and subsequently stores energy in high-energy phosphate compounds. Although a dis-

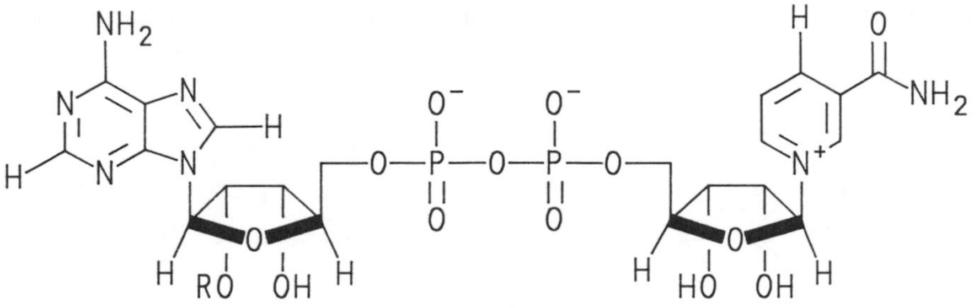

$NAD(P)^+$

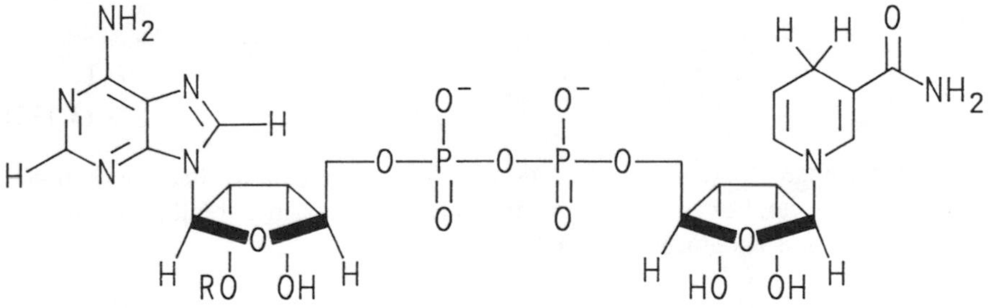

NAD(P)H

Figure 3.3 Structures of oxidized and reduced forms of nicotinamide adenine dinucleotide (NAD^+ and NADH) and nicotinamide adenine dinucleotide phosphate ($NADP^+$ and NADPH). $R = PO_3^{2-}$.

cussion of these reactions falls outside the scope of this book, there is an important class of compounds that participates in metabolic transformations and that also has been shown to be effective substrates in photoinduced electron transfer. These are the 1,4-dihydropyridine derivatives, which include the coenzymes nicotinamide adenine dinucleotide (NAD^+) and nicotinamide adenine dinucleotide phosphate ($NADP^+$), and their reduced forms, NADH and NADPH, respectively (Fig. 3.3).

The redox potential of NADH in water is 0.72 V vs. SCE. This value implies that NADH should be a good electron acceptor. The overall mechanism in the reduction of an organic substrate such as a ketone by NADH or NADPH appears to involve a hydride-transfer. There are some unresolved issues about the details of this reduction (e.g., does the hydride transfer in one step or is it actually two one-electron transfers followed by a protein transfer?). Ignoring these issues, we may summarize the reduction (or oxidation) of a carbonyl substrate as shown in 3.135:

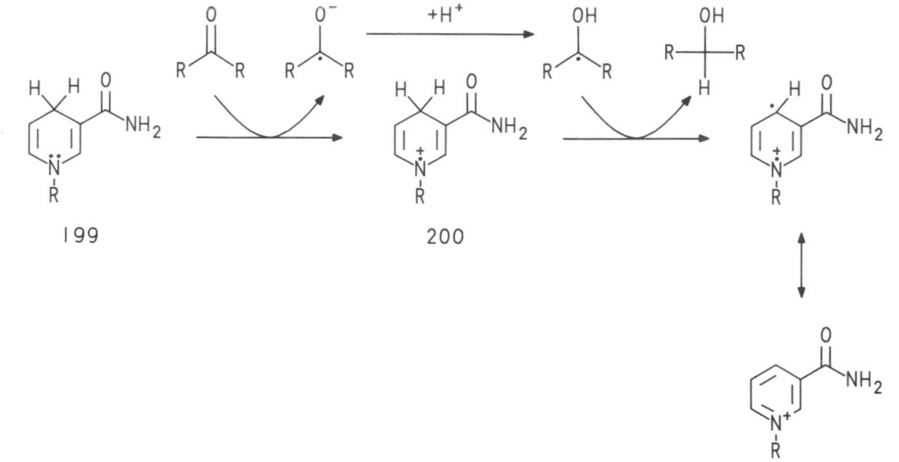

199 200 201

(3.135)

In the presence of excited electron donors, it has been shown that some model pyridinium ions can be reduced.[84] For example, **202** quenches the fluorescence of singlet ZnTPP and is reduced to **203**[85]:

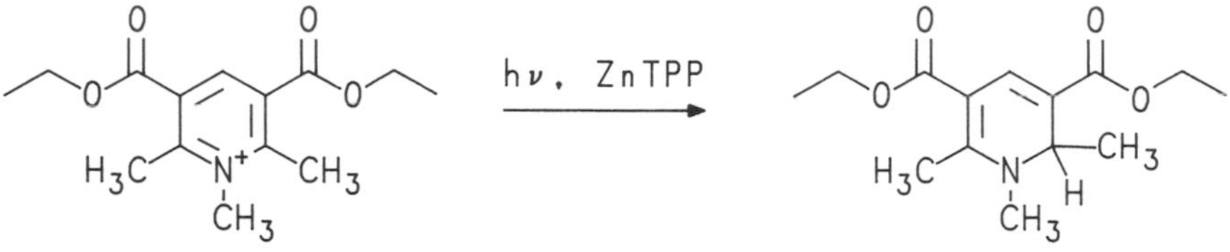

hν, ZnTPP

202 203 (3.136)

It has also been demonstrated that $Ru(bpy)_3^{2+}$ can photosensitize coupling and reduction pathways of certain substrates in the presence of NADH model 1,4-dihydropyridine compounds. One such reaction is the reduction of carbonyl compounds: irradiation of $Ru(bpy)_3^{2+}$ and 1-benzyl-1,4-dihydronicotinamide (**205**) in the presence of **204** yields substantial amounts of alcohol product (**206**)[86,87]:

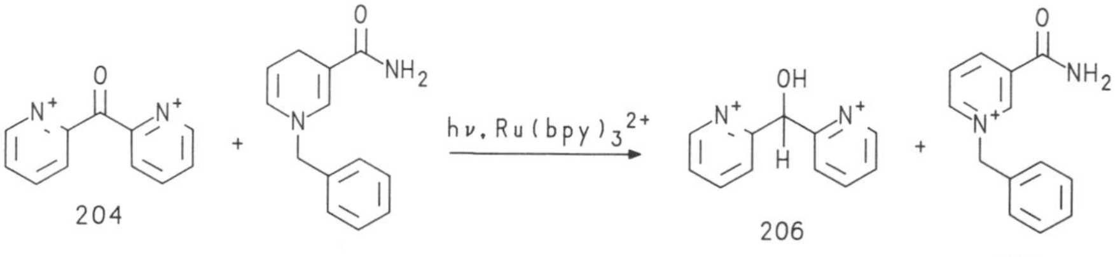

hν, $Ru(bpy)_3^{2+}$

204 205 206 207

(3.137)

206 is presumed to arise by a cosensitization pathway. Thus, the emission of $[Ru(bpy)_3^{2+}]^*$ is quenched by **205** resulting in the latter's oxidation. In the meantime, as $Ru(bpy)_3^+$ returns to $Ru(bpy)_3^{2+}$, **204** is reduced. The remaining steps consist of several electron and proton transfers to yield, finally, a cross-coupled product:

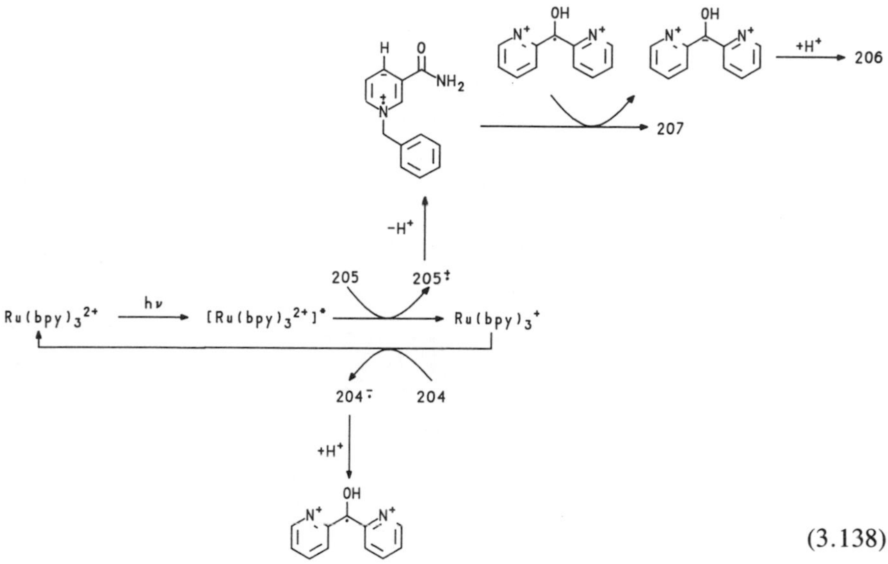

(3.138)

A critical step is electron transfer from the 1-benzyl-1,4-dihydronicotinamide radical to the ketyl radical. For this step to proceed spontaneously, the redox potential of the ketyl radical must be sufficiently positive to accept an electron. This appears to be true for the radical of di-2-pyridyl ketone, **204.** If this condition is not satisfied, then coupling products may be obtained:

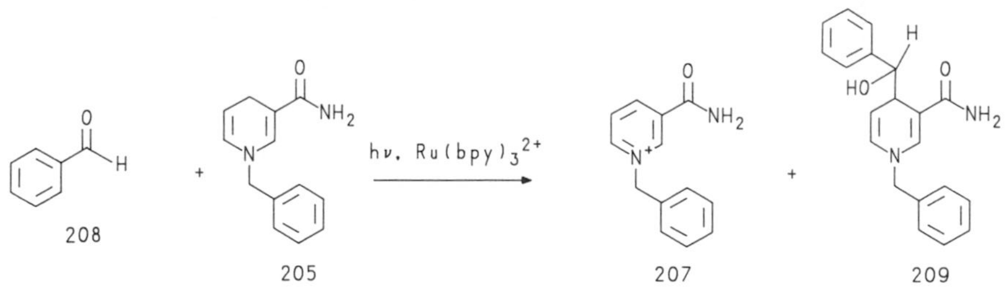

(3.139)

Similar pathways have been postulated for the reduction of certain olefins. An example is the photosensitized reduction of certain C—C double bonds by **205** in the presence of $Ru(bpy)_3^{2+}$ [88]:

$$\mathbf{210} + \mathbf{205} \xrightarrow[Mg^{2+}/CH_3OH]{h\nu,\ Ru(bpy)_3^{2+}} \mathbf{207} + \mathbf{211} \qquad (3.140)$$

Mg^{2+} was shown to catalyze the reaction.

Strategies to reduce ketones and acids using enzyme catalysis coupled with photosensitization have been investigated extensively by Willner and co-workers.[89] The reduction of 2-butanone (**212**) takes place within a multicomponent system consisting of $Ru(bpy)_3^{2+}$, $NADP^+$, methyl viologen (MV^{2+}), $(NH_4)_3EDTA$ as an electron donor, and ferredoxin/$NADP^+$/reductase (FDR) coupled with an alcohol dehydrogenase:

$$\underset{\mathbf{212}}{H_3C-CH_2-CO-CH_3} \xrightarrow[MV^{2+},\ NADP^+,\ (NH_4)_3EDTA,\ \text{enzymes}]{h\nu,\ Ru(bpy)_3^{2+}} \underset{\mathbf{213}}{H_3C-CH_2-CH(OH)-CH_3} \qquad (3.141)$$

On illumination with light, the sensitizer reduces MV^{2+}, which acts as an electron carrier. Subsequently, $MV^{+\cdot}$ reduces $NADP^+$ to NADPH, which then transfers an electron to the ketone. $Ru(bpy)_3^{3+}$ returns to $Ru(bpy)_3^{2+}$ by accepting an electron from $(NH_4)_3EDTA$:

(3.142)

Scheme 3.142 is an example of a multielectron relay that occurs in a micelle. More examples of photoinduced electron-transfer reactions occurring in micelles and other organized media are presented in Chapter 5.

Suggestions for Further Reading

The Pathways of Photoinduced Electron-Transfer Reactions

Kochi, J. *Adv. Free Radical Chem.* **1990,** *1,* 53. Discusses the chemistry of radical cations, which play an important role in electron-transfer photochemistry.

Lewis, F. D. In *Photoinduced Electron Transfer. Part C. Photoinduced Electron Transfer Reactions: Organic Substrates,* Fox, M. A., and Chanon, M., eds. Elsevier, Amsterdam, 1988, Part C, Chapter 4.1. Covers reactions through the late 1980s.

Mattes, S. L., and Farid, S. In *Organic Photochemistry,* Vol. 6, Padwa, A., ed. Marcel Dekker, New York, 1983, Chapter 4. An excellent review of photoinduced electron-transfer reactions through the earlier 1980s.

Photosensitized Electron-Transfer Isomerizations and Rearrangements

Böhm A., and Müllen, K. *Tetrahedron Lett.* **1992,** *33,* 611. The valence isomerization of 1,2-distyrylbenzene under electron-transfer conditions is discussed in this paper.

Coyle, J. D. *Introduction to Organic Photochemistry.* Wiley, Chichester, England, 1986, Chapter 2.

Photosensitized C–C Bond Cleavages

Maslak, P., and Chapman, W., Jr. *J. Org. Chem.* **1990,** *55,* 6334.

Saeva, F. D. *Top. Curr. Chem.* **1990,** *156,* 59.

Photoadditions via Electron Transfer

March, J. *Advanced Organic Chemistry,* 3rd ed. Wiley, New York, 1985, Chapters 10 through 13 contain much background information relevant to the material in this chapter.

Mattes, S. L., and Farid, S. *Science* **1984,** *226,* 917. Distinguishes between exciplex and radical ion reactions.

Müller, F., and Mattay, J. *Angew. Chem.* **1991,** *30,* 1336. This paper presents an example of a sensitized 1,3-dipolar cycloaddition reaction.

Simmons, H. E., and Bunnett, J. F., eds. *Orbital Symmetry Papers.* American Chemical Society, Washington, D. C., 1974. A compendium of articles on the role of symmetry in organic chemistry. Good background reading for the serious student.

Turro, N. J. *Modern Molecular Photochemistry.* Benjamin Cummings, Menlo Park, CA, 1978, Chapters 10 and 11. These chapters serve as good introductions to photoaddition and photocycloaddition reactions.

Woodward, R. B., and Hoffmann, R. *The Conservation of Orbital Symmetry.* Verlag Chemie, Deerfield Beach, FL, 1981. A classic monograph—required reading for the serious student.

Photoadditions between Aromatic Compounds and Amines

Görner, H., Elisei, F., and Aloisi, G. G. *J. Chem. Soc., Faraday Trans.* **1992,** *88,* 29. The quenching of excited styrylanthracene derivatives with a variety of electron donors and acceptors is discussed.

Lewis, F. D. *Acc. Chem. Res.* **1986,** *19,* 401.

Photadditions of Carbonyl Compounds

Coyle, J. D. *Pure Appl. Chem.* **1988,** *60,* 941.

Hoshino, M., and Shizuka, H. In *Photoinduced Electron Transfer. Part C. Photoinduced Electron Transfer Reactions: Organic Substrates,* Fox, M. A., and Chanon, M., eds. Elsevier, Amsterdam, 1988, Part C, Chapter 4.5.

Kanaoka, Y. *Acc. Chem. Res.* **1978,** *11,* 407.

Wagner, P. *Top. Curr. Chem.* **1976,** *66,* 1. General discussion on photochemistry of carbonyl compounds.

Photooxygenations

Lopez, L., *Top. Curr. Chem.* **1990,** *156,* 117.

Photoinduced Polymerizations

Timpe, H.-J. *Top. Curr. Chem.* **1990,** *156,* 167.

Photoinduced Electron-Transfer Reactions of Bioorganic Substrates

Lablache-Combier, A. In *Photoinduced Electron Transfer. Part C. Photoinduced Electron Transfer Reactions: Organic Substrates,* Fox, M. A., and Chanon, M., eds. Elsevier, Amsterdam, 1988, Part C, Chapter 4.4.

Smith, K. C. In *In Science of Photobiology,* Smith, K. C., ed. Plenum, New York, 1977, Chapter. 5.

The Photosensitized Reactions of NAD(P)H and $NAD(P)^+$

Fukuzumi, S., and Tanaka, T. In *Photoinduced Electron Transfer. Part C. Photoinduced Electron Transfer Reactions: Organic Substrates,* Fox, M. A., and Chanon, M., eds. Elsevier, Amsterdam, 1988, Part C, Chapter 4.10.

References

1. Tagaki, K., and Ogata, Y. *J. Org. Chem.* **1992,** *47,* 1409.
2. Kuriyama, Y., Arai, T., Sakuragi, H., and Tokumaru, K. *Chem. Lett.* **1988,** 1193.

3. Gutierrez, A. R., Meyer, T. J., and Whitten, D. G. *Mol. Photochem.* **1976,** *7,* 349.

4. Wong, P. C., and Arnold, D. R. *Tetrahedron Lett.* **1979,** 2101.

5. Hub, W., Klüter, U., Schneider, S., Dörr, F., Oxman, J. D., and Lewis, F. D. *J. Phys. Chem.* **1984,** *88,* 2308.

6. Gotoh, T., Kato, M., Yamamoto, M., and Nishijima, Y. *J. Chem. Soc., Chem. Commun.* **1981,** 90.

7. Majima, T., Pac, C., and Sakurai, H. *Chem. Lett.* **1979,** 1133.

8. Barber, R. A., de Mayo, P., Okada, K., and Wong S. K. *J. Am. Chem. Soc.* **1982,** *104,* 4995.

9. Roth, H. D., and Schilling, M. L. M. *J. Am. Chem. Soc.* **1981,** *103,* 1246.

10. Roth, H. D., and Schilling, M. L. M. *J. Am. Chem. Soc.* **1981,** *103,* 7210.

11. Evans, T. R., Wake, R. W., and Sifain, M. M. *Tetrahedron Lett.* **1973,** 701.

12. Roth, H. D., Schilling, M. L. M., and Raghavachari, K. *J. Am. Chem. Soc.* **1984,** *106,* 253.

13. Okada, K., Hisamitsu, K., and Mukai, T. *Tetrahedron Lett.* **1981,** *22,* 1251.

14. Miyashi, T., Takahashi, Y., Konno, A., Mukai, T., Roth, H. D., Schilling, M. L., and Abelt, C. J. *J. Org. Chem.* **1989,** *54,* 1445.

15. Hasegawa, E., Mukai, T., and Yanaki, K. *J. Org. Chem.* **1989,** *54,* 2053.

16. Jones, G., II, Schwarz, W., and Malba, V. *J. Phys. Chem.* **1982,** *86,* 2286.

17. Schwarz, W., Dangel, K.-M., Jones, G., II, and Bargon, J. *J. Am. Chem. Soc.* **1982,** *104,* 5686.

18. Arnold, D. R., and Humphreys, R. W. R. *J. Am. Chem. Soc.* **1979,** *101,* 2743.

19. Yamashita, Y., Yaegashi, H., and Mukai, T. *Tetrahedron Lett.* **1985,** 3579.

20. Okamoto, A., Snow, M. S., and Arnold, D. R. *Tetrahedron* **1986,** *42,* 6175.

21. Albini, A., and Mella, M. *Tetrahedron* **1986,** *42,* 6219.

22. Davis, H. F., Das, P. K., Reichel, L. W., and Griffin, G. W. *J. Am. Chem. Soc.* **1984,** *106,* 6968.

23. Majima, T., Pac, C., and Sakurai, H. *J. Chem. Soc., Perkin Trans. I* **1980,** 2705.

24. Turro, N. J., Tung, C.-H., Gould, I. R., Griffin, G. W., and Manmade, A. *J. Photochem.* **1984,** *24,* 265.

25. Das, P. K., Muller, A. J., Griffin, G. W., Gould, I. R., Tung, C.-H., and Turro, N. J. *Photochem. Photobio.* **1984,** *39,* 281.

26. Ledwith, A. *Acc. Chem. Res.* **1972** 5, 133.

27. Farid, S., and Shealer, S. E. *J. Chem. Soc., Chem. Commun.* **1973,** 677.

28. Mizuno, K., Kaji, R., and Otsuji, Y. *Chem. Lett.* **1977,** 1027.

29. Neuenteufel, R. A., and Arnold, D. R. *J. Am. Chem. Soc.* **1972,** *95,* 4080.

30. Mattes, S. L., and Farid, S. *J. Am. Chem. Soc.* **1983,** *105,* 1386.

31. Mizuno, K., Hashizume, T., and Otsuji, Y. *J. Chem. Soc., Chem. Commun.* **1983,** 722.

32. Bellville, D. J., Wirth, D. D., and Bauld, N. L. *J. Am. Chem. Soc.* **1981,** *103,* 718.

33. Mattay, J., Trampe, G., and Runsink, *J. Chem. Ber.* **1988,** *121,* 1991.

34. Shigemitsu, Y., and Arnold D. R. *J. Chem. Soc., Chem. Commun.* **1975,** 407.

35. Maroulis, A. J., Shigemitsu, Y., and Arnold, D. R. *J. Am. Chem. Soc.* **1978,** *100,* 535.

36. Majima, T., Pac, C., Nakasone, A., and Sakurai, H. *J. Am. Chem. Soc.* **1981,** *103,* 4499.

37. Gassman, P. G. In *Photoinduced Electron Transfer. Part C. Photoinduced Electron Transfer Reactions: Organic Substrates,* Fox, M. A., and Chanon, M., eds. Elsevier, Amsterdam, 1988, Chapter 4.2.

38. Gassman, P. G., Olson, K. D., Walter, L., and Yamaguchi, R. *J. Am. Chem. Soc.* **1981,** *103,* 4977.

39. Dinnocenzo, J. P., Farid, S., Goodman, J. L., Gould, I. R., Todd, W. P., and Mattes, S. L. *J. Am. Chem. Soc.* **1989,** *111,* 8973.

40. Mattes, S. L., and Farid, S. *Acc. Chem. Res.* **1982,** *15,* 80.

41. Caldwell, R. A., Mizuno, K., Hansen, P. E., Vo, L. P., Frentrup, M., and Ho, C. D. *J. Am. Chem. Soc.* **1981** *103,* 7263.

42. Arnold, D. R., Wong, P. C., Maroulis, A. J., and Cameron, T. S. *Pure Appl. Chem.* **1980,** *52,* 2609.

43. Borg, R. M., Arnold, D. R., and Cameron, T. S. *Can. J. Chem.* **1984,** *62,* 1785.

44. Albini, A., Fasani, E., and Mella, M. *J. Am. Chem. Soc.* **1986,** *108,* 4119.

45. Mattay, J. *Angew. Chem.* **1987,** *26,* 825.

46. Hub, W., Schneider, S., Dörr, F., Oxman, J. D., and Lewis, F. D. *J. Am. Chem. Soc.* **1984,** *106,* 708.

47. Lewis, F. D., and Ho, T.-I. *J. Am. Chem. Soc.* **1980,** *102,* 1751.

48. Lewis, F. D., Ho, T.-I., and Simpson, J. T. *J. Am. Chem. Soc.* **1982** *102,* 1924.

49. Lewis, F. D., and Correa, P. E. *J. Am. Chem. Soc.* **1984,** *106,* 194.

50. Gersdorf, J., Mattay, J., and Görner, H. *J. Am. Chem. Soc.* **1987,** *109,* 1203.

51. Wagner, P. J., Truman, R. J., Puchalski, A. E., and Wake, R. *J. Am. Chem. Soc.* **1986,** *108,* 7727.

52. Cohen, S. G., Parola, A., and Parsons, G., Jr. *Chem. Rev.* **1973,** *73,* 141.

53. Mazzocchi, P. H., Wilson, P., Khachik, F., Klingler, L., and Minamikawa, S. *J. Org. Chem.* **1983,** *48,* 2981.

54. Mazzocchi, P. H., Minamikawa, S., and Wilson, P. *J. Org. Chem.* **1985,** *50,* 2681.

55. Mazzocchi, P. H., and Fritz, G. *J. Am. Chem. Soc.* **1986,** *108,* 5362.

56. Mariano, P. S. In *Photoinduced Electron Transfer. Part C. Photoinduced Electron Transfer Reactions: Organic Substrates,* Fox, M. A., and Chanon, M., eds. Elsevier, Amsterdam, 1988, Part C, Chapter 4.6.

57. Cho, I.-S., Tu, C.-L., and Mariano, P. S. *J. Am. Chem. Soc.* **1990,** *112,* 3594.

58. Ohga, K., and Mariano, P. S. *J. Am. Chem. Soc.* **1982,** *104,* 617.

59. Ohga, K., Yoon, P. S., and Mariano, P. S. *J. Org. Chem.* **1984,** *49,* 213.

60. Borg, R. M., and Mariano, P. S. *Tetrahedron Lett.* **1986,** 2821.

61. Lan, A. J. Y., Quillen, S. L., Heuckeroth, R. O., and Mariano, P. S. *J. Am. Chem. Soc.* **1984,** *106,* 6439.

62. Borg, R. M., Heuckeroth, R. O., Lan, A. J. Y., Quillen, S. L., and Mariano, P. S. *J. Am. Chem. Soc.* **1987,** *109,* 2728.

63. Gorman, A. A., and Rodgers, M. A. J. In *CRC Handbook of Organic Photochemistry,* Scaiano, J. C., ed. CRC Press, Boca Raton, FL, 1989, Vol, 11, p. 229.

64. Cox, G. S., Whitten, D. G., and Giannotti, C. *Chem. Phys. Lett.* **1979,** *67,* 511.

65. Hopf, F. R., and Whitten, D. G. In *Porphyrins and Metalloporphyrins,* Smith, K. M., ed. Elsevier, Amsterdam, 1975, p. 161.

66. Eriksen, J., and Foote, C. S. *J. Am. Chem. Soc.* **1980,** *102,* 6083.

67. Spada, L. T., and Foote, C. S. *J. Am. Chem. Soc.* **1980,** *102,* 391.

68. Manring, L. E., Eriksen, J., and Foote, C. S. *J. Am. Chem. Soc.* **1980,** *102,* 4275.

69. Jiang, Z. Q., and Foote, C. S. *Tetrahedron Lett.* **1983,** *24,* 461.

70. Wilczak, W. A., and Schuster, D. I. *Tetrahedron Lett.* **1986,** *27,* 5331.

71. Reichel, L. W., Griffin, G. W., Muller, A. J., Das, P. K., and Ege, S. N. *Can. J. Chem.* **1984,** *62,* 424.

72. Schaap, A. P., Lopez, L., and Gagnon, S. D. *J. Am. Chem. Soc.* **1983,** *105,* 633.

73. Schaap, A. P., Siddiqui, S., Prasad, G., Palomino, E., and Lopez, L. *J. Photochem.* **1984,** *25,* 167.

74. Davidson, R. S., Goodwin, D., and Pratt, J. E. *J. Photochem.* **1985,** *30,* 167.

75. Akasaka, T., and Ando, W. *J. Am. Chem. Soc.* **1987,** *109,* 1260.

76. Shim, S. C., and Lee, H. J. *Bull. Korean Chem. Soc.* **1988,** *9,* 68.

77. Shim, S. C., and Lee, H. J. *Bull. Korean Chem. Soc.* **1988,** *9,* 112.

78. Davidson, R. S., and Steiner, P. R. *Chem. Commun.* **1971,** 1115.

79. Davidson, R. S., and Steiner, P. R. *J. Chem. Soc., Perkin 2* **1972,** 1357.

80. Okada, K., Okamoto, K., and Oda, M. *J. Am. Chem. Soc.* **1988,** *110,* 8736.

81. Yasuda, M., Yamashita, T., Matsumoto, T., Shima, K., and Pac, C. *J. Org. Chem.* **1985,** *50,* 3667.

82. Kubota, H., and Ogiwara, Y. *J. Appl. Polym. Sci.* **1982,** *27,* 2683.

83. Segawa, H., Shimidzu, T., and Honda, K. *J. Chem. Soc., Chem. Commun.* **1989,** 132.

84. Hedstrand, D. M., Kruizinga, W. H., and Kellog, R. M. *Tetrahedron Lett.* **1978,** 1255.

85. Ogata, Y., Takagi, K., and Tanabe, Y. *J. Chem. Soc., Perkin Trans. 2* **1979,** 1069.

86. Ishitani, O., Pac, C., and Sakurai, H. *J. Org. Chem.* **1983,** *48,* 2941.

87. Ishitani, O., Yanagida, S., Takamuku, S., and Pac, C. *J. Org. Chem.* **1987,** 52, 2790.

88. Ishitani, O., Ihama, M., Miyauchi, Y., and Pac, C. *J. Chem. Soc., Perkin Trans 1* **1985,** 1527.

89. Mandler, D., and Willner, I. *J. Chem. Soc., Perkin Trans. 2* **1986,** 805.

Problems

1. Rewrite Schemes 3.1 to 3.4 for cases where the primary sensitizer is an electron acceptor.
2. Consider reaction 3.30. It was shown in CIDNP studies of this system that sensitized isomerization with phenanthrene as sensitizer resulted in $\Phi_{isom} = 0.16$

in marked contrast to sensitization with anthracene where $\Phi_{isom} = 0.005$. Use an energy diagram to explain this result.

3. Calculate ΔG_{el} for reaction 3.7 given that the redox potential of **1** in acetonitrile is $E^0(D^{\dot{+}}/D) = -0.74$ vs. SCE, using ZnTPP and $Ru(bpy)_3^{2+}$ as sensitizers. Which of these two sensitizers is more effective in electron transfer and why? Which sensitizer is more likely to sensitize isomerization of **1** by an energy-transfer mechanism?
4. Using resonance-type structures, show that α-protonation in diallylamine (see reaction 3.93) is favorable.
5. Write out mechanisms for reactions 3.118 and 3.119.
6. Write out mechanisms for the following reactions:

(a)

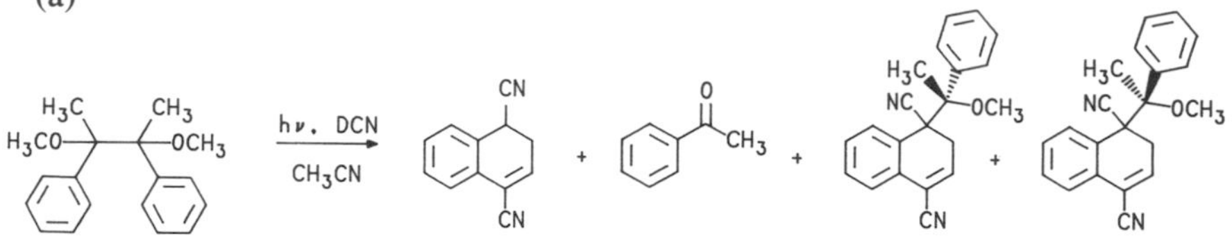

(Albini, A., and Mella, M. *Tetrahedron* **1986,** *42,* 6219.)

(b)

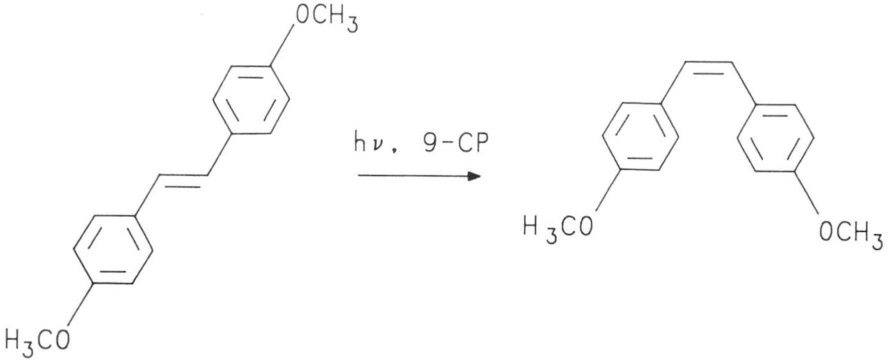

(Roth, H. D., and Schilling, M. L. M. *J. Am. Chem. Soc.* **1980,** *102,* 4303.)

(c)

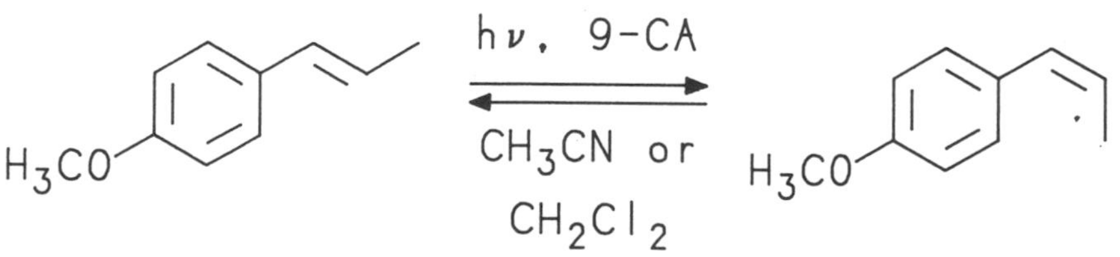

(Lewis, F. D., and Kojima, M. *J. Am. Chem. Soc.* **1988,** *110,* 8664.)

(d)

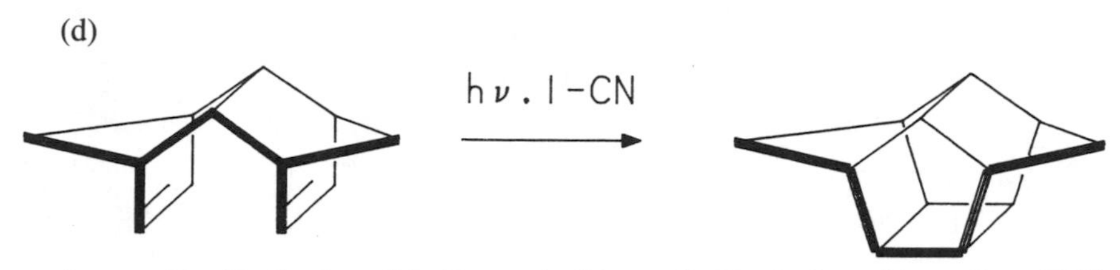

(Jones, G., II, Becker, W. G., and Chiang, S.-H. *J. Am. Chem. Soc.* **1983,** *105,* 1269.)

(e)

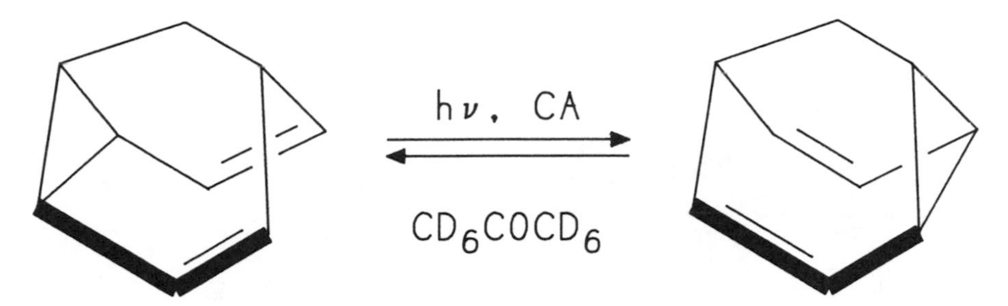

(Roth, H. D., and Abelt, C. J. *J. Am. Chem. Soc.* **1986,** *108,* 2013.)

(f)

$(CH_2)_n$ — hν, DCA / CH_3CN →

n = 1,5,6,7,12,20

(Mizuno, K., Kagano, H., and Otsuji, Y. *Tetrahedron Lett.* **1983,** *24,* 3849.)

(g)

hν, DCA / $CH_3CN/CH_3CN/C_6H_{12}$ →

(Watanabe, H., Kato, M., Tabei, E., Kuwabara, H., Hirai, N., Sato, T., and Nagai, Y. *J. Chem. Soc., Chem. Commun.* **1986,** 1662.)

(h)

(Albini, A., and Mella, M. *Tetrahedron* **1986,** *42,* 6219.)

(i)

(Gollnick, K., and Weber, M. *Tetrahedron Lett.* **1980,** *31,* 4585.)

(j)

(Tomioka, H., and Inoue, O. *Bull. Chem. Soc. Jpn.* **1988,** *61,* 1404.)

(k)

(Mizuno, K., Hashizume, T., and Otsuji, Y. *J. Chem. Soc., Chem. Commun.* **1983,** 977.)

(l)

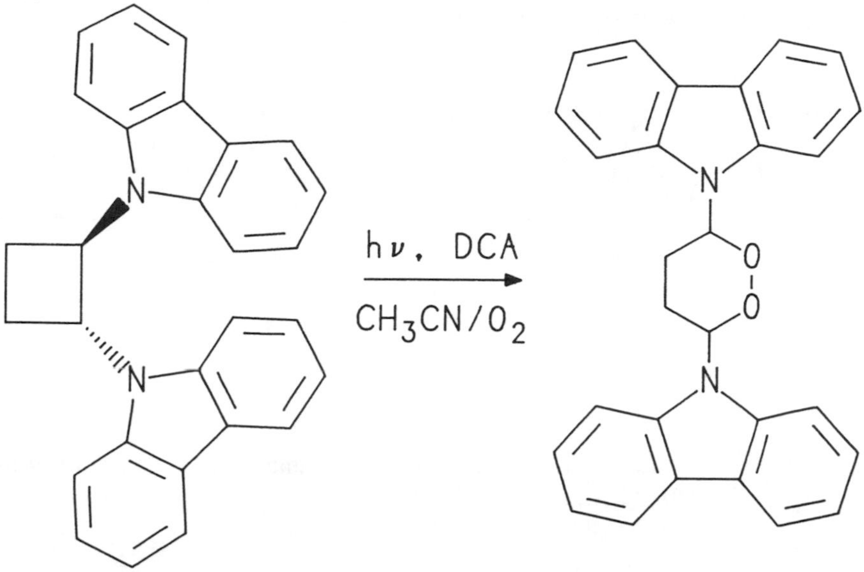

$h\nu$, DCA

CH_3CN/O_2

(Mizuno, K., Murakami, K., Kamiyama, N., and Otsuji, Y. *J. Chem. Soc., Chem. Commun.* **1983,** 462.)

7. Cellulose, a major constituent of plants and wood, is a natural polymer built from monomer units of D-glucose. The use of solar radiation to degrade biological materials such as cellulose for commercial purposes has attracted the interest of photochemists. Suggest a mechanism for the photosensitized cleavage of aryl glycosides in the presence of 1,4-dicyanonaphthalene:

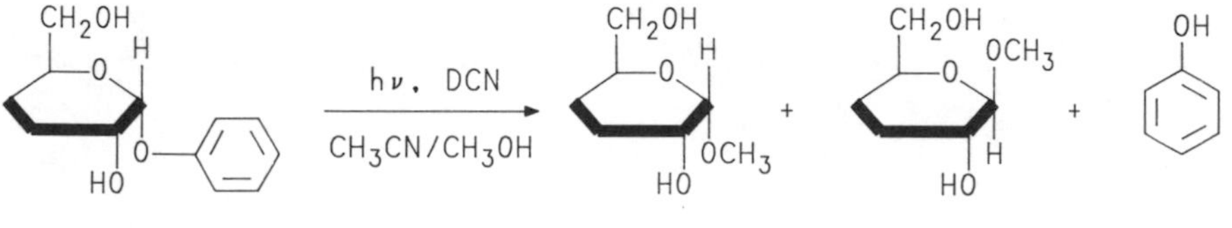

$h\nu$, DCN

CH_3CN/CH_3OH

(Timpa, J. D., Legendre, M. G., Griffin, G. W., and Das, P. K. *Carbohydrate Res.* **1983,** *117,* 69.)

8. Ultraviolet light is known to damage human DNA (deoxyribonucleic acid). Such damage may lead to harmful effects in human skin, including aging and cancer. One of the ways by which ultraviolet light interacts with DNA is to induce the formation of cyclobutane-like dimers between chains of the nucleic acid. It is known that cyclobutane dimers in DNA can be cleaved in the presence of certain repair enzymes on irradiation with near UV or visible light. A study of the

anthraquinone-2-sulfonate (AQS) sensitized splitting of pyrimidine dimer supports the formation of the dimer radical cation:

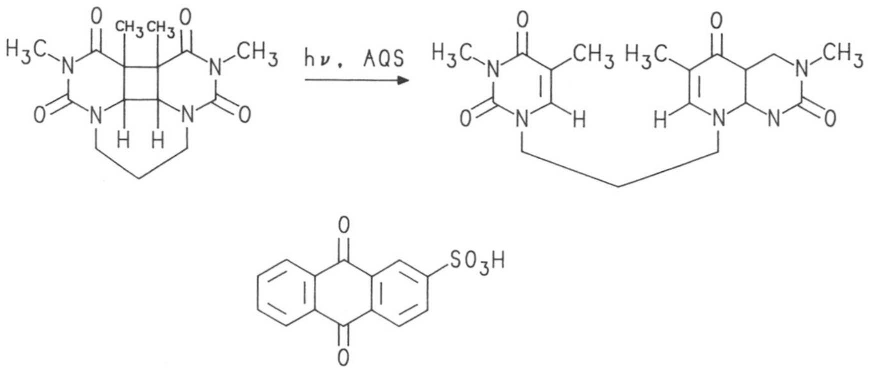

$\xrightarrow{h\nu,\ AQS}$

Suggest a mechanism. (Young, T., Nieman, R., and Rose, S. D., *Photochem. Photobiol.* **1990,** *52,* 661.)

CHAPTER

4

Intramolecular Photoinduced Electron Transfer

4.1. Introduction

We now turn to a fascinating area of photoinduced electron transfer—reactions between donor and acceptor molecules linked together by flexible and rigid spacer molecules. Intramolecular photoinduced electron transfer is a complex phenomenon presenting the chemist with interesting problems and opportunities. One of the challenges is to understand the role of the spacer or bridge group on the rates and efficiencies of electron transfer. We will restrict the discussion to excited-state electron transfer in systems where the donor and acceptor are connected by a molecular linking group.

For our purposes, we shall adapt the practice of classifying molecular spacers as *rigid* or *flexible*. A rigid spacer maintains a fixed distance and orientation between the donor and acceptor. Consequently, molecular motions are severely restricted. In many rigid systems, the separation distance and orientation can be determined with some degree of certainty, say by X-ray crystallographic or NMR studies. In contrast, a donor–acceptor pair attached via a flexible chain can interact by electron transfer at many distances and orientation since, depending on its structure, the chain can move more or less freely in solution. This changes the separation distance between the donor and acceptor. The donor and acceptor may be attached directly by a single carbon–carbon bond, or by a longer, more wobbly aliphatic hydrocarbon chain. Since the motions of flexible chains are influenced by the nature and length of the chain as well as the solvent viscosity and temperature, interpretations of experimental results obtained in studies on flexible systems frequently are complicated. In solution, electron transfer can take place at maximum distances where the

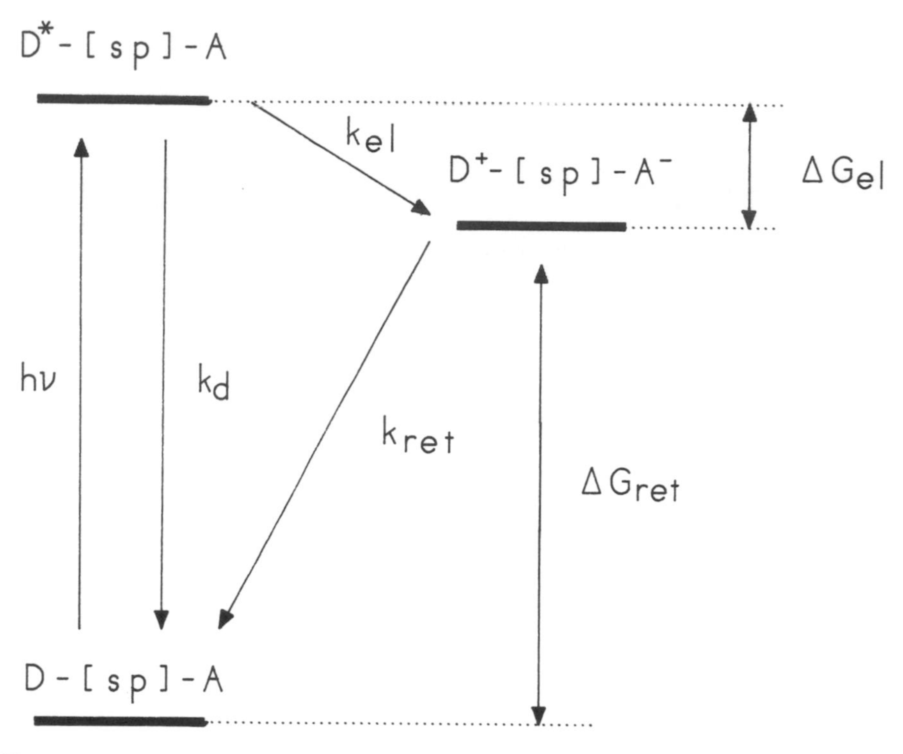

Figure 4.1 Energy diagram showing electron transfer and electron return in an intramolecular system consisting of an electron donor and acceptor molecule linked by a spacer molecule.

chain is fully extended or at closer distances where the donor and acceptor adopt a sandwich-like geometry. There are also many intermediate donor–bridge–acceptor conformations between the extended and folded conformations. Electron transfer is usually more rapid for systems where favorable chain conformations bring the donor and acceptor to close distances. In the folded conformation, there is the possibility of exciplex formation between donor–acceptor pairs linked together by long chains. In fact, exciplex formation often precedes the appearance of ion pairs. This process is mediated by solvation effects as well as the nature of the donor, acceptor, and chain. A major theme in this chapter will be to understand how structure, chain length, and solvent influence the competition between direct electron transfer and exciplex formation.

The energy diagram Fig. 4.1 will serve as an operational model for this chapter. This figure shows the energy changes accompanying photoinduced electron transfer between an excited donor and acceptor connected by a rigid or flexible spacer. The electron transfer step between the photoexcited group and the quencher is sometimes referred to as *forward* electron transfer or *charge separation*.

The quantum efficiency for generation of the ion pair, ϕ_{IP}, within an intramolecular system can be written as

$$\phi_{IP} = \frac{k_{el}}{k_{el} + k_d} \tag{4.1}$$

where k_{el} is the rate of (forward) electron transfer and k_d is the rate constant of radiationless and emissive decay of the excited state (we ignore other photochemical pathways involving the excited state). In attempting to maximize ϕ_{IP}, the challenge is to maximize k_{el} and minimize k_d by making electron transfer more competitive with the other pathways. One also wishes to maximize τ_{IP} so as to make the ion-pair state available for some useful application (Eq. 1.8).

At this point, we present several propositions without proof:

1. The energy difference between states plays a decisive role in determining the magnitude of the rate constant of electron-transfer processes.
2. A relatively small energy difference usually separates the reactant and the ion-pair states. In this cases, k_{el} increases with an increase in the energy difference.
3. A large energy difference separates the ion pair and ground states. An increase in the energy difference usually leads to a decrease in k_{ret}.
4. The rate of electron transfer decreases with increasing separation distance between donor and acceptor.

(The reader will learn more about the theory behind these generalizations in Chapter 6.) Applying these "rules" allows one to "tune" a reactant system to give a desired ϕ_{IP} and τ_{IP}. This shall be demonstrated in the examples presented later in this chapter. For now, we accept the validity of these rules and proceed on.

4.2. Through-Space and Through-Bond Coupling

Photoinduced electron transfer in rigid or flexible intramolecular systems can take place via *through-space* or *through-bond coupling*. By through-space electron transfer, we imply that the electron migrates through the space separating the donor from the acceptor. Here electron transfer depends on orbital overlap, which in turn is determined by the separation distance and relative orientation of the reactant orbitals. The orbital interactions may be strong (i.e., as in exciplex formation), or quite weak (i.e., as in electron transfer over large distances). In general, the rate of electron transfer, k_{el}, is predicted to be proportional to orbital overlap.

In through-bond electron transfer, electron transfer is influenced by the nature and number of bonds in the molecular bridge through which the electron "travels." We may envision that the electron "hops" through the orbitals of the spacer molecule. The through-bond mechanism holds particular interest in that the spacer group can be visualized as a conductor of electronic charge. Consider the illustration in Fig. 4.2 depicting an electron migrating from D over the LUMOs of the spacer, finally

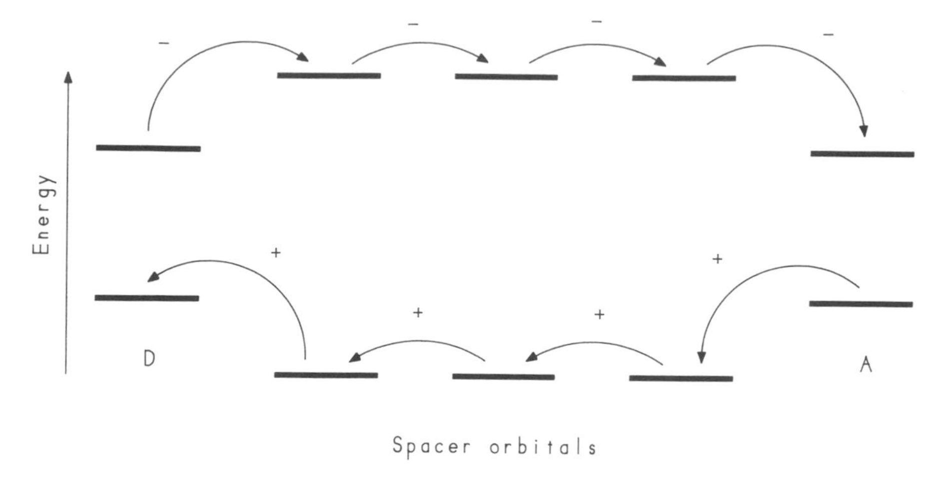

Figure 4.2 Electron and hole transfer through a molecular bridge. This figure shows the participation of the molecular link (spacer) in electron and hole transfer. (Adapted with permission from Kavarnos, G. *Top. Curr. Chem.* **1990,** *156,* 21. Copyright 1990 Springer-Verlag.)

ending up on the LUMO of the acceptor. We can also visualize a positive hole migrating from the acceptor along the spacer HOMOs until it reaches an empty donor orbital. The electron or hole may actually "hop" via successive jumps from the adjoining molecules to occupied or unoccupied orbitals of the bridge. Alternatively, we can regard the pathway as a resonance interaction through the barrier set up by the molecular bridge. This interaction is sometimes called a *superexchange*. What is meant by this terminology is that the molecular bridge provides the electronic coupling between the donor and acceptor. The magnitude of the electronic interaction in through-bond electron transfer is a function of the distance between the donor and acceptor, which in turn is determined by the number of bonds through which the electron migrates. Examples of several common molecular bridges include saturated cyclohexane and norbornane structures. It is likely that hydrocarbon bridges possessing different orbital topologies result in different electron-transfer rates.[1,2]

According to current theories, the migration of a hole through the valence orbitals, or electron through antibonding orbitals, also depends on the energy spacings between the donor and acceptor and spacer orbitals. Thus, the energy gap between the donor orbital and the nearest LU molecular orbital of the spacer, or between the acceptor orbital and nearest HO molecular of the spacer, can determine whether the electron moves along the LU molecular orbitals or whether the hole travels across the HO orbitals. π-orbitals are more effective than σ-orbitals in propagating an electron or hole, since the π-orbitals of the unsaturated spacer normally are positioned close in energy to the orbitals of the donor and acceptor.

Ultimately, the precise physical picture of an electron moving through a complex molecule will be achieved only by careful measurements using ultrafast techniques such as femtosecond laser spectroscopy. At this stage, although the exact pathway of an electron remains elusive, it appears that the bridge *does* mediate the *rate* of electron or hole migration. In this connection, an important point is whether a through-bond pathway prevails over a through-space pathway and, if so, under what conditions. This question can be examined only if we know the precise details of the molecular structure of the system and have a way to map the path of the electron. It is likely that the through-space pathway dominates in systems where there is effective orbital overlap between the orbitals of the donor and acceptor. Consequently, through-space coupling should be favorable for close distances and sandwich-type orientations. In rigid systems, the through-bond pathway probably dominates if the donor and acceptor are well separated. However, even in these systems, it is possible for the donor and acceptor to be oriented favorably for a through-space interaction. Perhaps the through-space and through-bond pathways coexist in some systems. These matters are important considerations in biological electron transfer where protein and DNA macromolecules can mediate electron transfers.

4.3. Photoinduced Electron Transfer in Bridged Metal–Ligand Systems

Studies on photoinduced electron transfers between metal complexes linked by π-delocalized ligands and saturated linkages have sparked much interest among inorganic chemists. Much of this work is closely connected with the pioneering contributions of Henry Taube and collaborators, who studied electron transfer and exchange in metal complexes. The terms *intervalence transfer, metal–metal charge transfer, and metal–ligand charge transfer* have become part of the language used by chemists to describe electronic transitions in metal complexes called *polynuclear complexes*. This terminology refers to inorganic complexes containing two or more metal centers connected by bridging ligands. When these compounds are exposed to visible or near ultraviolet light, they often display low-energy absorption bands. These spectral bands can be attributed to transitions involving the movement of an electron from one orbital to the other within the molecule. The reader may now wish to review Section 1.2.3 before proceeding.

If we adopt the following representation for a *binuclear* metal complex, i.e.,

$$L_a–M_a–B–M_b–L_b$$

where L represents a ligand, M a metal center, and B a bridging ligand, then we can classify the types of electronic transitions in these compounds. The simplest transition between the metal centers M_a and M_b is known as intervalence transfer or IT. Thus, the transfer of an electron from M_a to M_b can be depicted as shown in 4.2:

$$L_a–M_a–B–M_b–L_b \xrightarrow{h\nu} L_a–M_a{}^{+}–B–M_b{}^{-}–L_b \tag{4.2}$$

The charge-transfer state, $L_aM_a{}^+–B–M_b{}^-–L_b$, can be regarded as an IT state. Likewise, absorption of light can lead to the transfer of an electron from the metal to the ligand. The resulting state is called the metal-to-ligand charge-transfer state, or MLCT:

$$L_a–M_a–B–M_b–L_b \xrightarrow{h\nu} L_a{}^-–M_a{}^+–B–M_b–L_b \tag{4.3}$$

A metal-to-bridge charge transfer, or MBCT, is shown below:

$$L_a–M_a–B–M_b–L_b \xrightarrow{h\nu} L_a–M_a{}^+–B^-–M_b–L_b \tag{4.4}$$

The formation and decay of these short-lived states may be monitored by time-resolved spectroscopy.

Before exploring several examples, however, it may be worthwhile to discuss several general points about electron transfer in polynuclear metal complexes. First, the reader will note that in the literature the terminology *optical* electron transfer is often used to describe electron transfer accompanying electronic excitation of one of the metal centers. Optical electron transfer should not be confused with *photoinduced* electron where electron transfer *follows* excitation of the reactant, e.g.:

Optical electron transfer:

$$D + A \xrightarrow{h\nu} D^+ + A^- \tag{4.5}$$

Photoinduced electron transfer:

$$D + A \xrightarrow{h\nu} D^* + A \rightarrow D^+ + A^- \tag{4.6}$$

Note the difference: in 4.5 electron transfer occurs *during* photoexitation whereas in 4.6, it occurs *after*.

A second point—one that is quite important—is that one metal center can be excited independently of the other. It is assumed, therefore, that electronic delocalization between the centers is minimal. In this case, the system can be treated in terms of each separate, independent component. These complexes are classified as class I mixed-valence ions. In contrast, there may be "communication" between the metal centers in the sense that there is considerable electronic interaction between them (class III). Here the components of the molecule cannot be separated into discrete components, and the electrons are quite delocalized. In between these two extremes are metal complexes classified as class II where there is some degree of interaction between the metal centers. (In Chapter 6, a useful physical model that helps to describe the difference between weakly and strongly interacting metal center will be introduced. We shall not discuss this theoretical model here except to alert the reader to the existence of a large variety of polynuclear metal complexes where varying degrees of electronic interactions between the metal centers exist.)

The third point is that the bridge may or may not participate in the optical electron transfer. This is usually dictated by the molecular structure of the bridge. The bridge can be an unsaturated π- or saturated σ-ligand. As shown in Fig. 4.2, the electronic transition can involve an electron or hole transfer. The bridge participates as an electron or hole acceptor. In either case, we can visualize the following pathways:

$$M_a\text{–}B\text{–}M_b \xrightarrow{h\nu} M_a^{+}\text{–}B^{-}\text{–}M_b \rightarrow M_a^{+}\text{–}B\text{–}M_b^{-} \tag{4.7}$$

$$M_a\text{–}B\text{–}M_b \xrightarrow{h\nu} M_a\text{–}B^{+}\text{–}M_b^{-} \rightarrow M_a^{+}\text{–}B\text{–}M_b^{-} \tag{4.8}$$

Reaction 4.7 shows the bridge as an electron acceptor. In 4.8 the bridge is an electron donor; consequently, this pathway represents a hole transfer. As noted earlier, the transfer can be regarded as a superexchange or as a series of hops resulting in possibly discrete intermediates. Occasionally, the oxidized or reduced bridge may be detected by spectroscopic methods.

One of the classic examples of an optical electron transfer was noted in the following mixed-valence complex (**1**), which is often referred to as the Creutz-Taube ion[3,4]:

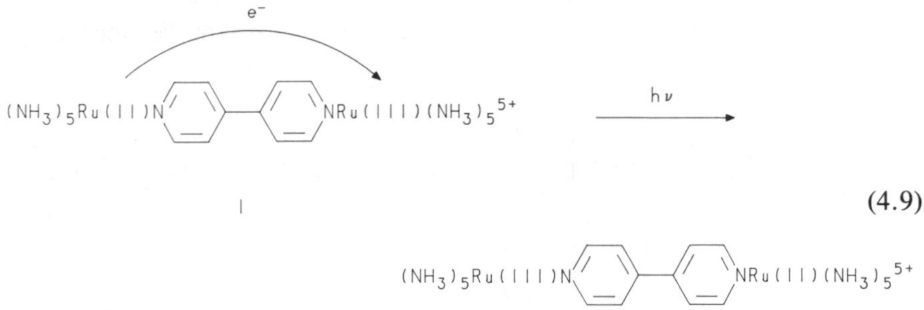

e^-

$(NH_3)_5Ru(II)N$ [4,4'-bipyridine] $NRu(III)(NH_3)_5^{5+}$ $\xrightarrow{h\nu}$

I

(4.9)

$(NH_3)_5Ru(III)N$ [4,4'-bipyridine] $NRu(II)(NH_3)_5^{5+}$

In this complex, the ruthenium centers are held together by the pyrazine bridging ligand. On photoexcitation, an electron is transferred from the reduced to oxidized metal. This is an example of an IT transition in which the unsaturated bridge interacts strongly with the metal centers.

In $[(NH_3)_5Ru(II)\text{–}pz\text{–}Ru(III)\text{-}EDTA]^+$ (**2**) where pz = pyrazine, a metal-to-bridge electronic transition was observed employing picosecond laser photolysis[5]:

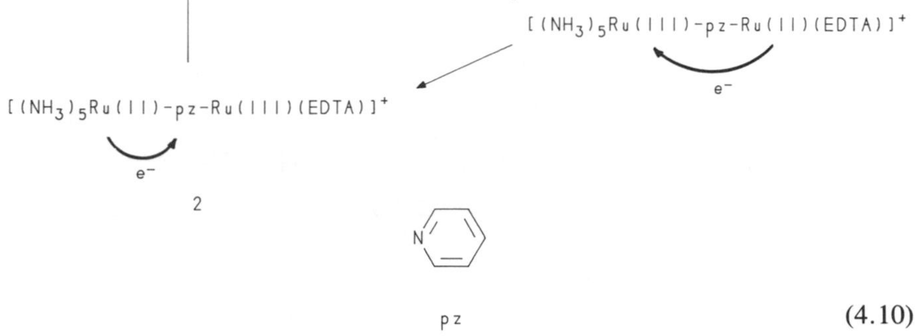

e^-

$[(NH_3)_5Ru(III)\text{-}pz^{\bar{\cdot}}\text{-}Ru(III)(EDTA)]^+$

$h\nu$

$[(NH_3)_5Ru(III)\text{-}pz\text{-}Ru(II)(EDTA)]^+$

e^-

$[(NH_3)_5Ru(II)\text{-}pz\text{-}Ru(III)(EDTA)]^+$

e^-

2

N

pz

(4.10)

Spectral analysis of this system showed a visible Ru(II) → pz MBCT transition. Evidently, photoexcitation of the complex leads to formation of the MBCT state. Deactivation of this state is accompanied by a bridge-to-metal electron transfer resulting in the formation of a short-lived IT state. The IT state decays finally to the ground state.

In the previous example, the bridging ligand plays an important role in acting to transfer the electron between the ruthenium centers. In other systems, however, the bridge can be pictured as a passive observer. The bridge does not participate in the electron transfer in the sense that it becomes oxidized or reduced. For example, in $[(NH_3)_5Ru(II)–bpa–Ru(II)(bpy)_2Cl]^{2+}$ (**3**) the Ru → bpy MLCT absorption band is quite prominent.[6] The bridge is saturated and does not participate in the formation of the MLCT state. Emission from the MLCT state is similar to the emission from $Ru(bpy)_2Cl–bpa^+$. It is concluded that the emission is not quenched by an intramolecular process. Accordingly, the following pathway was proposed for this system:

e^-

$$[(NH_3)_5Ru(II)–bpa–Ru(III)(bpy^{\bar{\cdot}})(bpy)Cl]^{2+}$$

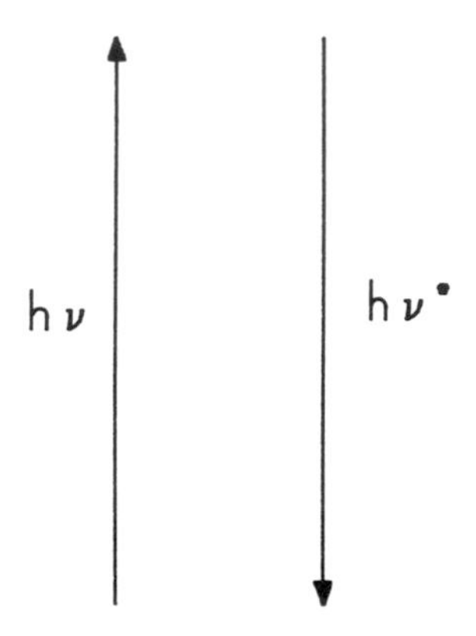

$$[(NH_3)_5Ru(II)–bpa–Ru(II)(bpy)_2Cl]^{2+}$$

e^-

3

N⟨pyridine⟩–CH₂–CH₂–⟨pyridine⟩N

bpa (4.11)

In contrast, in $[(NH_3)_5Ru(II)\text{–bpe–}Ru(II)(bpy)_2Cl]^{2+}$ (**4**) where bpe is an unsaturated analogue of the "insulating" bpa bridge, the emission lifetime is substantially reduced. In this case, the bridge apparently interacts via electron delocalization with the metal and ligand orbitals:

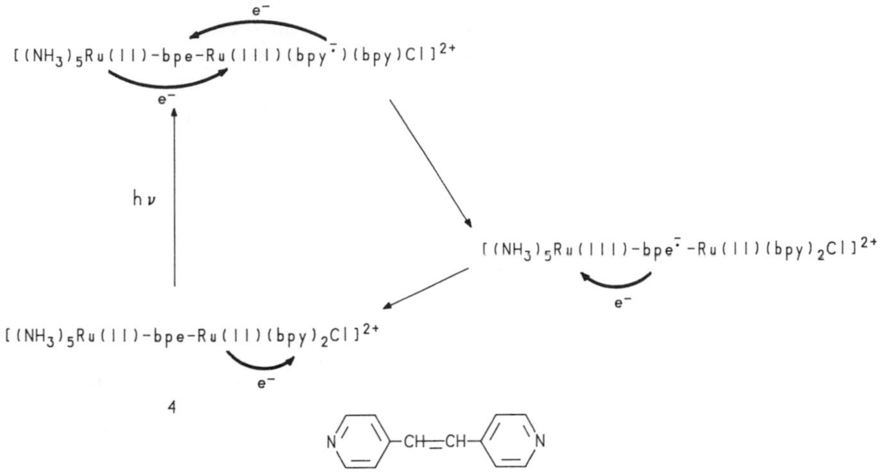

$[(NH_3)_5Ru(II)\text{-bpe-}Ru(III)(bpy^{\cdot -})(bpy)Cl]^{2+}$

e^- e^-

$h\nu$

$[(NH_3)_5Ru(III)\text{-bpe}^{\cdot -}\text{-}Ru(II)(bpy)_2Cl]^{2+}$

e^-

$[(NH_3)_5Ru(II)\text{-bpe-}Ru(II)(bpy)_2Cl]^{2+}$

e^-

4

N⟨⟩–CH=CH–⟨⟩N

bpe (4.12)

Intramolecular quenching of the MLCT state actually occurs by two electron transfers. The MBCT state, which has a lifetime of less than 30 ps, deactivates by a single electron transfer.

A particularly striking example of optical electron transfer has been documented in $[py(NH_3)_4Ru\text{–CN–}Ru(bpy)_2\text{–CN–}Ru(NH_3)_5{}^{5+}]$ (**5**) whose structure is shown below[7]:

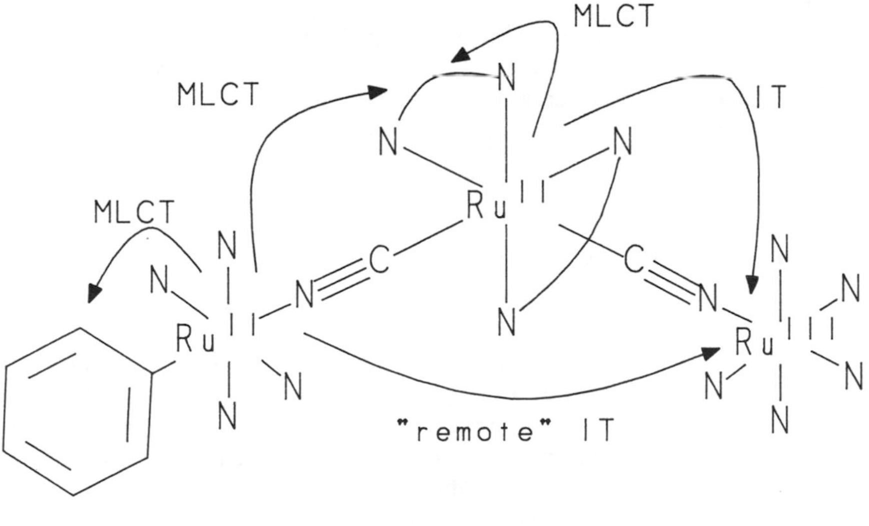

MLCT MLCT MLCT IT

Ru^{II} Ru^{II} Ru^{III}

"remote" IT

5

The bridging unit in this trinuclear complex is the cyanide group. The absorption spectrum of this complex suggests a large number of electronic transitions. The reader will note that several electron transitions can be excited separately. These transitions are indicated by the labels accompanying the structure shown above. Of special interest is the appearance of "remote" MLCT and IT transitions. These transitions involve electron transfers between well-separated (remote) centers. Decay processes subsequently follow the formation of the MLCT, MBCT, and IT states.

Studies of long-range electron transfer in excited Co(III)–Cu(I) binuclear complexes linked by bicyclooctene rings (**6** and **7**) have revealed interesting distance effects[8]:

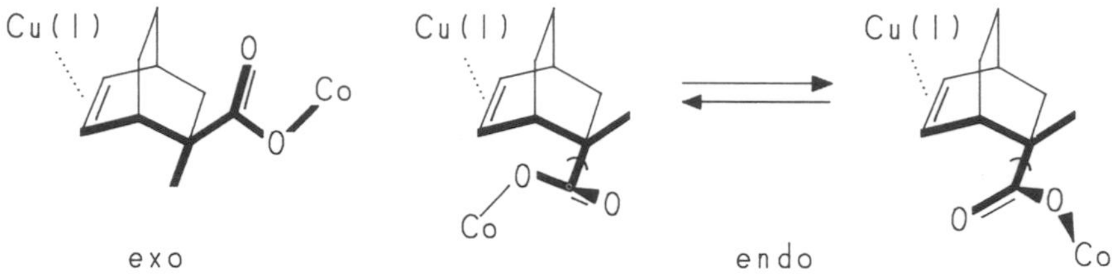

exo

6

endo

7

In these structures, Cu(I) is coordinated with the unsaturated bond. On photoexcitation of **6** or **7**, electron transfer occurs from Cu(I) to the Co(III) ion coordinated with the carboxylate group. The *endo* isomer (**7**) is found to be more reactive. It is thought that in the *endo* isomer, Co(III) can be brought much nearer to Cu(I) to distances approaching ~4 Å. In the extended *exo* isomer (**6**), a greater separation of 7.1 to 7.9 Å is maintained between the ions so that in this case, electron transfer is slower.

Numerous optical electron transfers have been documented and will not be dealt with here. The reader, who is urged to refer to the excellent reviews listed at the end of this chapter, will be impressed by the rich tapestry of optical electron transfers in inorganic complexes.

4.4. Photoinduced Electron Transfer in Flexible Organic Systems

Chemists have long been fascinated by the behavior of long hydrocarbon chains. By developing methods to probe the complex dynamics of chain motions, they have been able to learn much about the functions and properties of long-chain flexible polymers, which play an important role in many practical and commercial applications. In short flexible chains consisting of polymethylene groups, certain functional groups can be attached at the ends. Subsequently, these systems can undergo

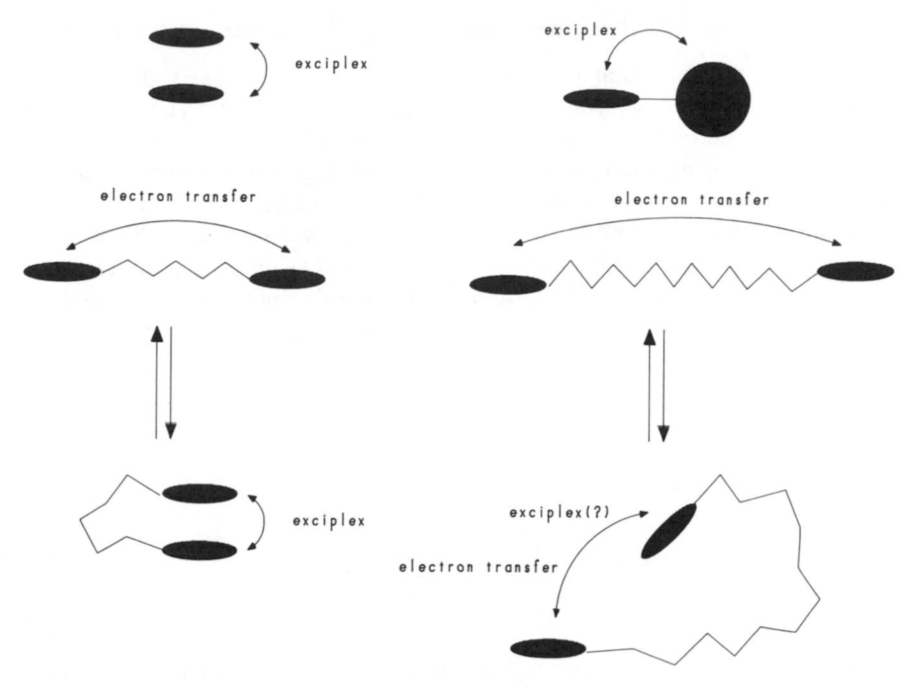

Figure 4.3 Effects of size of chain on electron transfer *vs.* exciplex formation. Unconnected donor and acceptor molecules can form sandwich-like exciplexes (upper left). Donor and acceptor connected by medium-size flexible chain can form exciplexes at close distances and can engage in electron transfer in the extended form (lower left). Donor and acceptor connected by large flexible chain can display electron transfer at extended and intermediate distances, but whether the donor and acceptor can form an exciplex at close distances depends on the dynamics of the chain (lower right). Donor and acceptor molecules connected by a single chain can form twisted exciplexes (top right).

ring closure reactions. The kinetics of these processes depend to a great extent on chain length and the dynamics of chain conformations. These conformational changes can sometimes be studied by an intramolecular reaction such as exciplex formation involving an end-to-end interaction of the linked donors and acceptor. In general, for exciplex formation to occur, the intramolecular donor or acceptor should be able to approach closely to one another. If the pair cannot move to close distances required for exciplex formation, electron transfer may dominate. This is illustrated schematically in Fig. 4.3.

Thus, under certain conditions, electron transfer involving the end groups can be observed in competition with exciplex formation. We shall review several examples in this section with emphasis on donor–acceptor molecules linked by flexible polymethylene chains. We start with a discussion of donor–acceptors attached by a single bond and then continue to examine more complex systems where several methylene groups connect the donor and acceptor.

4.4.1. Twisted Intramolecular Charge-Transfer Complexes

When a donor molecule is attached an acceptor by a single σ-bond, the intramolecular donor–acceptor molecule may display two kinds of fluorescence. This dual or "anomalous" fluoresence arises because photoexcitation produces two emitting states, each possessing a unique geometry. The nature of this fluorescence and its solvent dependence have been the subject of intense interest. For example, it is known that *p*-cyano-*N,N*-dimethylaniline (**8**) fluoresces when exposed to light. With increasing solvent polarity a red-shift fluorescence band begins to appear. The dual fluorescence is thought to arise because rotation about the single bond in **8** can lead to formation of a highly polar twisted conformation[9]:

and 4

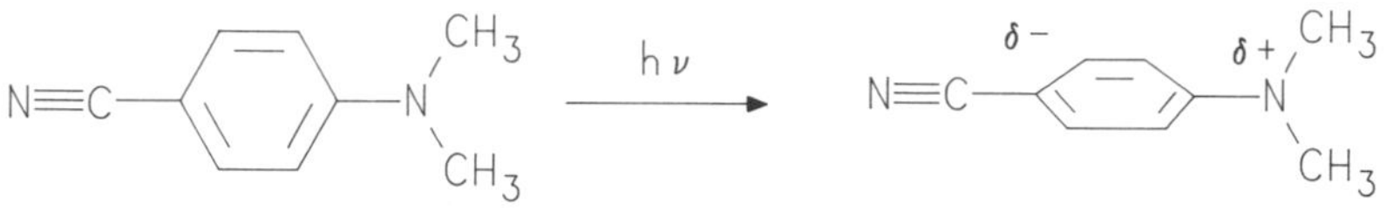

N≡C–⟨ring⟩–N(CH_3)(CH_3) $\xrightarrow{h\nu}$ N≡C–⟨ring⟩($\delta-$)–N($\delta+$)(CH_3)(CH_3)

8

Twisted charge-transfer state

(4.13)

Photoexcitation of the planar ground state leads to a locally excited singlet state. An electron then migrates from the dimethyl amino nitrogen into an orbital of the aromatic ring to give a planar radical ion pair. Fluorescence can subsequently occur from this planar complex. In polar solvents, the red-shifted fluorescence arises from a twisted radical ion pair state (TICT). In the twisted state, the π-electrons are localized over $D^{\dot{+}}$ and $A^{\dot{-}}$, and there is no sharing of charge between these ions. The following pathway suggests a possible route to explain dual fluorescence:

$$\begin{array}{ccccc} \text{D–A} & \xrightarrow{h\nu} & \text{D*–A} & \rightarrow & \text{TICT} \\ & & \downarrow h\nu_f & & \downarrow h\nu_f' \\ & & \text{D–A} & & \text{D–A} \end{array} \qquad (4.14)$$

The energy diagram in Fig. 4.4 will help the reader understand the changes in energy accompanying the formation of the twisted structure. A case is shown where the donor and acceptor are free to rotate in their ground state. The red-shifted fluorescence originates from the polar twisted excited state. Note from Fig. 4.4 that $h\nu_f > h\nu_f'$. If solvent relaxation is rapid, fluorescence band broadening may be noticeable due to transitions from states that differ in the manner that solvent molecules are oriented about the donor–acceptor excited state. In some cases, the fluorescence decay may display nonexponential behavior. The TICT state can be visualized as an intramolecular exciplex. Since formation of the twisted structure involves a bond rotation, the energy barrier should depend on steric effects and solvent interactions. The transition from the excited structure to ground state may actually pass through a Franck–Condon state, which then rapidly relaxes to a relatively nonpolar twisted ground state.

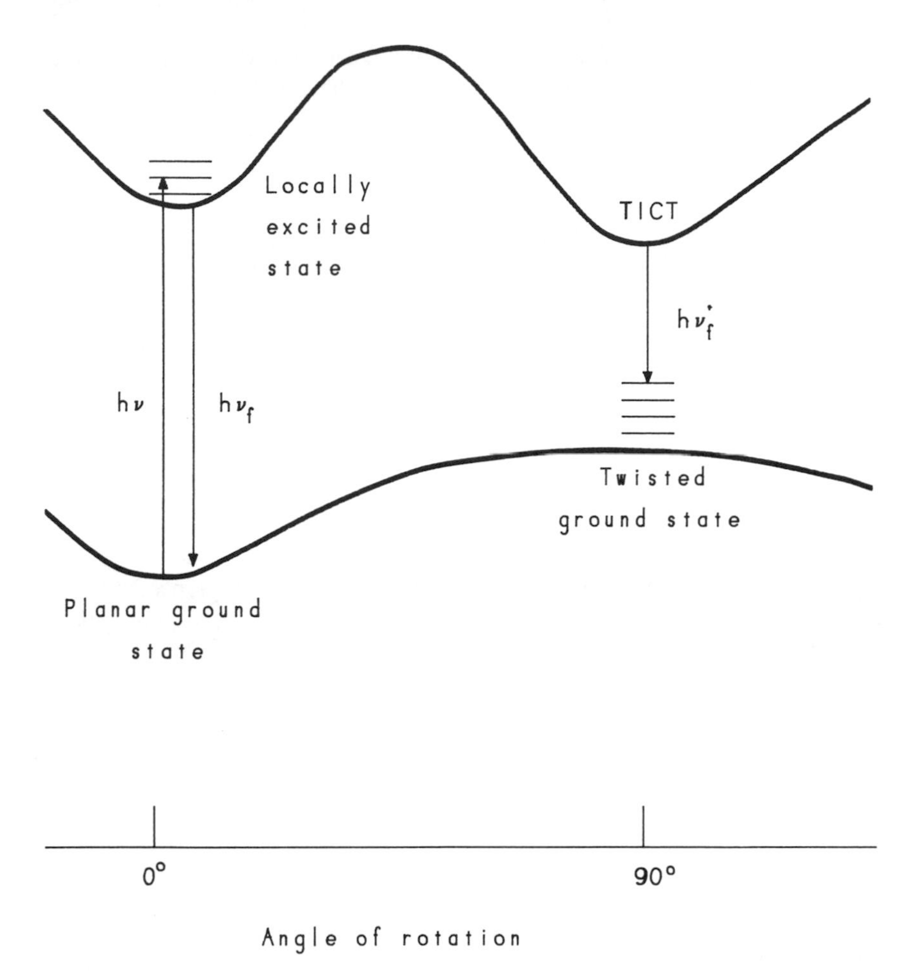

Figure 4.4 Formation of the "twisted" charge-transfer complex. Photoexcitation of a planar ground state results in formation of a locally-excited state. The latter may fluoresce or undergo conformational changes to a twisted charge-transfer state (TICT). The fluorescence from TICT (normally shifted to the red) leads to formation of a twisted ground state, which subsequently relaxes to the planar ground state.

Electron delocalization is related to the degree of rotation about the single bond. In a 90° fully twisted conformation, the mobile π-electrons are localized on D and A. Hence, this polar structure possesses a relatively large dipole moment; in a planar structure, however, the electrons are more delocalized, and consequently, the dipole moment is smaller.

Flexible donor–acceptors—molecules that are free to rotate between planar and perpendicular structures such as **8**—should produce dual fluorescence. However, molecules with restrictions to rotation may not display the same fluorescence patterns of flexible donor–acceptor molecules. For example, since 5-cyano-*N*-methylindoline

(**9**) is "frozen" into a coplanar geometry, this molecule displays only the higher wavenumber fluorescence since rotation to a perpendicular structure is not allowed:

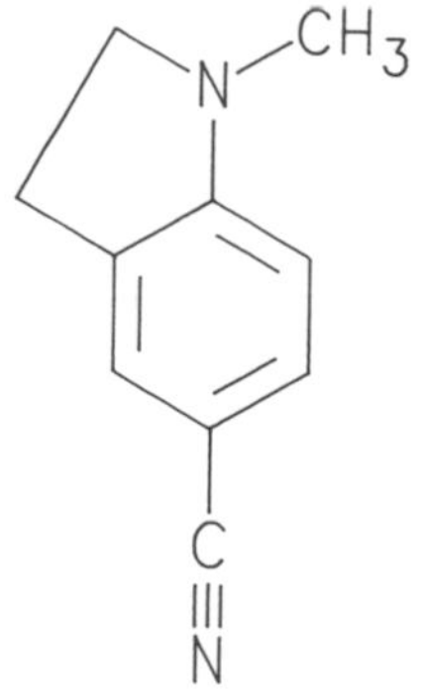

9

In contrast, 6-cyanobenzquinuclidine, **10,** which is fixed in a perpendicular conformation and therefore is not free to rotate to a planar structure, displays only red-shifted fluorescence:

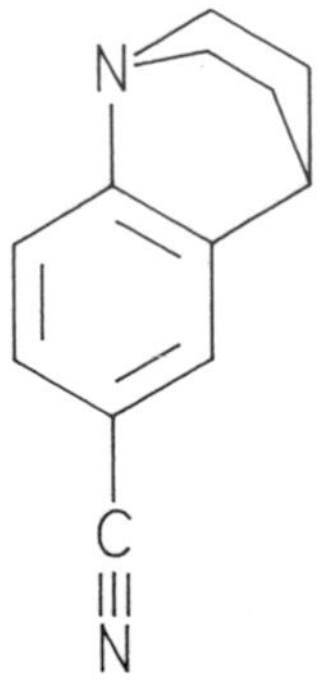

10

Other investigators have reported similar observations for donor–acceptor complexes with restricted and unrestricted geometries.[10,11] These results support the TICT hypothesis. In fact, the observation of red-shifted fluorescence in restricted, perpendicular geometries implies the existence of electron transfer in these molecules.

There have been many attempts to clarify the role of the solvent in the formation of twisted charge-transfer states. It appears that viscosity and solvent motion can play a role by influencing the twisting motion of the connecting single bond. The solvent motion may be strongly coupled with changes in the geometries of the reactants, so that the rate is determined largely by the time it takes for the solvent molecules to relax following electron transfer. Solvent molecules can also impede molecular motion. This frictional force is more important for viscous solvents so that we might predict that the ability of solvent molecules to equilibrate the charge-transfer products of electron transfer is closely related to the viscosity of the solvent. These considerations are particularly pertinent for the formation of a twisted charge-transfer state in which the surrounding polar solvent molecules must respond to the electronic changes in the charge-transfer structure.

The concept of *solvent-induced* electron transfer also applies to photoinduced electron transfer in donor–acceptor molecules connected by longer flexible chains, which we consider in the next section.

4.4.2. Photoinduced Electron Transfer in Flexible Molecular Systems: Exciplex vs. Ion Pair Formation

The dynamic motions of flexible intramolecular systems are fairly complex. Here we must consider what restrictions, if any, the connecting chain imposes on the overall rate of electron transfer. These effects may be steric and dynamic. In a polymethylene chain, —$(CH_2)_n$—, the approach of a donor and acceptor attached to a chain depends on the energy barriers due to rotations of C—C bonds. These energy barriers are influenced by the structure and length of the flexible chain and as well as by interactions between the reactants and surrounding solvent molecules. Typically, the rate of C—C bond rotation in an alkyl chain in solution is $\sim 10^9$ s^{-1}. The reader may wonder if is there a preferred chain conformation and distance, or does electron transfer occur over a wide range of separation distances and conformations? Actually, theory tells us that electron transfer is possible over a range of distances, the actual rate varying in an exponential manner with distance (Chapter 6). Thus, electron transfer at large distances should be relatively slow and less observable as compared to electron transfer at short separation distances.

Another complication in flexible systems is that exciplex formation often competes with direct electron transfer. This competition is dependent on solvent characteristics such as polarity and viscosity as well as structural features of the donor–bridge–acceptor. Because exciplex formation in flexible intramolecular systems is a complex phenomenon, it is helpful to resort to a simplified picture. In Fig. 4.5, a donor molecule connected to an acceptor via a flexible polymethylene chain is shown. Two idealized ground-state geometries, an extended and end-to-end conformation, are illustrated. The conformations and geometries of these chains are dictated by the size of the chain as well as entropy and internal energies accompanying bond rotations. Numerous "intermediate" geometries are possible but for expedience we do not explicitly consider these structures. Assuming that at any given temperature there is an equilibrated mixture of possible ground-state conformations, we envision that photoexcitation generates an excited chain in the same conformation as the absorbing ground state structure. Within the short lifetime of the excited state, the chain continues to engage dynamic rotations. In any one of a number of intermediate structures, electron transfer may take place in competition with radiative and nonradiative decay processes. If the donor and acceptor move

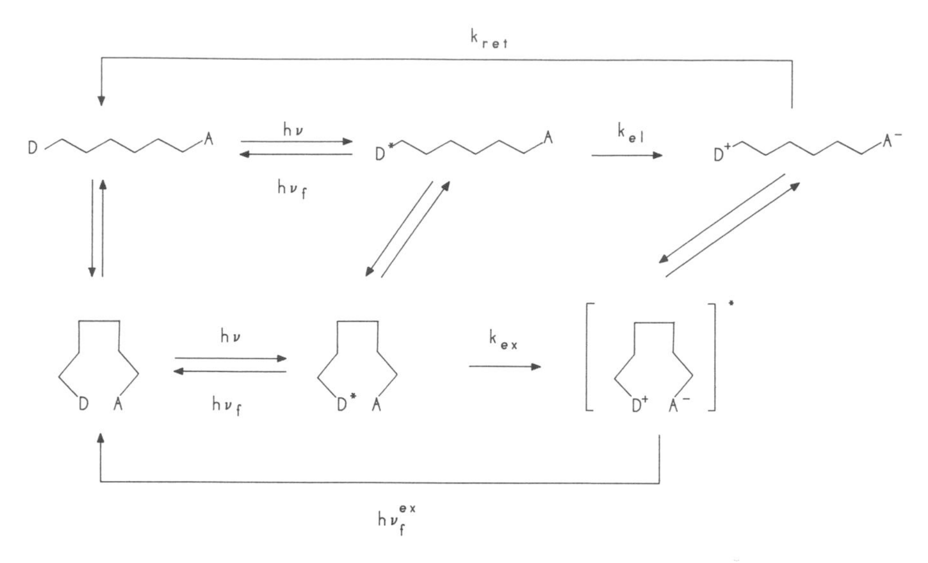

Figure 4.5 Idealized representations of photoinduced electron transfer in donor-acceptor pairs linked by flexible polymethylene chains.

into a folded conformation, the donor and acceptor are favored to generate an exciplex.

In the following examples, there is an emphasis on the effect of chain length. Although it is not possible to establish formal rules for determining whether exciplex formation or direct electron transfer predominates in the quenching step, we will note some trends in flexible systems. In general, for $n = 0$, twisted exciplexes are favored. In exciplexes separated by only a few methylene groups, nonplanar (perpendicular) structures are favored. With longer chains, there is the possibility of forming planar, sandwich-like exciplexes. If n is larger, say between 2 and 6, the possibility of planar, sandwich-like exciplexes exists. For very large n (perhaps greater than 6), the possibility of forming planar exciplexes becomes less likely since the chain must now undergo numerous rate-limiting conformational motions to form the exciplex.

These considerations probably hold true in nonpolar solvents where the mutual orientation of donor and acceptor plays an important role. In polar solvents, these restrictions are relaxed because direct electron transfer in these media does not demand as strict an orientation as does formation of an exciplex. Usually, electron transfer to give ion pairs rather than exciplexes prevails in polar solvents because of energetics (Chapter 2). However, even in polar solvents, exciplex may be observed because of solvent-assisted stabilization of the emitting state. Perhaps the best way to appreciate the complex interplay of chain length and solvent is to examine a few examples.

Davidson and co-workers, for instance, have considered the question of exciplex formation and electron transfer in a series of flexible ω-(1-naphthyl)-*N*-alkylpyrroles.[12] Using fluorescence quantum yield and lifetime measurements, they investigated the effect of solvent on the fluorescence yields, lifetimes, and maximum wavelength of exciplex emission in **11, 12,** and **13** in nonpolar and polar solvents:

H_3C

$(CH_2)_n$

H_3C

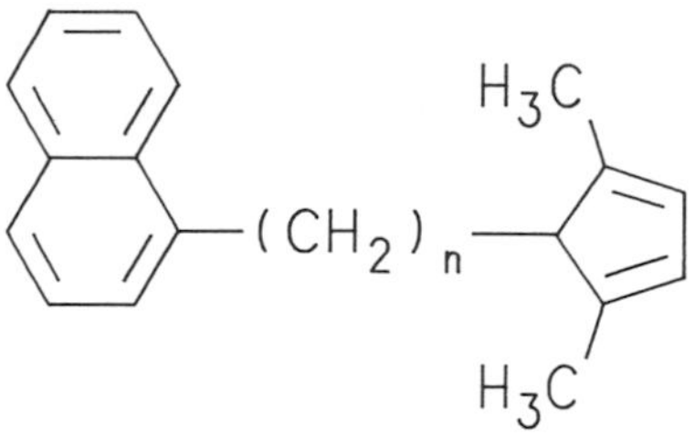

11 n = 0

12 n = 1

13 n = 2

Some results are summarized in Table 4.1. It is of particular interest that in all three cases (n = 0,1,2), exciplex fluorescence emission is observed in both nonpolar and polar solvents. If we consider the idealized exciplex structures of **11, 12,** and

Table 4.1 FLUORESCENCE QUANTUM YIELDS, LIFETIMES, AND MAXIMUM EXCIPLEX WAVELENGTHS FOR **11, 12,** AND **13**[a]

	Φ_f^{EX}	τ_f^{EX} (ns)	λ_{max} (nm)
	11 (n = 0)		
Cyclohexane	0.03	7.2	405
Benzene	0.02	5.0	440
1-Propanol	0.015	7.6	470
Acetonitrile	0.01	6.9	505
	12 (n = 1)		
Cyclohexane	0.03	15.8	398
Benzene	0.03	10.2	412
1-Propanol	0.02	14.3	429
Acetonitrile	0.02	14.1	490
	13 (n = 2)		
Cyclohexane	0.36	36.1	396
Benzene	0.28	75.6	414
1-Propanol	0.14	0.61	455
Acetonitrile	0.12	41.0	485

[a] Adapted from reference 12; in degassed solutions.

13 shown in Fig. 4.6, we note that the exciplex of **11** can form a twisted charge-transfer state whereas **13** is favored to form a sandwich-like conformation. The exciplex of **12** probably lies somewhere between these structures. An important piece of experimental evidence is that structures **11** and **12** give rise to low fluorescence quantum yields, Φ_f^{EX}, and short fluorescence lifetimes, τ_f^{EX}, when compared to the values for **13.** Since the latter is held by the longest chain length, the geometry of its exciplex probably comes closest to a planar, face-to-face structure, resembling an intermolecular pair of naphthalene and pyrrole. Therefore, for **13,** exciplex formation prevails in all solvents.

Let us consider another piece of evidence that further substantiates the proposed

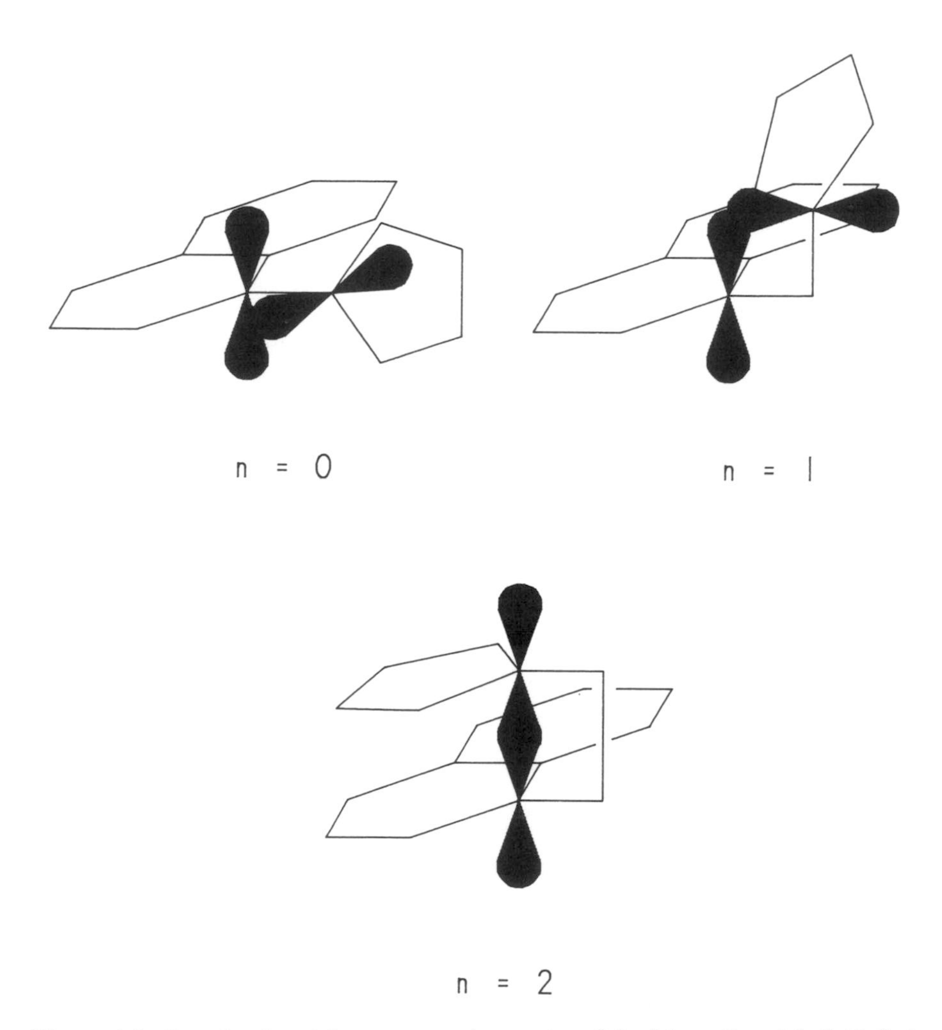

Figure 4.6 Postulated exciplex structures in a series of flexible ω-(1-naphthyl)-N-alkylpyrroles. There are usually significant conformational changes accompanying exciplex formation. For n = 0 and 1, the exciplexes are in mutually perpendicular orientations in contrast to the planar, sandwich-like structure for n = 2.

model. The risetime in exciplex emission of compounds **11, 12,** and **13,** which is defined as the time required to form the exciplex structure, is found to be sensitive to the distance separating the donor and acceptor. In **11** and **12,** the risetime is more rapid than in **13.** This should not be surprising since in the latter there must be numerous chain rotations to bring the donor and acceptor into a suitable orientation for exciplex formation. The close proximity of donor and acceptor in **11** and **12** results in a faster appearance of the exciplex, even though Φ_f^{EX} is low. Thus, exciplex formation in **13** is more sensitive to dynamic chain rotations, which eventually bring the donor and acceptor into the face-to-face orientation.

Dynamic competition between direct electron transfer and exciplex formation was postulated by Eisenthal and co-workers, who examined the photoinduced reactions of **14**[13,14]:

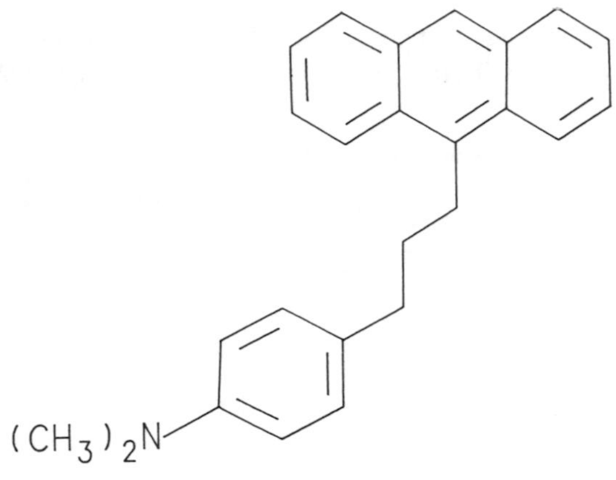

$(CH_3)_2N$

14

In acetonitrile, a "folded" ground-state conformation and "extended" conformation rapidly interconvert. The ion pairs resulting from both conformations were observed spectroscopically. The following pathway was propose to account for the results observed in polar solvents:

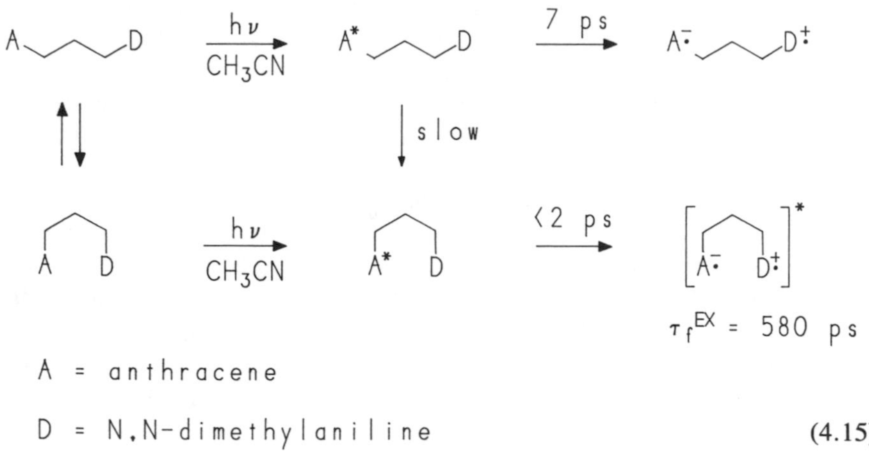

A—D $\xrightarrow[CH_3CN]{h\nu}$ A*—D $\xrightarrow{7\ ps}$ $A^{\bar{\cdot}}$—$D^{\dot{+}}$

⇅ (left); slow ↓ (middle)

A D (folded) $\xrightarrow[CH_3CN]{h\nu}$ A* D (folded) $\xrightarrow{<2\ ps}$ $[A^{\bar{\cdot}}\ D^{\dot{+}}]^*$

$\tau_f^{EX} = 580\ ps$

A = anthracene

D = N,N-dimethylaniline (4.15)

In contrast, in a nonpolar solvent such as 2-methylbutane, exciplex formation prevails:

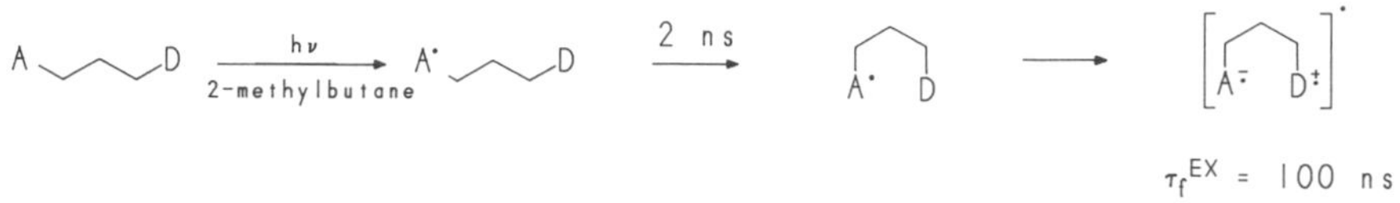

$$\mathrm{A{\sim}D} \xrightarrow[\text{2-methylbutane}]{h\nu} \mathrm{A^{\bullet}{\sim}D} \xrightarrow{2\ \mathrm{ns}} \mathrm{A^{\bullet}\ D} \longrightarrow [\mathrm{A^{\overline{\bullet}}\ D^{\dagger}}]^{\bullet} \qquad \tau_f^{EX} = 100\ \mathrm{ns} \tag{4.16}$$

The exciplex formed in 4.16 is much longer lived and stabilized than that formed in acetonitrile. It is likely that the exciplexes formed in 4.15 and 4.16 differ with respect to their structures.

In a related study, Mataga and co-workers documented intramolecular exciplex formation and competitive electron transfer in 1-pyrenyl-$(CH_2)_n$-*N,N*-dimethylaniline pairs (**15** and **16**)[15]:

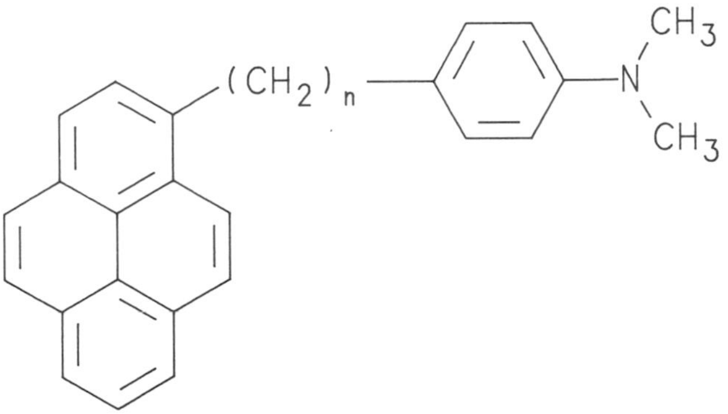

15 n = 1

16 n = 3

These workers studied the behavior of compounds **15** and **16** in acetonitrile, *n*-hexane, and 2-propanol. In **15,** quenching of pyrene fluorescence in acetonitrile occurs within several picoseconds and matches the "growing-in" of charge-transfer absorbance. Electron transfer is not favored in **15** in nonpolar solvents. In **16,** exciplex formation is observed in both nonpolar and polar solvents, although in *n*-hexane exciplex formation takes a longer time (nanoseconds) to grow in, so it is likely that the exciplex takes on the sandwich structure in *n*-hexane. In acetonitrile, the formation of the charge-transfer state is rapid, within a few picoseconds, indicating that formation of this state does not involve extensive conformational rotations. In 2-propanol, which is a more viscous solvent than *n*-hexane or acetonitrile, transient measurements indicate that electron transfer takes place within ~350 ps, followed by a slow formation of an exciplex (~1 ns). The results in 2-propanol suggest that this solvent slows down motions of the connecting methylene link.

Yang and collaborators have studied the effects of chain length on electron transfer and exciplex formation between the amino group and the excited anthryl group in **17, 18, 19,** and **20**[16,17]:

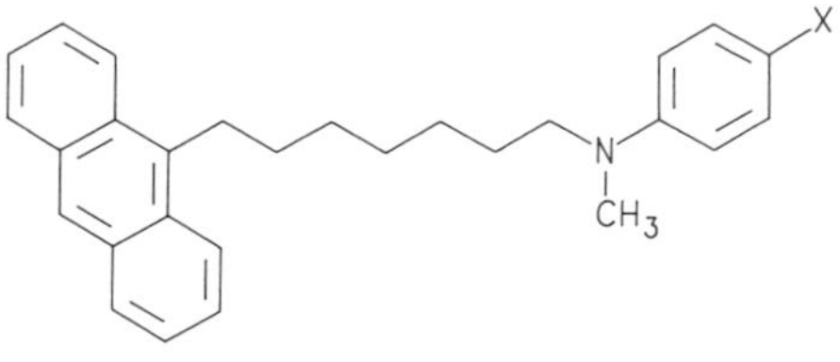

17 X = H
18 X = OCH_3

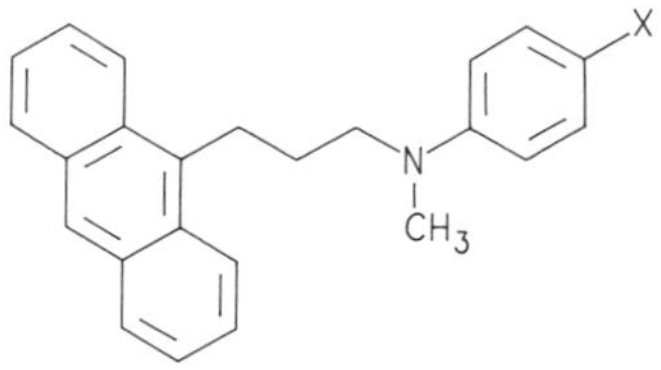

19 X = H
20 X = OCH_3

Their results indicated that exciplex formation is unfavorable for $n = 7$ as compared to efficient exciplex formation for $n = 3$. Apparently, with $n = 7$, the donor and acceptor cannot be positioned too easily into a sandwich-like conformation. The dynamic chain motions required to bring the donor and acceptor together occur on such a long time scale that the excited state decays before exciplex formation.

Let us now turn to an intramolecular system containing a ketone group. As pointed out in earlier chapters, there is much experimental evidence supporting the concept of electron-transfer quenching of triplet ketones by electron-rich quenchers. Recall from the earlier chapters that quenching of excited ketones by aliphatic and aromatic amines proceeds quite rapidly and probably involves the intermediacy of amine radical cations and carbonyl radical anions. The available evidence suggests that in order for quenching to be efficient, there must be effective orbital overlap between the electron-rich quencher and the carbonyl molecular orbitals. In flexible intramolecular ketone–amine systems, conformational motions are often rate limiting. This is demonstrated in the reaction of **21** and **22**[18]:

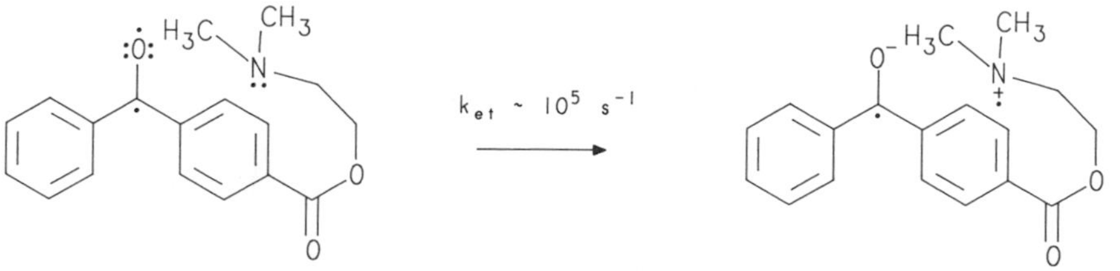

$k_{et} \sim 10^5\ s^{-1}$

$^3n,\pi^*$

21 (4.17)

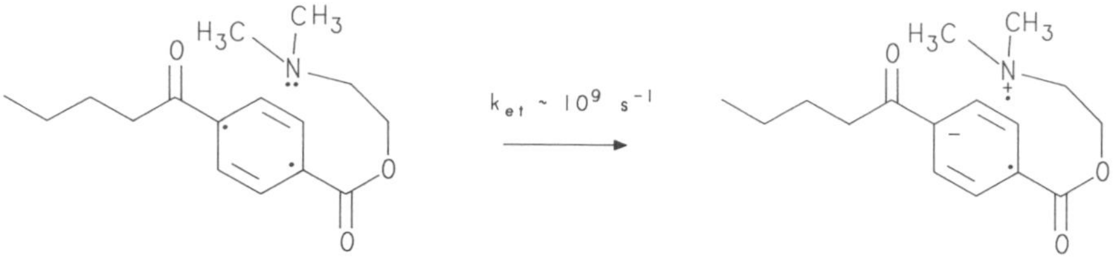

$k_{et} \sim 10^9\ s^{-1}$

$^3\pi,\pi^*$

22 (4.18)

The rate of electron transfer for **21** was found to be markedly lower than for **22.** In **21,** which is an n,π^* triplet, presumably the flexible chain is unable to bring the nitrogen orbital sufficiently close to the carbonyl group for effective orbital overlap. However, in the π,π^* triplet of **22** where excitation energy is partly localized in the phenyl group, the half-vacant, π-orbital of the carbonyl can be positioned over the aromatic group. The resulting increase in overlap favors electron transfer.

Other studies also lend support to the role of chain motions in ketone systems. For example, in **23,** ketone phosphorescence was found to be quenched efficiently for $n \geqslant 8$[19,20]:

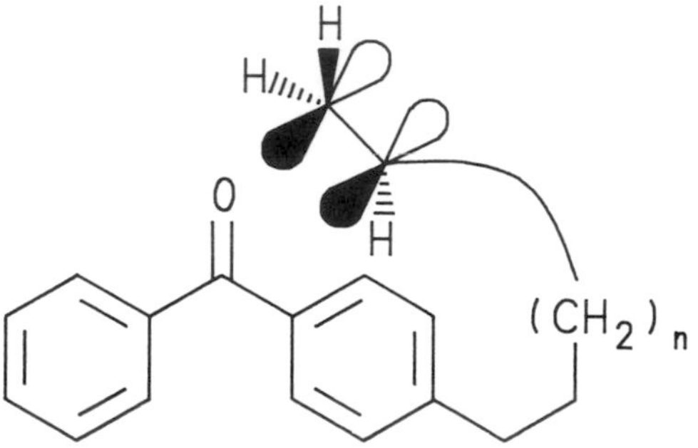

$(CH_2)_n$

23

In contrast, efficient quenching is not observed for $n < 8$. This is taken as evidence that the chain must be sufficiently long and flexible to wrap around to allow the π-electrons of the olefin to interact with the *n*-orbital on the ketone oxygen.

4.4.3. Photoinduced Electron Transfer in Ternary Intramolecular Systems

In some organic systems, three active groups may be connected together by a molecular link. These systems are referred to as ternary intramolecular compounds.

Each component acts independently, and one group functions as the light absorber. The remaining groups may interact with the photoexcited component to form an exciplex or ion pair. These interactions are usually sequential processes.

For example, exciplex formation has been suggested in **24** shown below[21]:

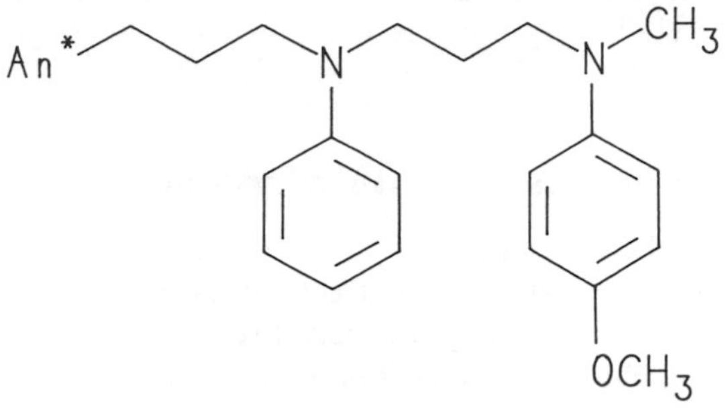

An* CH3 N N OCH3

24

Analysis of the effects of solvent polarity on the fluorescence emission displayed by **24** on exposure to light suggested the intermediacy of a polar exciplex with the participation of both amino groups:

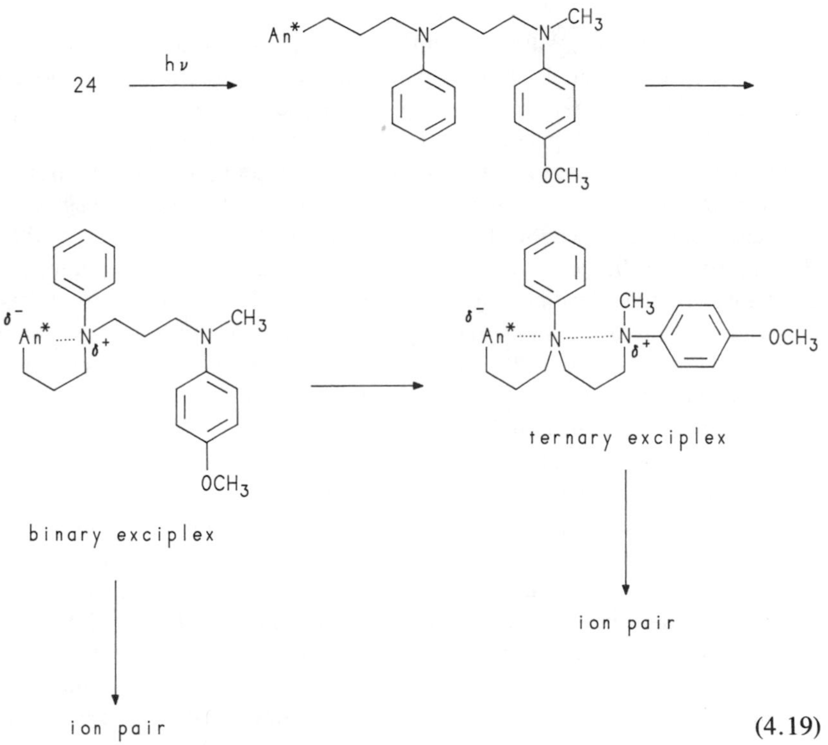

24 hν An* N N CH3 OCH3

δ⁻ An* N δ+ N CH3 OCH3

binary exciplex

ion pair

δ⁻ An* N CH3 N δ+ OCH3

ternary exciplex

ion pair

(4.19)

In the step-wise mechanism shown in 4.19, **24** initially forms a binary exciplex (an exciplex with two components). Subsequently, the third component becomes involved, and we have the creation of a ternary exciplex. In the ternary exciplex, there is a greater degree of charge separation so that the dipole moment in this form is larger than in the binary exciplex. In polar solvents, therefore, formation of the ternary exciplex is favored with subsequent formation of ion pairs.

4.4.4. Spin Dynamics in Flexible Systems

Another approach to the study of chain dynamics in intramolecular donor–acceptor systems has involved the magnetic field effect. At this point, the reader may want to review the section on electronic spin effects (Section 2.6). A study of the effects of magnetic fields on photoinduced electron transfer in a series of pyrene-$(CH_2)_n$-*N,N*-dimethylaniline molecules where $n = 4$–16 has yielded some interesting facets of chain dynamics. The following scheme summarizes the electron-transfer pathways in these systems[22]:

$$\begin{array}{ccccccccc} {}^1A^*\text{-sp-D} & \leftrightarrows & {}^1[{}^1A^*\text{-sp-D}] & \rightarrow & {}^1[{}^2A^{\bar{\cdot}}\text{-sp-}{}^2D^{\dot{+}}] & \leftrightarrows & {}^3[{}^2A^{\bar{\cdot}}\text{-sp-}{}^2D^{\dot{+}}] & \rightarrow & {}^3A^*\text{-sp-D} \\ & & & & \downarrow & & \downarrow & & \\ & & & & A\text{-sp-D} & & {}^2A^{\bar{\cdot}}\text{-sp-}{}^2D^{\dot{+}} & & \end{array} \quad (4.20)$$

(A = pyrene, D = *N,N*-dimethylaniline, and sp = spacer). The initial electron transfer generates a singlet radical pair followed by electron return to ground-state reactants or intersystem crossing to produce a triplet radical ion pair. It is assumed that triplet pyrene formation within the triplet ions is exothermic. Triplet yield of pyrene, Φ_T, was monitored by transient absorption measurements in the presence of a changing external magnetic field. It was found that with decreasing chain length, Φ_T decreases: $\Phi_T \simeq 0.6$ ($n = 16$), $\Phi_T \simeq 0.5$ ($n = 11$), and $\Phi_T \simeq 0.02$ ($n = 4$). For a given chain length, there are dynamic conformational motions leading to a distribution of end-to-end distances. These distances range from a minimum and effective distance for electron transfer, which we designate as d_{eff}, to the longest distance of the extended chain conformation, d_{long}, where electron transfer is not as favorable. In the longest chains, $E_{hc} > 2J_{eff} > 2J_{long}$ (recall from Section 2.6 that J, which measures the spin exchange interaction, is related to separation distance). We can visualize that the average time that the donor and acceptor are close together is similar to that when the donor and acceptor are not linked together (i.e., they are independent and freely diffusing). The observed values of Φ_T imply that hyperfine coupling between degenerate S, T_+, T_-, and T_0 states is the predominant interaction in two-component systems connected by long chains (Fig. 4.7). Increasing the magnetic field leads to a decrease in Φ_T because of the increasing energy gap between S and T_+ and T_-. In chains of intermediate length ($6 < n < 12$), E_{hc} lies somewhere between the maximum and minimum J so that hyperfine coupling is no longer a dominant interaction. In

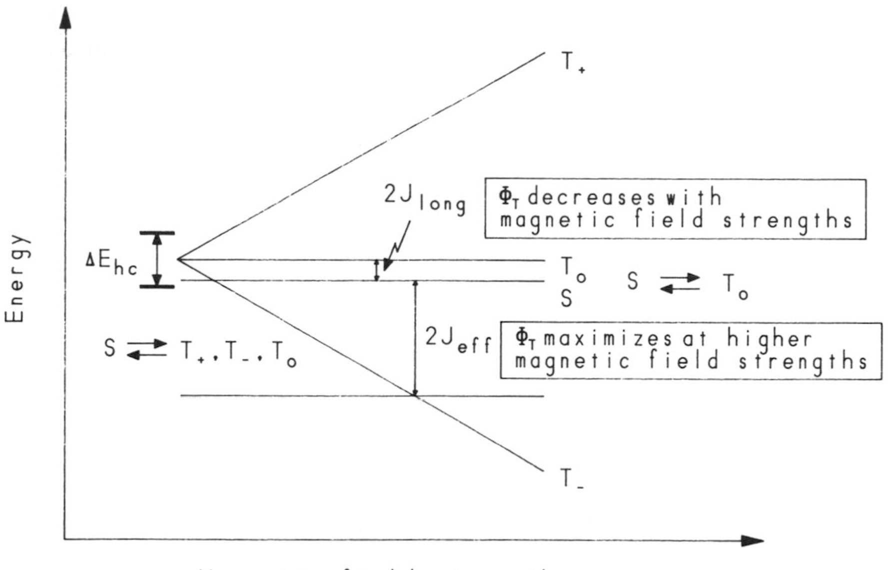

Figure 4.7 The effects of magnetic fields on photoinduced electron transfer in a series pyrene-$(CH_2)_n$-N,N-dimethylaniline compounds. The Zeeman interaction results in splitting of T_+, T_o, and T_{-1} energy levels of these radical pairs. Singlet-triplet splitting can result from distance-dependent spin exchange interactions, J. These interactions are considered relative to hyperfine interactions, E_{hc}. Singlet-triplet splitting is stronger at smaller distances (J_{eff}) than at larger distances.

these systems, the donor and acceptor are closer together so that the electronic exchange interaction is enhanced. Here, an increasing magnetic field leads to a crossing between the S and T_- states, and results in a maximum Φ_T at higher fields. In short chains ($n \sim 6$), the close proximity of the ends leads to a large singlet–triplet energy interaction. Now $J > E_{hc}$, and hyperfine coupling is inoperative. Under these conditions, intersystem crossing is dominated by spin-orbital coupling.

These magnetic field effects are consistent with the theme that the dynamic motions of chains in flexible donor–acceptor systems can control the rates and efficiencies of electron-transfer processes.

4.4.5. Photoinduced Electron Transfer between Donors and Acceptors Linked by Peptide Spacers

In examining many of the previous examples, the reader should now be aware of the difficulty in dealing with donor–acceptor systems connected by flexible chains.

To overcome this difficulty, the use of certain synthetic peptide spacers offers one solution. For instance, oligoproline spacers can adopt only certain conformations, which slowly interconvert. In emission studies of electron transfer between $[(bpy)Re(I)(CO)_3(pyr)]^+$ and 4-(*N*,*N*-dimethylamino)benzoate linked together by two L-proline spacers, two significantly different electron-transfer rates were measured[23]:

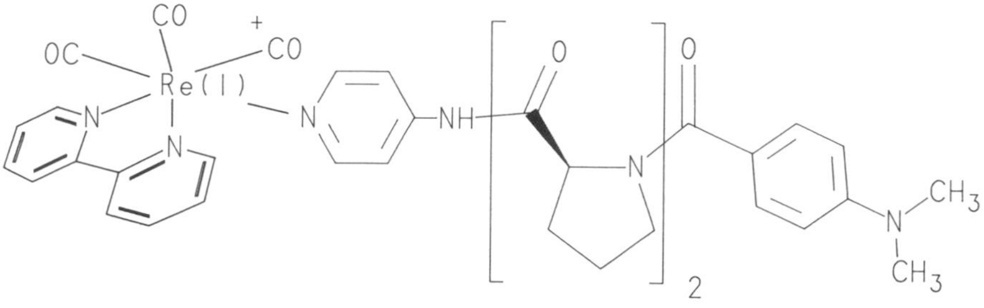

25

The Re(I) complex displays a MLCT band, and the excited state is luminescent, long-lived, and a good oxidant. It was determined that the molecule exists in two interconverting conformations, and that photoinduced electron transfer can occur involving both of these conformations. Since the rate of interconversion appeared to be relatively slow, it was suggested that the dependence of electron transfer on conformation implies that the electron transfer occurs "through-space."

4.5. Photoinduced Electron Transfer in Rigid Systems

Photoinduced electron transfer between two functional groups attached together by one or more *rigid* spacer molecules has attracted the interest of chemists, physicists, and biologists. One of the reasons for this attention is that covalently linked donor–acceptor pairs serve as heuristic models of biological electron transfer and photosynthesis (Chapters 5 and 6). A particular advantage of a rigid molecule is that the distance and orientation of electron acceptor and donor remain "frozen" during the course of the electron transfer. Here the complications introduced by a flexible chain are completely absent.

By investigating a series of systems where the number of rigid spacer groups linking the electron donor and acceptor is varied, one can test the distance dependence of the rate of electron transfer. Similarly, the effect of orientation may be studied by selecting rigid bridges that restrict the donor and acceptor to known geometries.

Photoinduced electron transfer in a rigid system normally does not involve exciplex intermediates, since the donor and acceptor are usually too far apart. The electron transfer may be a through-space or through-bond interaction. Unlike flexible chain systems where the reactants can approach to close distances, rigid groups maintain a fixed distance and geometry between the reactants. In this section, we will limit the subject matter to examples of photoinduced electron transfer in rigid systems consisting of organic donors and acceptors linked by saturated bridges.

The first example is shown in **26** where a methoxybenzene group is attached to dicyanoethylene by several fused cycloalkane rings[24]:

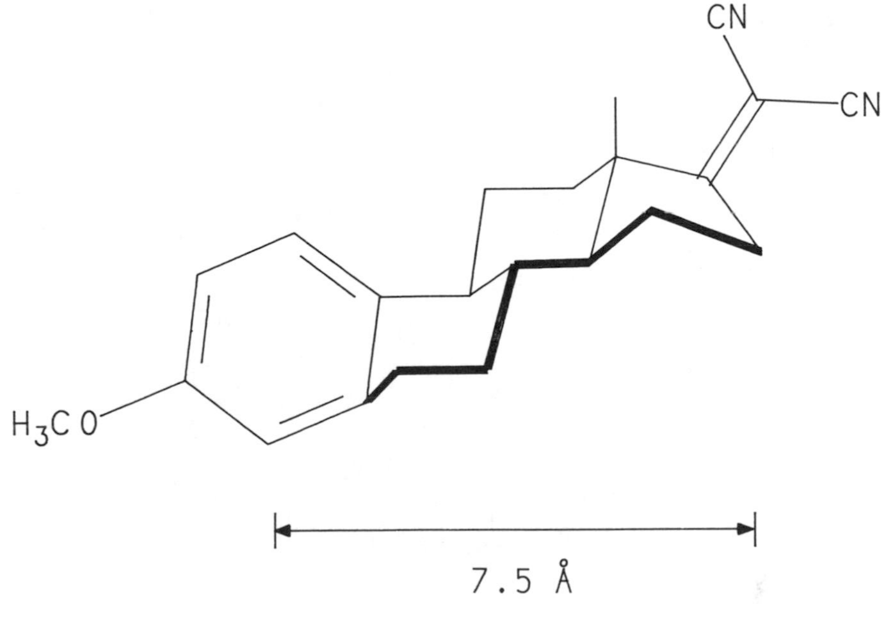

CN

CN

H_3CO

7.5 Å

26

Methoxybenzene is the light-absorbing species. On illumination with light, it readily donates an electron to the strong electron acceptor. The role of the cycloalkane spacer group is to maintain a rigid distance between methoxybenzene and dicyanoethylene.

A spectacular example of long-range electron transfer has been observed in a series of molecular assemblies (**27, 28, 29, 30,** and **31**). In this series, the dimethoxynaphthalene chromophore is separated from the dicyanoethylene acceptor by an increasing number of hydrocarbon bridges[25]:

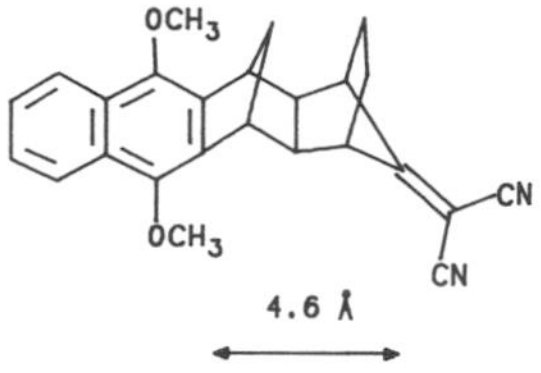

4.6 Å

27

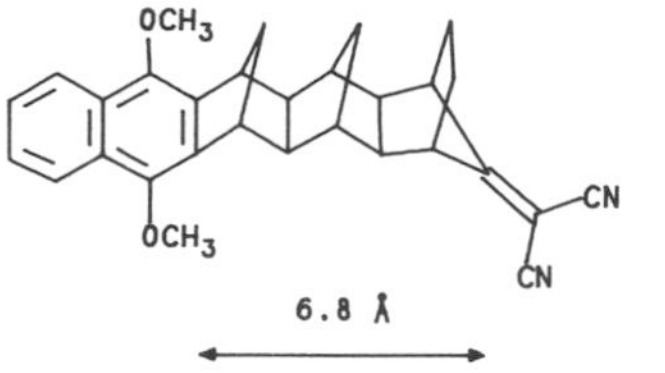

6.8 Å

28

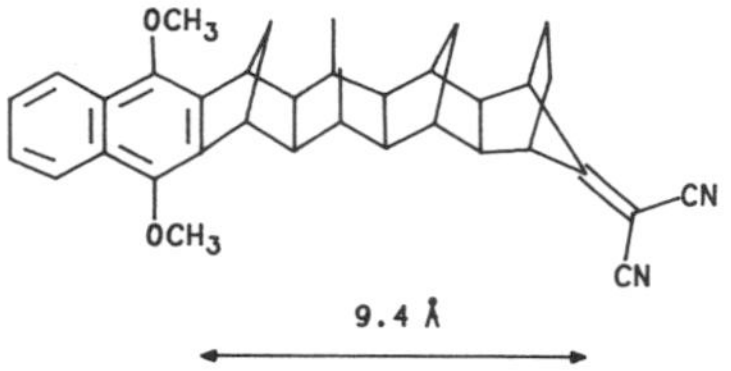

9.4 Å

29

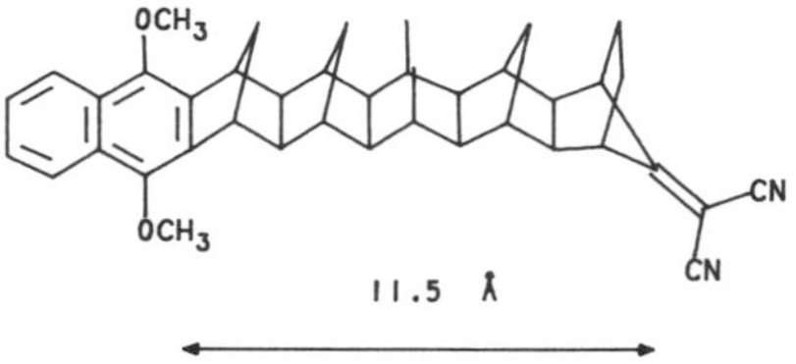

11.5 Å

30

13.5 Å

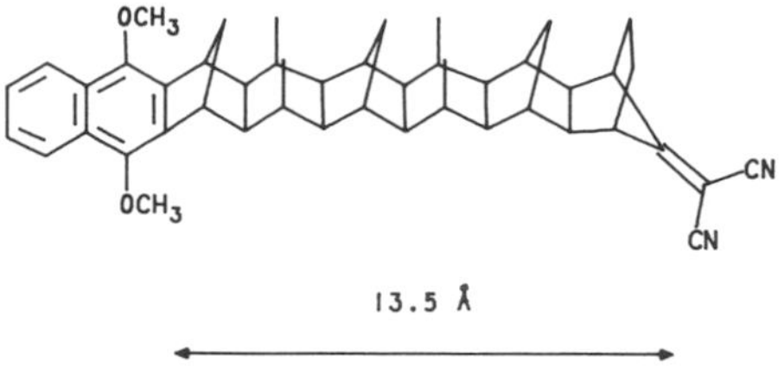

31

The approximate edge-to-edge separation distance, d_{ee}, between the naphthalene donor and ethylene acceptor is shown with each molecule. An electron transfer mechanism was assumed since previous studies had noted a correlation between fluorescence quenching of the donor and generation of charge-transfer bands.[26] Additional evidence was provided by time-resolved microwave conductivity, from which the dipole moments of the charge-transfer transients were estimated (the ion pairs were dubbed "giant" dipoles because of the large separation distances). The forward rates of electron transfer range between $\sim 10^9$ and $>10^{12}$ s^{-1} and decrease, as expected, with increasing separation distance. In contrast, the rates of electron return fall within $\sim 10^6$ to $>10^9$ s^{-1}. The low values of k_{ret} are indicative of the large energy gap between the ion-pair and ground states. The "slowness" of electron return favors formation of long-lived ion pairs (Eq. 1.8).

4.6. Models of Photosynthesis: Photoinduced Electron Transfer in Porphyrin–Electron Acceptor Pairs

As we have seen in the previous section, there are examples that demonstrate that photoinduced electron transfer in rigid systems can be influenced by the separation distance between the donor and acceptor, the geometry of the molecular bridge, the energy gap between the ion-pair state and ground state, and the nature of the solvent. Understanding these factors, qualitatively at least, is particularly useful in the quest for methods to enhance charge separation (i.e., to increase the efficiency of formation and the lifetimes of ionic intermediates). Control of electron return to ground state is desirable to maximize the lifetimes of ion pairs. In freely diffusing intermolecular systems, the close proximity of two ions, say $D^{\dot{+}}$ and $A^{\dot{-}}$, and the electrostatic attractive forces between them may impede the formation of a long-lived ion pair by favoring rapid electron return. This is particularly true of singlet ion-pair states where electron return is spin-allowed but less true of spin-forbidden electron return in triplet ion pairs. When donor and acceptor are connected to a spacer molecule, electron return can be controlled by varying the energy gap between the donor and acceptor, as was suggested in the previous section.

A desire to understand the intricacies of natural photosynthesis has motivated many chemists and molecular biologists to synthesize and study a wide variety of porphyrin–quinone pairs and other intramolecular systems. These molecules were intended to serve as models for photoinduced electron transfer in an attempt to simulate nature's technique in bringing about efficient charge separation (this is discussed in Chapters 5 and 6). Initially, attempts were made to investigate simple intramolecular porphyrin–quinone pairs. **32** is representative of an early system[27]:

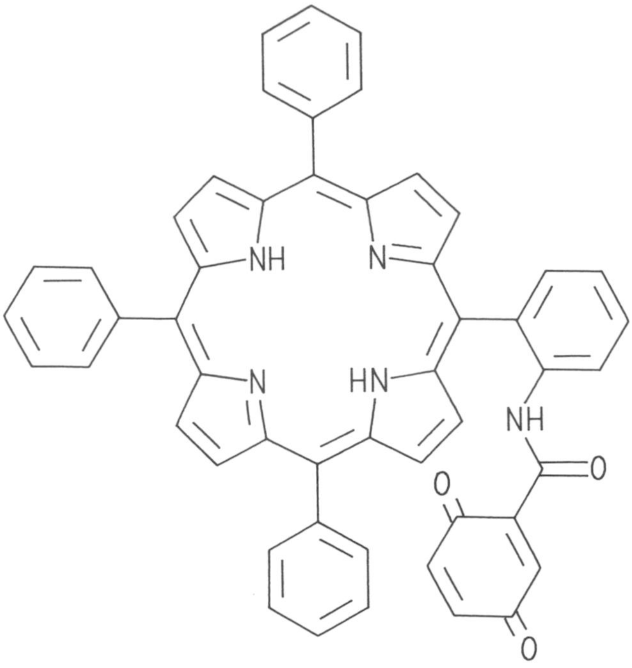

32

In **32** the porphyrin is linked to a quinone via an amide group. On exposure to light, the porphyrin is excited into a singlet state, which decays or undergoes intersystem crossing to the triplet state in competition with electron transfer. The quinone, a strong electron acceptor, readily removes an electron from the excited porphyrin. Electron transfer can occur from singlet or triplet porphyrin.

In porphyrin–quinone intramolecular systems, the rates of electron transfer and electron return can be controlled by the judicious choice of a spacer. The spacer may keep the reactants well apart so that once electron transfer takes place, the ion pair still keeps the original distance, thereby slowing down electron return.

One of the first direct observations of photoinduced electron transfer in porphyrin–quinone pairs separated by a flexible link was made in 1981.[28] This investigation was performed by measuring the transient absorbance of the porphyrin cation from 450 to 700 nm in acetonitrile. The presence of this intermediate was confirmed by ESR. In a study of porphyrin–quinone molecules linked by diamide groups, ESR measurements at low temperatures revealed signals for $n = 2$ to 4, which were assigned to the $P^{+\cdot}$–$Q^{-\cdot}$ ion pair.[29] Fluorescence studies on these systems suggested that the quinone can fold and complex with the singlet porphyrin prior to electron transfer.[30] The optimum chain length was found to be $n = 3$. With shorter chain lengths the folded structure cannot be attained; with longer chain lengths, dynamic conformations of the connecting link may prevent the porphyrin and quinone from maintaining a close distance.

The porphyrin can also be attached to several quenchers. For instance, photoinduced electron transfer was studied in a porphyrin molecule to which four viologen molecules were covalently attached (33)[31]:

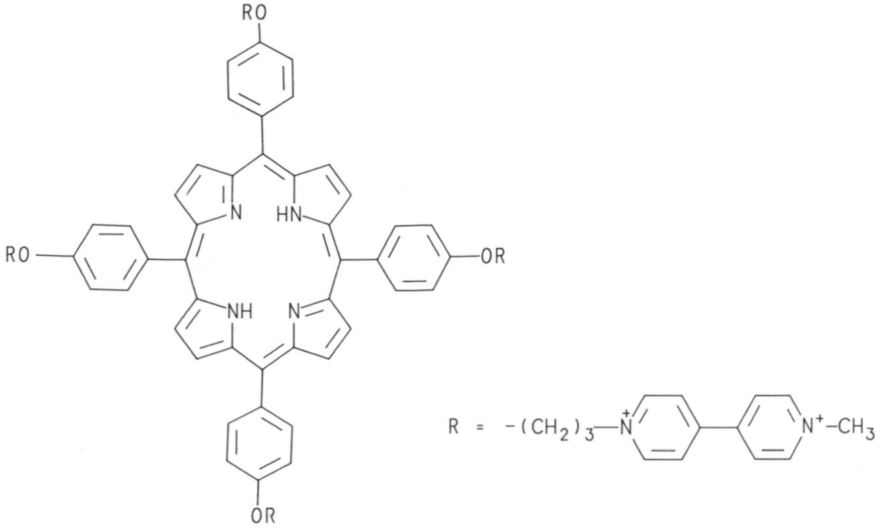

RO
N HN
RO OR
NH N
OR

$R = -(CH_2)_3-N^+ \langle\rangle - \langle\rangle N^+-CH_3$

33

With nanosecond and picosecond laser spectroscopy, the fluorescence of the excited singlet state of **33** was found to be quenched quite efficiently. Quenching of triplet **33,** formed by intersystem crossing, was also observed. The ion pairs resulting from quenching are fairly long-lived (~6 μs). An analysis of the time-resolved fluorescence decay curves confirmed that quenching, which is complete within several nanoseconds, occurs from two conformations. A fast quenching occurs while the porphyrin and viologens are arranged in an orthogonal geometry. In this structure, the porphyrin is about 11.1 Å from the viologen group. A slow quenching is postulated from a less orthogonal arrangement, at a porphyrin–viologen separation distance of 15.1 Å. These structures were deduced from computer modeling (Fig. 4.8). The two conformations probably interconvert on a time scale much slower than electron transfer, so that the quenching occurs from nonequilibrated geometries.

In an attempt to bypass some of the complexities due to the dynamic motions of the flexible link, numerous investigators have examined electron transfer between porphyrins and quinones separated by a rigid spacer. Some of this work has been directed to sorting out the effects of mutual orientation of donor and acceptor. This can be done quite elegantly by proper selection of the spacer group. Consider zinc tetraphenylporphyrin attached to an electron-accepting quinone via a triptycentyl structure (**34**)[32]:

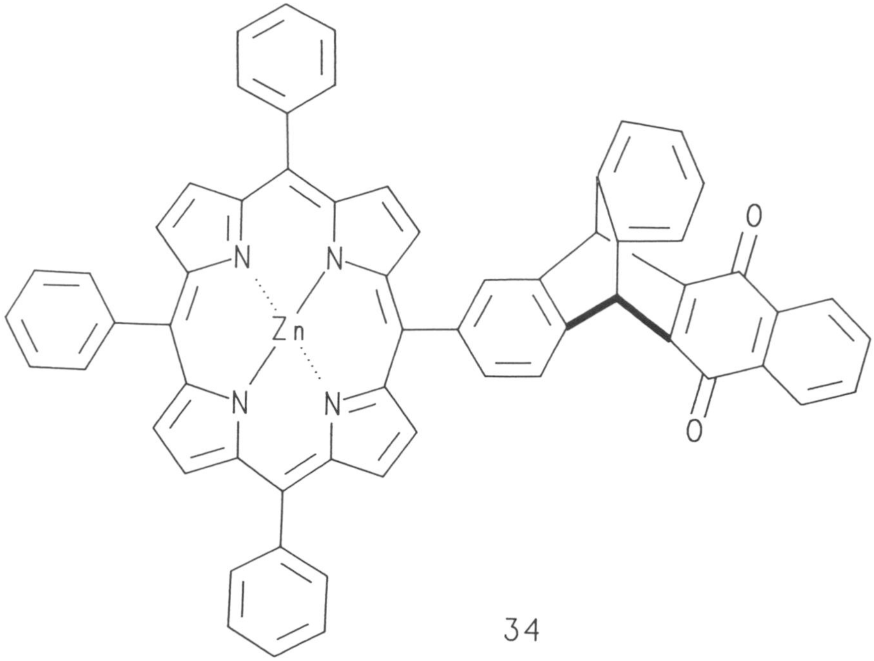

34

In **34,** the center-to-center distance has been estimated to range from 10 to 11 Å. Because of its rigid structure, the triptycentyl spacer prevents overlap between the π-orbitals of the porphyrin and quinone. As a consequence, there is little possibility of the face-to-face π-interaction of an exciplex. Instead, the zinc porphyrin's singlet state rapidly donates an electron to the quinone to generate a radical cation–anion pair. This work suggests that the spacer may be playing a key role in mediating superexchange and that through-bond interactions may be of importance. A similar conclusion was reached in a study involving **35** and **36**[33]:

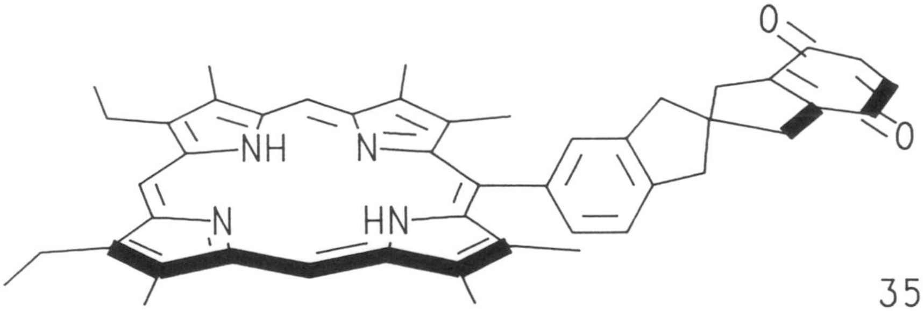

35

The center-to-center distance is about the same for both compounds (d_{cc} ~12.5 Å for **35** and d_{cc} ~12.9 Å for **36**). But the mutual orientations of each π-system are

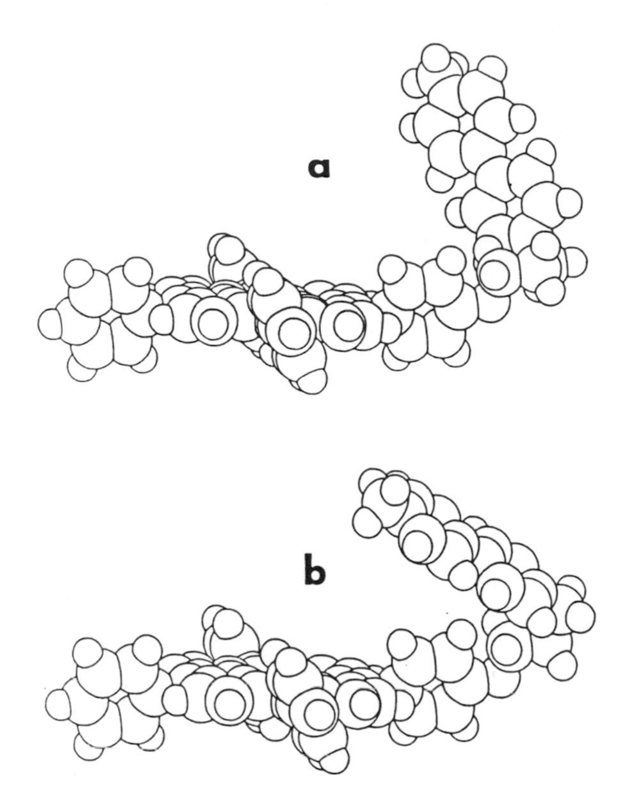

Figure 4.8 Structures of porphyrin-monoviologen complex deduced from computer modeling. (Reprinted with permission from Batteas, J. D., Harriman, A., Kanda, Y., Mataga, N., and Nowak, A. K. *J. Am. Chem. Soc.* **1990,** *112*, 126. Copyright 1990 American Chemical Society.)

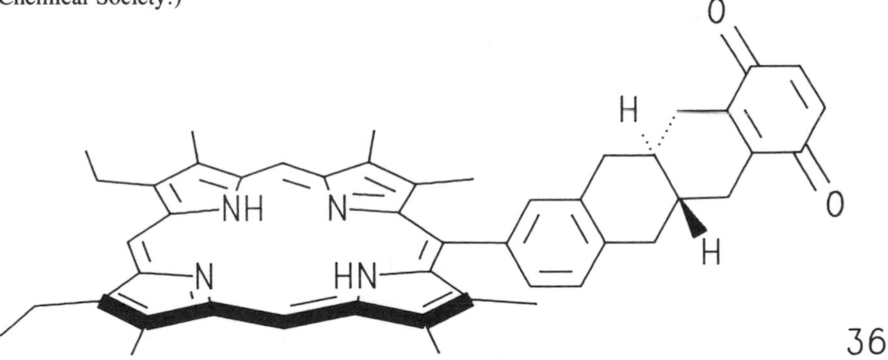

different. In **35** the dihedral angle between the planar rings is ~150°, whereas in **36** it is ~90°. Since the fluorescence of the porphyrin in **35** was found to be quenched faster than for **36,** and k_{el} ~3.3 × 10^8 s^{-1} (**35**) and k_{el} ~6.5 × 10^7 s^{-1} (**36**), the porphyrin–quinone pair appears to be better poised for electron transfer in **35.**

The decrease in k_{el} as the separation distance increases holds true in intramolecular systems when the number of bonds in the spacer group are increased. Rigid, intramolecular porphyrin–quinone systems are ideal models to test the effects of distance. For example, consider **37, 38,** and **39**[34]:

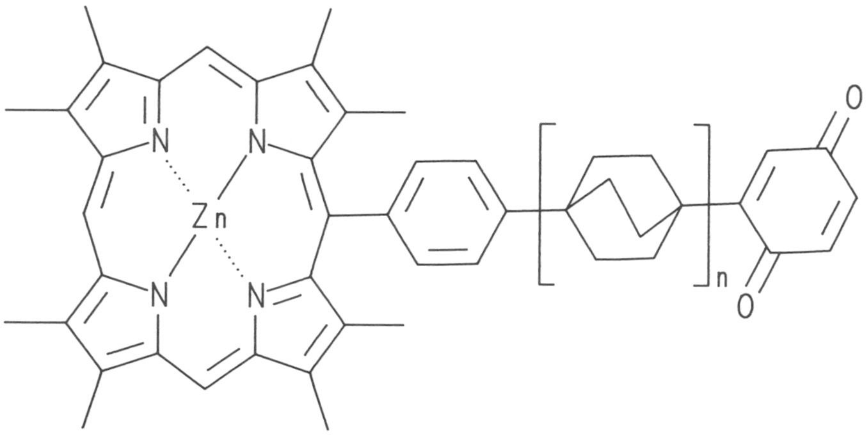

37 n = 0

38 n = 1

39 n = 2

k_{el} was measured, in order, 2.2×10^{11} (**37**), 1.8×10^{9} (**38**), and $\leq 2 \times 10^{7}$ (**39**) s^{-1}. This trend follows as predicted by theory, with increasing edge-to-edge separation distance from ~10 Å in **37** to ~14 Å in **39.**[35]

A complication in porphyrin-quinone systems has to do with the effects of energy separation between states. Recall that the rate of forward electron transfer is predicted to decrease with increasing separation distance and decreasing energy difference between the excited- and ion-pair states. However, the rates of electron return behaves differently. While it also decreases with separation distance, k_{ret} decreases with increasing energy separation. To sort out effects of distance and energy separation, it is desirable to design an experiment where one of these effects is invariant. For example, to study the effects of thermodynamics in a series of intramolecular donor–acceptor pairs, the distance separating the donor and acceptor can be held constant. This was precisely what was done in a series of porphyrin–quinone pairs which were synthesized so that the porphyrin–quinone separation distance is identical (**40, 41, 42, 43, 44,** and **45** on facing page).[36] The redox potentials of these compounds were measured and correlated with the rates of k_{el} and k_{ret}. It was found that k_{el} increases from $\sim 10^{10}$ to $\sim 10^{11}$ s^{-1} with increasing energy separation between the excited and ion-pair state but that k_{ret} decreases from $\sim 10^{11}$ to $\sim 10^{9}$ s^{-1} with energy separation between the ion-pair and ground states (the energies for several molecules of the series are shown in Table 4.2), just as expected.

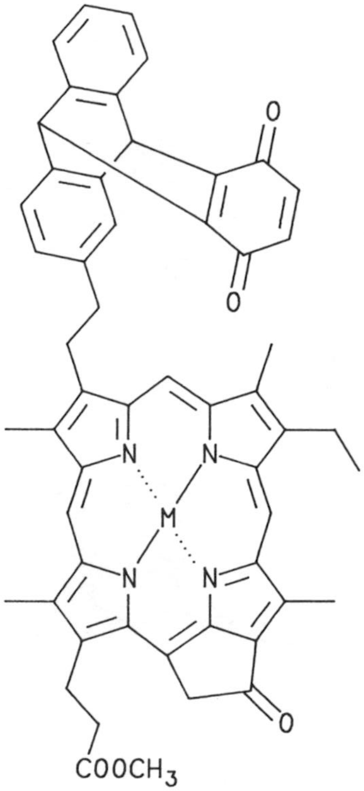

COOCH$_3$

40 M = H_2

41 M = Mg

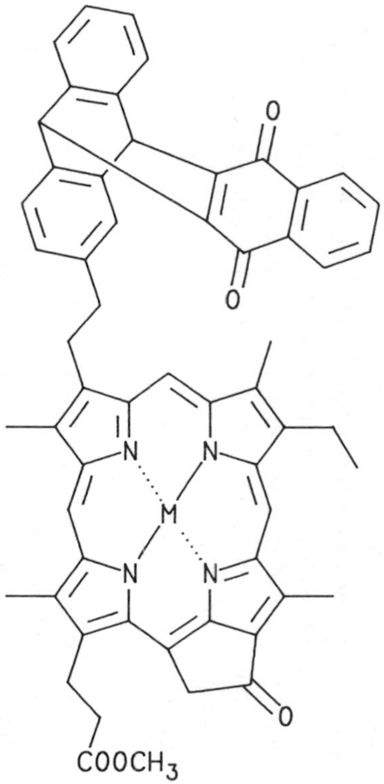

COOCH$_3$

42 M = H_2

43 M = Mg

COOCH$_3$

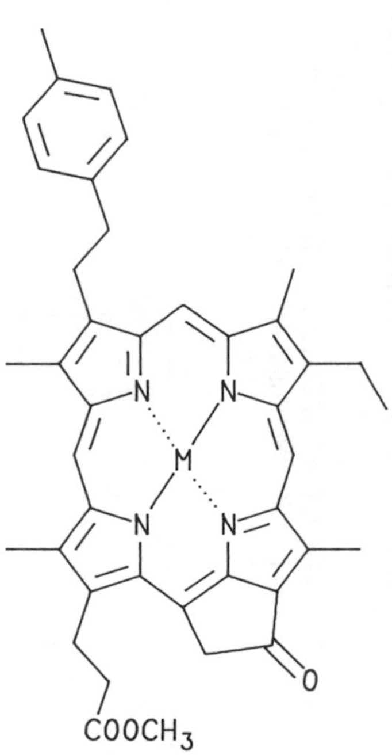

44 M = H_2

45 M = Mg

Table 4.2 KINETIC AND FREE-ENERGY DATA FOR QUENCHING OF CHLOROPHYLL–QUINONE PAIRS AT FIXED DISTANCES[a]

Molecule	Φ_f	k_{el} (s^{-1})	$-\Delta G_{el}$ (kcal mol^{-1})	k_{ret} (s^{-1})	$-\Delta G_{ret}$ (kcal mol^{-1})
40	0.012	8×10^{10}	8.3	8×10^{9}	34
41	0.009	4×10^{11}	16.4	1×10^{11}	26
42	0.026	2×10^{10}	5.1	2×10^{9}	37
43	0.0069	2×10^{11}	13.1	1×10^{11}	29.1

[a] Adapted from reference 36.

4.6.1. Photoinduced Electron Transfer in Molecular Triads, Tetrads, and Pentads

In porphyrin–quinone pairs separated by several rigid spacer groups, it is possible to generate long-lived ion pairs, but the forward rate is reduced as the number of spacer groups increases so that the efficiency of ion-pair formation decreases. Thus, these systems would appear to be unsuitable for obtaining long-lived ion pairs required for artificial photosynthesis. To work around this dilemma, it was proposed that effective and efficient charge separation might be achieved by attaching several acceptors to the porphyrin in such a way as to prevent rapid electron return. Suppose three different acceptors are attached to an electron donor:

$$\begin{aligned} D^*\text{–}A_1\text{–}A_2\text{–}A_3 &\rightarrow D^{\dot{+}}\text{–}A_1^{\dot{-}}\text{–}A_2\text{–}A_3 \rightarrow D^{\dot{+}}\text{–}A_1\text{–}A_2^{\dot{-}}\text{–}A_3 \\ &\rightarrow D^{\dot{+}}\text{–}A_1\text{–}A_2\text{–}A_3^{\dot{-}} \end{aligned} \tag{4.21}$$

Electron transfer is initially favored between D* and A_1 since these groups are closest. Electron hopping then occurs between closest neighbors and should be favored because of the close proximity of neighboring donors and acceptors. Now as the charge-separation distance between the ions increases, electron return becomes increasingly unfavorable resulting ultimately in long-lived ion pairs.

To exemplify this concept, we will consider several studies on molecular triads (three components), tetrads (four), and pentads (five).

Mataga and co-workers, for example, detected a long-lived ion-pair state in **46,** which consists of a porphyrin connected to two quinones (Q_A and Q_B)[37]:

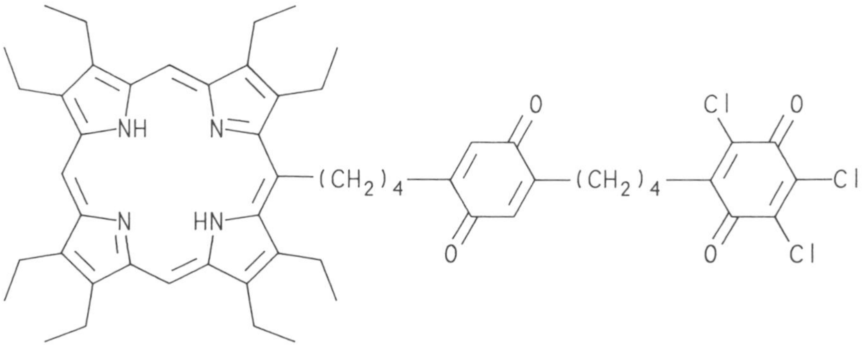

NH N N HN (CH2)4 O O (CH2)4 Cl O Cl O Cl

46

In this triad, photoexcitation of the porphyrin is followed by a sequential electron transfer:

$$P\text{–}Q_A\text{–}Q_B \xrightarrow{h\nu} {}^1P^*\text{–}Q_A\text{–}Q_B \rightarrow P^{\dot{+}}\text{–}Q_A^{\dot{-}}\text{–}Q_B \rightarrow P^{\dot{+}}\text{–}Q_A\text{–}Q_B^{\dot{-}} \qquad (4.22)$$

The lifetime of $P^{\dot{+}}\text{–}Q_A\text{–}Q_B^{\dot{-}}$ was found to be much longer than a simple dyad consisting of a porphyrin linked to a quinone by four methylene groups. This is probably due to the large separation distance between $P^{\dot{+}}$ and $Q_B^{\dot{-}}$ in 4.22.

Moore, Gust, Land, and collaborators synthesized and investigated several intramolecular porphyrin–quinone–carotenoid systems.[38–43] Initially, their models consisted of porphyrin–quinone–carotenoid triads (**47, 48, 49,** and **50**) shown below (C = carotenoid group):

47 n = 1
48 n = 2
49 n = 3
50 n = 4

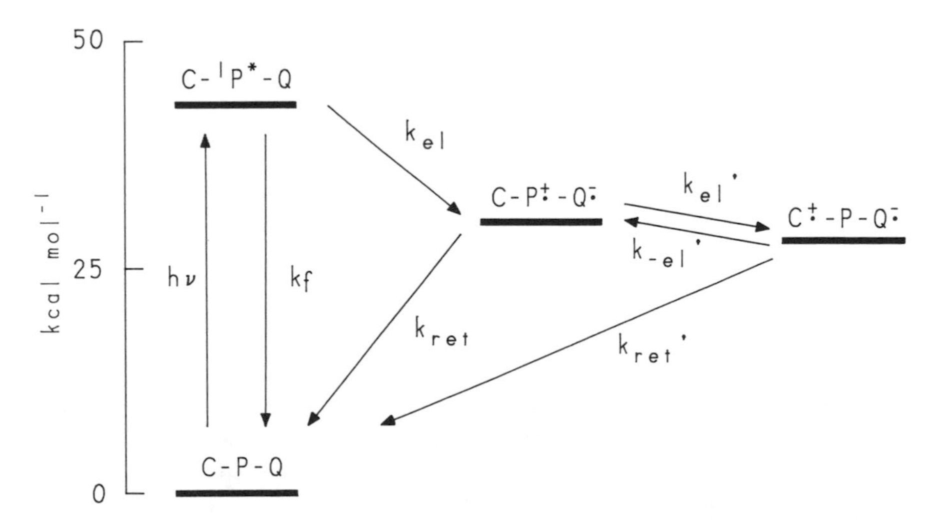

Figure 4.9 Photoinduced electron transfer in a porphyrin-quinone-carotenoid triad.

It was first established by high-resolution ^{1}H NMR that the carotenoid and quinone units do not fold across the porphyrin but instead adopt a linear conformation with respect to one another.[42] On the basis of electrochemical, nanosecond, and picosecond transient spectroscopy (absorption and fluorescence), a pathway of electron transfer was postulated (Fig. 4.9). Initially, photoexcitation of the porphyrin produces its singlet state. This step is followed by an electron transfer from singlet porphyrin to the electron-accepting quinone to generate $C–P^{+\cdot}–Q^{-\cdot}$. A second electron transfer was postulated leading to $C^{+\cdot}–P–Q^{-\cdot}$ (the long-lived carotenoid cation whose lifetime was estimated to be several hundred nanoseconds was identified by its absorption at 950 nm). A kinetic analysis undertaken as part of this study suggested the possibility of reversible electron transfer from $C^{+\cdot}–P–Q^{-\cdot}$ to $C–P^{+\cdot}–Q^{-\cdot}$.

This work was extended to include tetrad molecules containing one porphyrin, a carotenoid, and two quinones (**51**, **52**, and **53**):

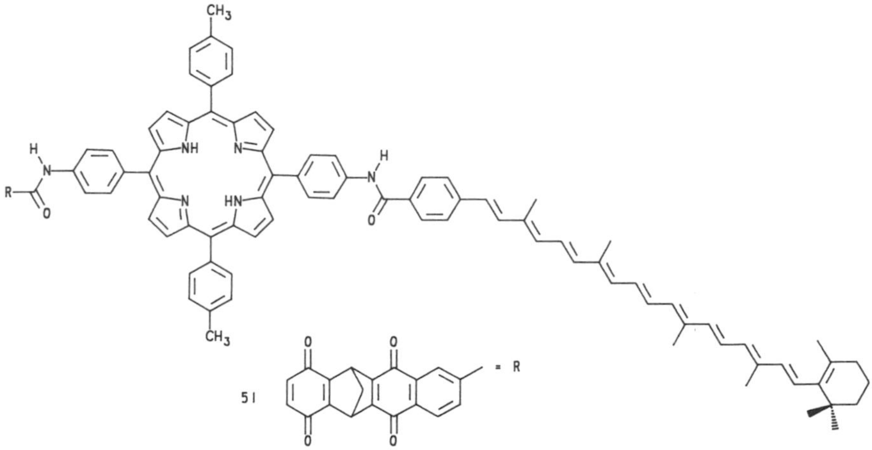

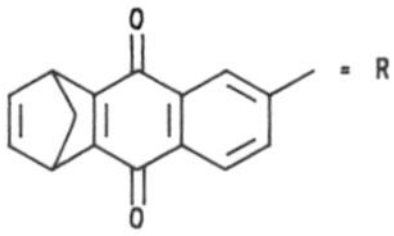

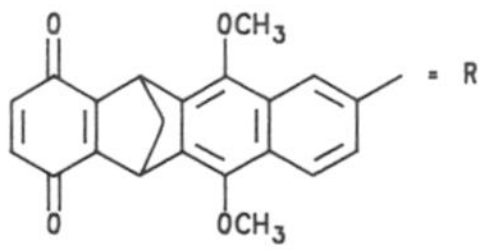

The measured lifetimes of the ion pairs were found to be considerably longer than those obtained for the triads. Once again the results indicated that electron transfers between nearest neighbors are favored. A suggested pathway is shown in Fig. 4.10. The long lifetime of $C^{+\cdot}–P–Q_A–Q_B^{-\cdot}$ was ascribed to a large separation distance.

A tetrad containing two porphyrins covalently attached to carotenoid and a naphthoquinone was synthesized and studied (**54** on page 224). $C^{+\cdot}–P_A–P_B–Q^{-\cdot}$ was

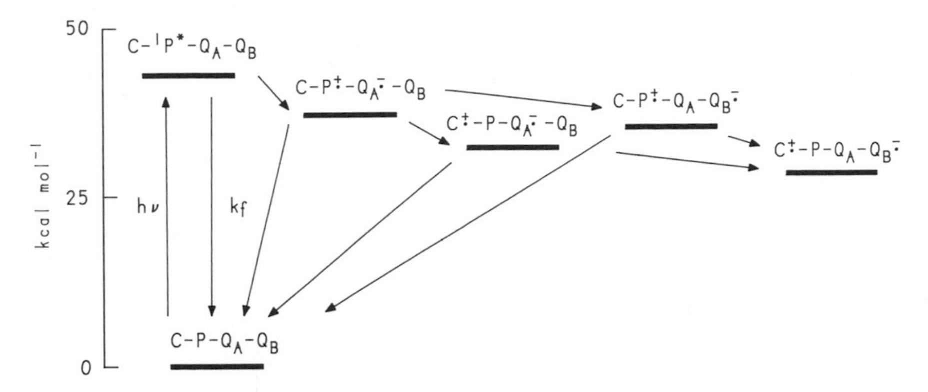

Figure 4.10 Photoinduced electron transfer in a tetrad molecule containing one porphyrin, a carotenoid, and two quinones.

identified by its absorption at $\lambda = 980$ nm. The lifetime of this intermediate was measured as $\tau_{IP} = 2.9$ μs. The initial pathway was postulated to be singlet energy transfer between the porphyrin groups (Fig. 4.11). Subsequent forward electron transfers lead to the final ion-pair state with an efficiency of $\Phi_{IP} = 0.25$. Again, the long-lifetime of this intermediate is due to the large separation distance between $C^{\dot{+}}$ and $Q^{\dot{-}}$. Interestingly, porphyrin triplets, which are also generated in this series, are quickly quenched by the carotenoid group by triplet energy transfer. In this respect, this series provides a fine example of the multi-functional aspects in natural photosynthesis (Chapter 6).

Efficient photoinduced electron transfer has also been documented in a molecular pentad (**55** on page 224).[43] Transient absorption studies revealed that on photoexcitation $C^{\dot{+}}–P_A–P_B–Q_A–Q_B^{\dot{-}}$ is generated (Fig. 4.12). This intermediate is produced with a lifetime of $\tau_{IP} = 0.55$ μs with an overall quantum efficiency of 0.83.

In summary, the intramolecular triads, tetrads, and pentads presented in this

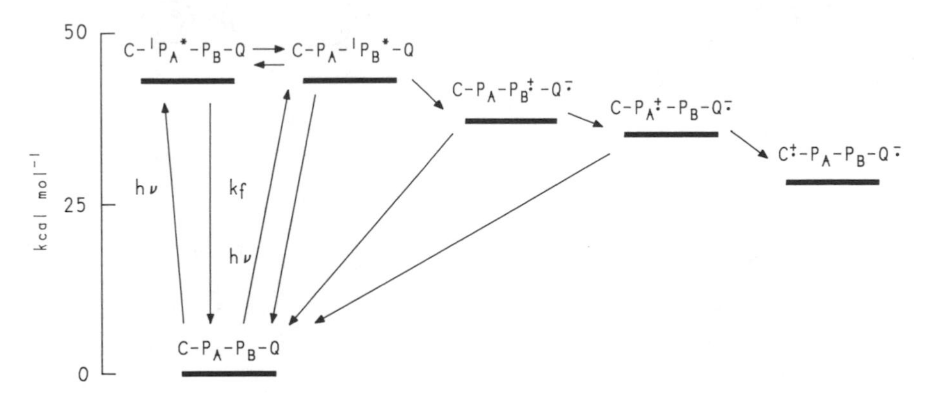

Figure 4.11 Photoinduced electron transfer in a tetrad molecule containing two porphyrins, a carotenoid, and one quinone.

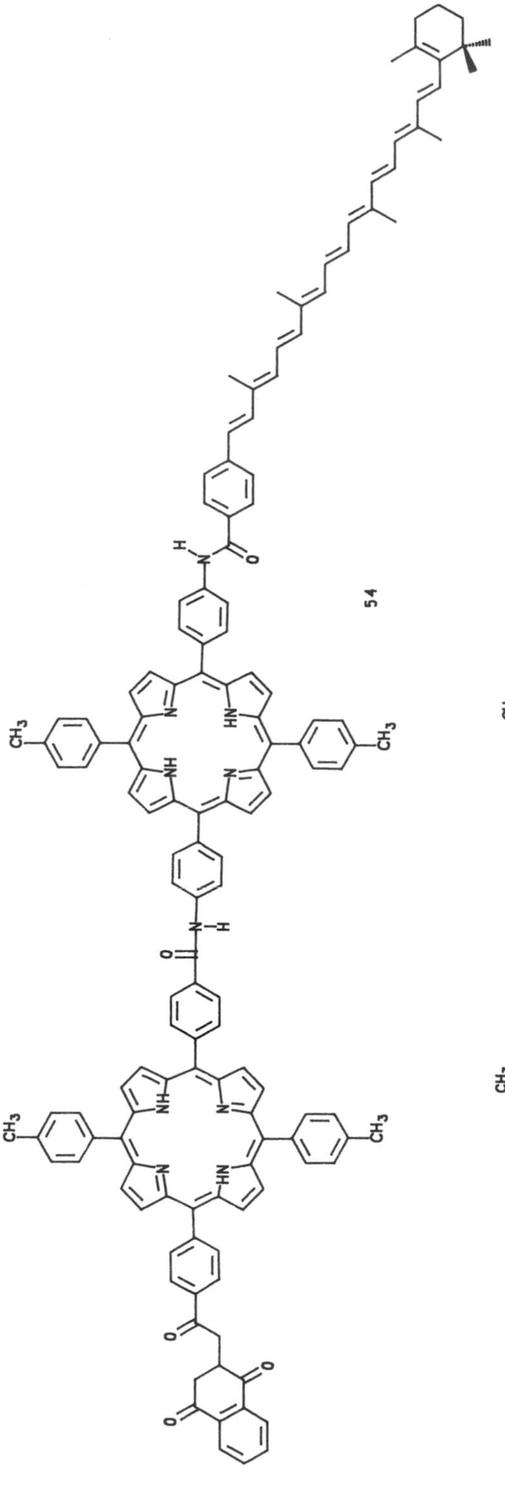

54

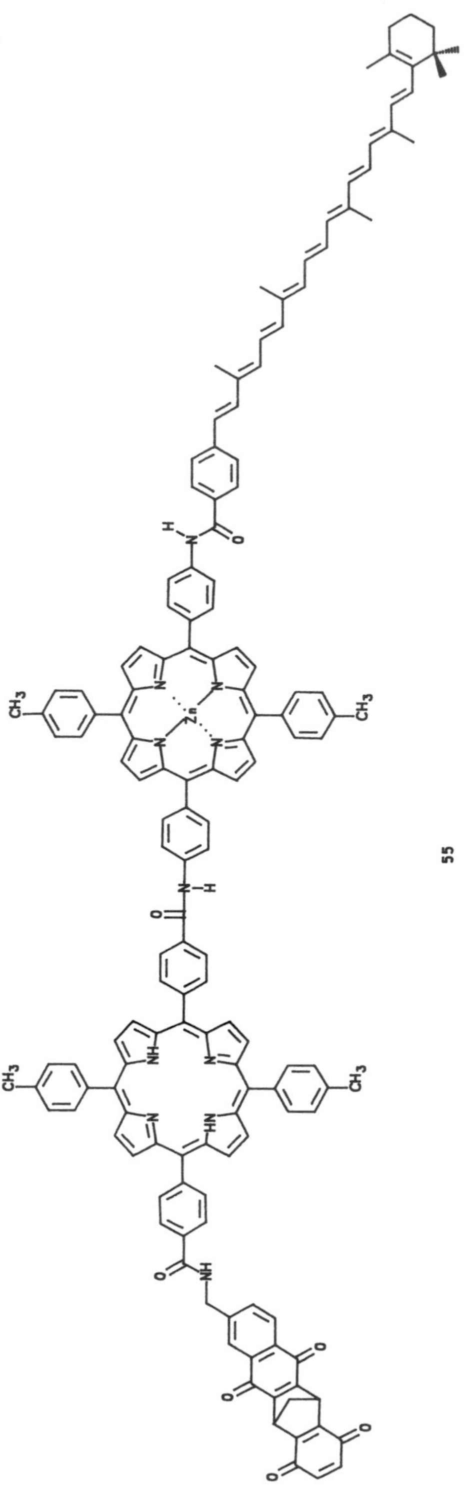

55

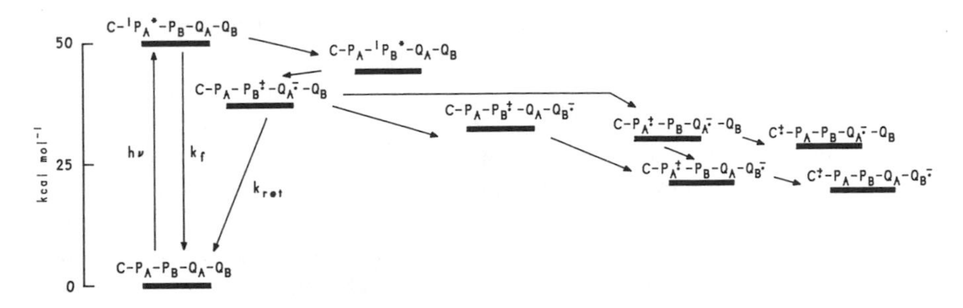

Figure 4.12 Photoinduced electron transfer in a pentad molecule containing two porphyrins, a carotenoid, and two quinones.

section were studied in part because they serve as elegant examples that sequential electron transfers can be exploited to give long-lived ion-pair states. But also they bear close resemblance to molecules that participate in natural photosynthesis. Although natural photosynthesis does not involve true intramolecular species (in fact, the molecular assemblies in photosynthesis are intermolecular species embedded in a medium of lipid bilayer membrane and surrounding proteins), the study of triads, tetrads, and pentads has yielded useful information. Most importantly, the results obtained in these systems emphasize the significance of energetics and distance.

4.7. Supramolecular Photoinduced Electron Transfer

The examples previously discussed in this chapter are representative of complex systems made up of smaller molecular subunits bound covalently. At least one of these components has to be a light-sensitive molecule capable of capturing a desired portion of the spectrum. In addition, there may be bound to this photosensitizer one or more electron donors or acceptors. These examples lead naturally to consideration of a new and exciting branch of chemistry called *supramolecular chemistry*. The Latin adverb *supra,* which means "above" or "over," gives a hint of the definition of *supramolecular*. A supramolecular system can be regarded as a complex molecular ensemble consisting of smaller molecular components somewhat like the examples considered in this chapter. When isolated, each of the molecular subunits is characterized by its own distinct properties. The properties of the supramolecular system, however, may be "above" and "over" the properties of the individual components. Supramolecular chemistry is a rapidly evolving field requiring an interdisciplinary knowledge of chemistry, physics, biology, and material science. The applications are "futuristic" in the sense that this field is only in its infancy. Nonetheless, the ramifications of supramolecular chemistry are indeed staggering if one wishes to apply imaginative solutions to serious questions in such diverse areas as energy conservation, information processing, medicine, and the environment.

Supramolecular systems are *molecular devices* capable of performing some function. For example, a photochemical supramolecule contains a light-sensitive subunit that transmits the excitation energy to some other component in the system. This may involve some photochemical process such as photoisomerization, energy transfer, or electron transfer, to name several possibilities. Keeping in mind the wide scope of supramolecular photochemistry, we shall consider only photoinduced electron-transfer processes taking place in supramolecular systems.

A photoinduced electron-transfer supramolecular ensemble thus contains one subunit, which we designate as the light-sensitive component. Other components in the system may be capable of relaying electrons between various electron donors and acceptors in the structure and the photosensitizer. These components may be covalently linked by additional components—molecular "connectors." These molecular devices can be designed and tailored for specific applications as, for example, artificial photosynthesis (Chapter 5) and signal generation by the transduction of optical into electronic energy (Section 4.8).

Supramolecular photochemical devices based on the concept of photoinduced electron transfer can make use of any number of configurations. We may envision supramolecules with two, three, four, five, etc., components designed to achieve long-lived charge-separated species (note that a one-component system is not suitable for this purpose because of very rapid electron return). The intramolecular systems introduced throughout this chapter can be regarded as supramolecular systems. In fact, the bridged polynuclear complex **5** is a typical example. The $Ru(bpy)_2$ is the group that captures light energy; the other Ru centers participate in subsequent photoinduced electron transfer processes. The porphyrin–carotenoid–quinone intramolecular systems are also good examples. In a molecular triad, tetrad, or pentad, for example, effective charge separation is achieved by carefully maintaining a suitable connectivity between the individual components.

The binding of the individual components does not necessarily have to be covalent. The molecular subunits can be bound together by van der Waals forces and electrostatic interactions. The photoinduced electron-transfer processes taking place in the macrocycle **56** provide an interesting example.[44] This macrocycle contains several molecular subunits including the zinc porphyrin as the light-sensitive component and the [18]-N_2O_4 groups. The fluorescence quantum yield and excited singlet lifetime are about 0.039 and 2 ns, respectively. However, the addition of silver triflate reduces these values sharply. It was proposed that silver strongly complexes with the [18]-N_2O_4 group forming a supramolecular system. After exposure to light, quenching of both singlet and triplet porphyrin proceeds by electron donation from the excited porphyrin to Ag^+ leading to Ag^0, the latter being bound much more weakly. The justification for adding the silver ion is to induce the electron transfer and generate a long-lived porphyrin cation.

We shall have occasion to examine other supramolecular photochemical systems. In Chapter 5, there will be a discussion of supramolecular devices capable of capturing solar energy for artificial photosynthesis and energy conversion. However, before we consider the unique molecular architecture and functionality of these molecular photochemical devices, we will conclude this chapter with a possible application in the area of molecular electronics.

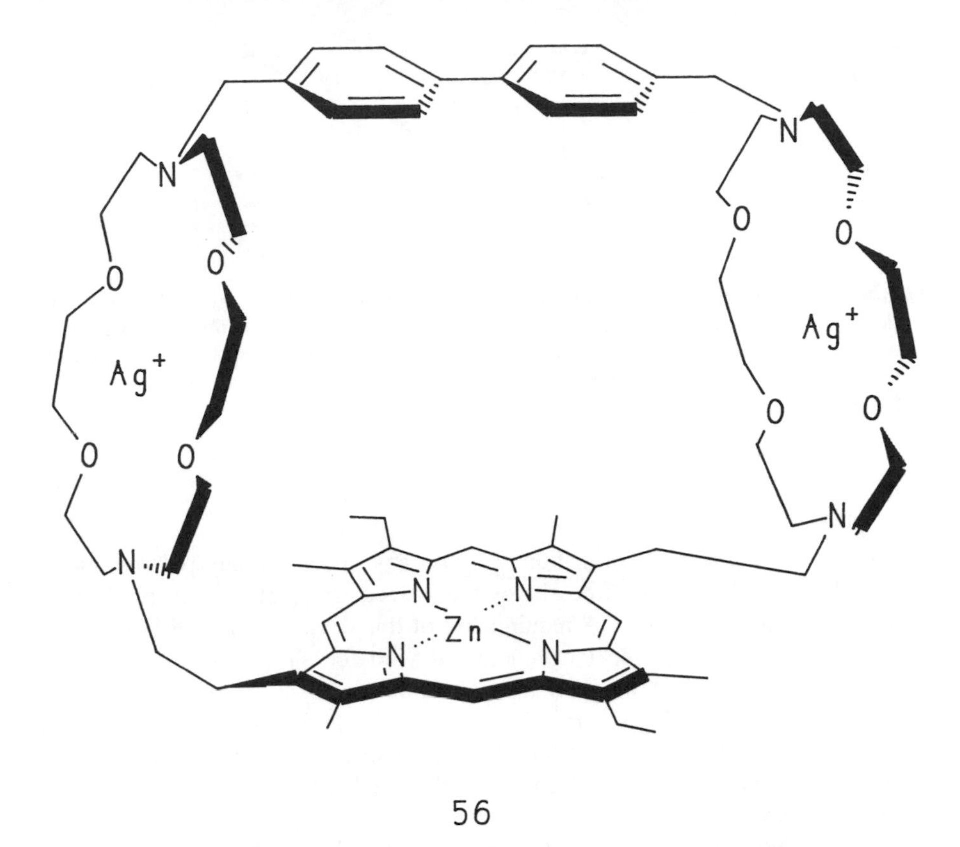

56

4.8. Photoinduced Electron Transfer in Molecular Electronic Devices

A novel suggestion has been advanced to use photoinduced electron transfer to construct a molecular shift register.[45] This possible application is based on the dynamics of intramolecular photoinduced electron transfer. A scheme is presented in Fig. 4.13. A polymer consisting of 600 repeating units—in this case each unit consists of a porphyrin attached to two quinones—is synthesized and covalently linked along with about 5000 other polymers to the ends of two electrodes. When a photon of light from a pulsed laser strikes a porphyrin, electron transfer is set in motion as shown in Fig. 4.13. Provided that the forward rate of electron transfer is faster than electron return, the electron "shifts" down the chain and finally to the electrode. This novel design suggests that information stored as bits on each chain (the "−" charge might be "1" and the "+" charge might be "0") can be retrieved as electrons and detected at the receiving electrode. Each molecular unit simulates a shift register, the contents of each shifting to the next adjacent register. With 5000 parallel chains containing the same information, it is therefore conceivable

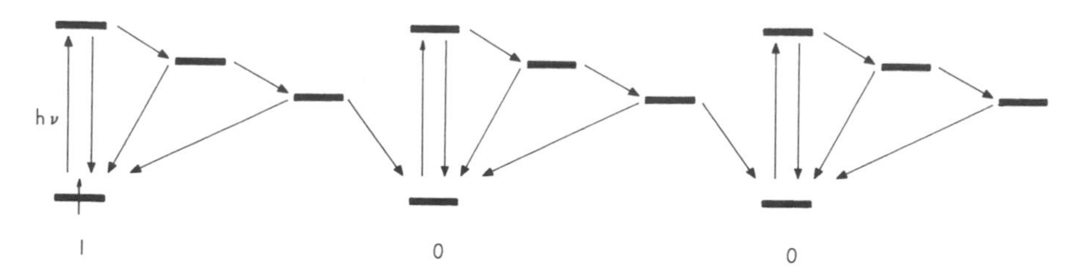

Figure 4.13 A possible scheme for photoinduced electron transfer in molecular electronic devices. A donor molecule containing one electron is excited with a photon. After photoexcitation, the electron can flow from left to right, but the efficiency of this flow is subject to the restrictions of return pathways. As the electron moves sequentially to each unit, the shift register changes from 100 to 010.

to protect the "electronic information," should any portion of any polymer experience chemical degradation (the chains are spaced far apart to prevent electron transfer from chain to chain). A requirement of this device is that the clock cycle of the pulsing laser must be long compared with the shift of one repeat unit. However, this system allows itself to be modified in many ways. For example, various functional groups can be substituted on the porphyrin and quinones, or both, to influence the rate of electron "shifting." Other chromophores can be used. Forking branched polymers ("molecular wires") that "amplify" the electron transfer into several directions can also be imagined.

The molecular "wires" presented in the preceding discussion are representative of futuristic molecular photochemical devices. These molecules represent an attempt to generate an electron flow photochemically by taking advantage of the multifunctional properties of large molecular assemblies.

Suggestions for Further Reading

Through-Space and Through-Bond Coupling

Beratan, D. N., Onuchic, J. N., and Hopfield, J. J. *J. Chem. Phys.* **1987,** *86,* 4488. An investigation of donor–acceptors connected by amino acid chains.

Chu, S.-Y., and Lee, T.-S., *Nouv. J. Chim.* **1982,** *6,* 155. An examination of electron "propagation" processes.

Closs, G. L., and Miller, J. R. *Science* **1988,** *240,* 440. Intramolecular electron transfer at large separation distances.

Halpern, J., and Orgel, L. E. *Faraday Discuss. Chem. Soc.* **1960,** *29,* 32.

Hoffman, R. *Acc. Chem. Res.* **1971,** *4,* 1. Professor Hoffman provides a lucid account of orbitals interacting through space and bonds.

Photoinduced Electron Transfer in Bridged Metal–Ligand Systems

Balzani, V., Barigelletti, F., and De Cola, L. *Top. Curr. Chem.* **1990,** *156,* 33. A survey of electron transfer in metal complexes.

Meyer, T. J. In *Mechanistic Aspects of Inorganic Reactions,* ACS Symposium Series 198, Rorabacher, D. B., and Endicott, J. F., eds. American Chemical Society, Washington, DC, 1982, Chapter 6.

Robin, M. B., and Day, P. *Adv. Inorg. Chem. Radiochem.* **1967,** *10,* 247.

Scandola, F., Indelli, M. T., Chiorboli, C., and Bignozi, C. A. *Top. Curr. Chem.* **1990,** *156,* 75.

Taube, H. *Angew. Chem.* **1984,** *23,* 329. The 1984 Nobel Lecture in Chemistry on electron transfer between metal complexes.

Twisted Intramolecular Charge–Transfer Complexes

Grabowski, Z. R., Rotkiewicz, K., Siemiarczuk, A., Cowley, D. J., and Baumann, W. *Nouv. J. Chim.* **1979,** *3,* 443.

Siemiarczuk A., Koput, J., and Pohorille, A. *Z. Naturforsch.* **1982,** *37a,* 598.

Visser, R. J., Varma, C. A. G. O., Konijnenberg, J., and Bergwerf, P. *J. Chem. Soc., Faraday Trans. 2* **1983,** *79,* 347. This paper examines the fluorescence of two aminobenzonitrile derivatives in glassy matrices and proposes an emitting solute–solvent exciplex in place of a twisted intramolecular charge-transfer state.

Photoinduced Electron Transfer in Flexible Molecular Systems: Exciplex vs. Ion Pair Formation

DeSchryver, F. C., Boens, N., and Put, J. *Adv. Photochem.* **1977,** *10,* 359. A review on intramolecular excitation.

Mataga, N., Migita, M., and Nishimura, T. *J. Mol. Struct.* **1978,** *47,* 199.

Vanderauwera, P., DeSchryver, F. C., Weller, A., Winnik, M. A., Zachariasse, K. A., *J. Phys. Chem.* **1984,** *88,* 2964.

Winnik, M. A. *Acc. Chem. Res.* **1977,** *10,* 173. Survey of flexible systems.

Winnik, M. A. *Chem. Rev.* **1981,** *81,* 491. An important review of chain dynamics.

Spin Dynamics in Flexible Systems

Weller, A., Staerk, H., and Treichel, R. *Faraday Discuss. Chem. Soc.* **1984,** *78,* 271.

Photoinduced Electron Transfer in Rigid Systems

Beecroft, R. A., Davidson, R. S., Goodwin, D., and Pratt, J. E. *Pure Appl. Chem.* **1982,** *54,* 1605. An interesting paper that includes a discussion of photoinduced electron transfer in rigid systems.

Connolly, J. S., and Hurley, J. K. In *Supramolecular Photochemistry,* Balzani, V., ed. Reidel, Dordrecht, 1987, p. 299.

Moore, T. A., Gust, D., Hatlevig, S., Moore, A. L., Makings, L. R., Pessiki, P. J., DeSchryver, F. C., Van der Auweraer, M., Lexa, D., Bensasson, R. V., and Rougée, M. *Israel J. Chem.* **1988,** *28,* 87. Discussion of manipulating the energy differences between excited- and ion-pair states and between ion and ground states to generate long-lived ion-pair states.

Supramolecular Photoinduced Electron Transfer

Balzani V., Moggi, L., and Scandola, F. In *Supramolecular Photochemistry,* Balzani, V., ed. Reidel, Dordrecht, 1988, p. 1.

Belser, P. *Chimia* **1990,** *44,* 226. The chemistry of ruthenium(II)–diimine complexes and how they might play a role in a supramolecular chemistry.

Lehn, J.-M. *Angew. Chem.* **1990,** *29,* 1304. An informative perspective on supramolecular chemistry.

Photoinduced Electron Transfer in Molecular Electronic Devices

Beratan, D. N. *Mol. Crystallogr. Liquid Crystallogr.* **1990,** *190,* 85. A discussion of controlled electron transfer for molecular electronics emphasizing a quantitative model that focuses attention on the interaction of the donor and acceptor connected by a number of interacting covalent bonds.

Miller, J. S. *Adv. Mater.* **1990,** *2,* 378, 495, and 601. Three articles on the future possibilities of molecular electronics.

References

1. Beratan, D. N., and Hopfield, J. J. *J. Am. Chem. Soc.* **1984,** *106,* 1584.
2. Onuchic, J. N., and Beratan, D. N. *J. Am. Chem. Soc.* **1987,** *106,* 6771.
3. Tom, G. M., Creutz, C., and Taube, H. *J. Am. Chem. Soc.* **1974,** *96,* 7827.
4. Curtis, J. C. and Meyer, T. J. *J. Am. Chem. Soc.* **1978,** *100,* 6284.
5. Creutz, C., Kroger, P., Matsubara, T., Netzel, T., and Sutin, N. *J. Am. Chem. Soc.* **1979,** 101, 5442.
6. Curtis, J. C., Bernstein, J. S., and Meyer, T. J. *Inorg. Chem.* **1985,** *24,* 385.
7. Scandola, F. In *Photochemical Energy Conversion,* Norris, J. R., Jr., and Meisel, D., eds. Elsevier, New York, 1989, pp. 60–73.
8. Norton, K. A., and Hurst, J. K. *J. Am. Chem. Soc.* **1982,** *104,* 5960.
9. Grabowski, Z., and Dobkowski, J. *Pure Appl. Chem.* **1983,** 245.
10. Hayashi, T., Mataga, N., Umemoto, T., Sakata, Y., and Misumi, S. *J. Phys. Chem.* **1977,** *81,* 424.
11. Becker, H. D., and Sandros, K. *Chem. Phys. Lett.* **1978,** *53,* 228.
12. Luo, X.-J., Beddard, G. S., Porter, G., Davidson, R. S., and Whelan, T. D. *J. Chem. Soc., Faraday Trans. 1* **1982,** *78,* 3467.

13. Crawford, M. K., Wang, Y., and Eisenthal, K. B. *Chem. Phys. Lett.* **1981,** *79, 529.*

14. Wang, Y., Crawford, M. C., and Eisenthal, K. B. *J. Am. Chem. Soc.* **1982,** *104,* 5874.

15. Mataga, N. *Pure Appl. Chem.* **1984,** *56,* 1255.

16. Yang, N. C., Neoh, S. B., Naito, T., Ng, L.-K., Chernoff, D. A., and McDonald, D. B. *J. Am. Chem. Soc.* **1980,** *102,* 2806.

17. Yang, N.-C., Minsek, D. W., Johnson, D. G., and Wasielewski, M. R. In *Photochemical Energy Conversion,* Norris, J. R., Jr., and Meisel, D., eds. Elsevier, New York, 1989, p. 111.

18. Wagner, P. J., and Siebert, E. J. *J. Am. Chem. Soc.* **1981,** *103,* 7335.

19. Winnik, M. A., and Hsiao, C. K. *Chem. Phys. Lett.* **1975,** *33,* 518.

20. Mar, A., Fraser, S., and Winnick, M. A. *J. Am. Chem. Soc.* **1981,** *103,* 4941.

21. Larson, J. R., Petrich, J. W., and Yang, N.-C. *J. Am. Chem. Soc.* **1982,** *104,* 5000.

22. Weller, A. In *Supramolecular Photochemistry,* Balzani, V., ed. Reidel, Dordrecht, 1987, p. 343.

23. Cabana, L. A., and Schanze, K. S. In *Electron Transfer in Biology and the Solid State,* Johnson, M. K., King, R. B., Kurtz, D. M. Jr., Kutal, C., Norton, M. L., and Scott, R. A., eds. Advances in Chemistry 226; American Chemical Society, Washington, DC, 1990, Chapter 5.

24. Pasman, P., Koper, N. W., and Verhoeven, J. W. *Recl. Trav. Chim. Pays-Bas* **1982,** *101,* 363.

25. Paddon-Row, M. N., Oliver, A. M., Warman, J. M., Smit, K. J., De Haas, M. P., Oevering, H., and Verhoeven, J. W. *J. Phys. Chem.* **1988,** *92,* 6958.

26. Hush, N. S., Paddon-Row, M. N., Cotsaris, E., Oevering, H., Verhoeven, J. W., and Heppener, M. *Chem. Phys. Lett.* **1985,** *117,* 8.

27. Tabushi, I., Koga, N., and Yanagita, M. *Tetrahedron Lett.* **1979,** 257.

28. Migita, M., Okada, T., Mataga, N., Nishitani, S., Kurata, N., Sakata, Y., and Misumi, S. *Chem. Phys. Lett.* **1981,** *84,* 263.

29. McIntosh, A. R., Siemiarczuk, A., Bolton, J. R., Stillman, M. J., Ho, T.-H., and Weedon, A. C. *J. Am. Chem. Soc.* **1983,** *105,* 7215.

30. Siemiarczuk, A., McIntosh, A. R., Ho, T.-H., Stillman, M. J., Roach, K. J., Weedon, A. C., Bolton, J. R., and Connolly, J. S. *J. Am. Chem. Soc.* **1981,** *105,* 7224.

31. Batteas, J. D., Harriman, A., Kanda, Y., Mataga, N., and Nowak, A. K. *J. Am. Chem. Soc.* **1990,** *112,* 126.

32. Wasielewski, M. R., and Niemczyk, M. P. *J. Am. Chem. Soc.* **1984,** *106,* 5043.

33. Sakata, Y., Nakashima, S., Goto, Y., Tatemitsu, H., and Misumi, S. *J. Am. Chem. Soc.* **1989,** *111,* 8979.

34. Closs, G. L., Piotrowiak, P., MacInnis, J. M., and Fleming, G. R. *J. Am. Chem. Soc.* **1988,** *110,* 2652.

35. Joran, A. D., Leland, B. A., Felker, P. M., Zewail, A. H., Hopfield, J. J., and Dervan, P. B. *Nature* (*London*) **1987,** *327,* 508.

36. Wasielewski, M. R., Johnson, D. G., and Svec, W. A. In *Supramolecular Photochemistry,* Balzani, V., ed. Reidel, Dordrecht, 1987, p. 255.

37. Mataga, N., Karen, A., Okada, T., Nishitani, S., Kurata, N., Sakata, Y., and Misumi, S. *J. Phys. Chem.* **1984,** *88,* 5138.

38. Moore, T. A., Gust, D., Mathis, P., Mialocq, J. C., Chachaty, C., Bensasson, R. V., Land, E. J., Doizi, D., Liddell, P. A., Nemeth, G. A., and Moore, A. L. *Nature* (*London*) **1984,** *307,* 630.

39. Gust, D., Moore, T. A., Bensasson, R. V., Mathis, P., Land, E. J., Chachaty, C., Moore, A. A., Liddell, P. A., and Nemeth, G. A. *J. Am. Chem. Soc.* **1986,** *107,* 3631.

40. Gust, D., and Moore, T. A. In *Supramolecular Photochemistry,* Balzani, V., ed. Reidel, Dordrecht, 1987, p. 267.

41. Gust, D., and Moore, T. A. *Science* **1989,** *244,* 35.

42. Gust, D., Moore, T. A., Liddell, P. A., Nemeth, G. A., Making, L. R., Moore, A. L., Barrett, D., Pessiki, P. J., Bensasson, R. V., Rougée, Chachaty, C., De Schryver, F. C., Van der Auweraer, M., Holzwarth, A. R., and Connolly, J. S. *J. Am. Chem. Soc.* **1987,** *107,* 846.

43. Gust, D., Moore, T. A., Moore, A. L., Lee, S.-J., Bittersmann, E., Kuttrull, D. K., Rehms, A. A., DeGraziano, J. M., Ma, X. C., Gao, F., Belford, R. E., and Trier, T. T. *Science* **1990,** *248,* 199.

44. Lehn, J.-M. In *Supramolecular Photochemistry,* Balzani, V., ed. Reidel, Dordrecht, 1987, p. 29.

45. Hopfield, J. J., Onuchic, J. N., and Beratan, D. N. *Science* **1988,** *241,* 817.

Problems

1. Optical electron transfer was observed in the complex, $[(bpy)(CO)_3Re(py\text{-}PTZ)]^+$, where py-PTZ has the structure shown below:

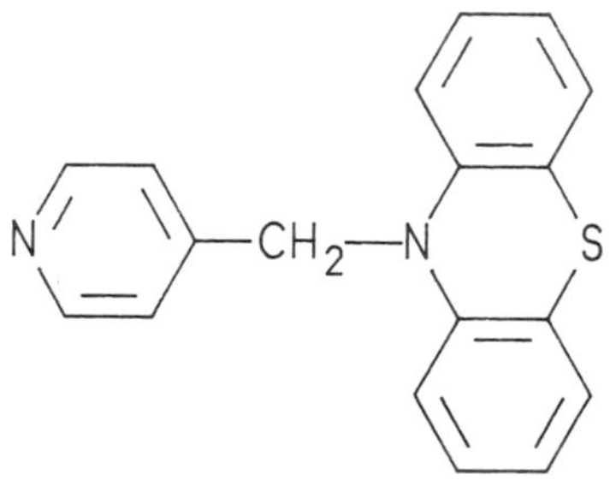

py-PTZ

At 77 K, an emissive MLCT band of the complex was observed in an ethanol: methanol glass. As the glass was warmed, the emission dramatically decreased. Transient spectroscopy revealed the appearance of a new transient with spectral features suggestive of PTZ^+. Suggest a mechanism that is in accord with these observations. (Chen, P., Westmoreland, T. D., Danielson, E., Schanze, K. S., Anthon, D., Neveux, P., Jr., and Meyer, T. J. *Inorg. Chem.* **1987,** *26,* 1116.)

2. The metal complex, $[(bpz)Re(I)(CO)_3\text{-ch-DTF}]^+$, where ch is the *trans*-1,4-cyclohexane spacer and DTF is the 1,3-benzodithiafulvene electron donor, was studied by emission and transient absorption spectroscopy:

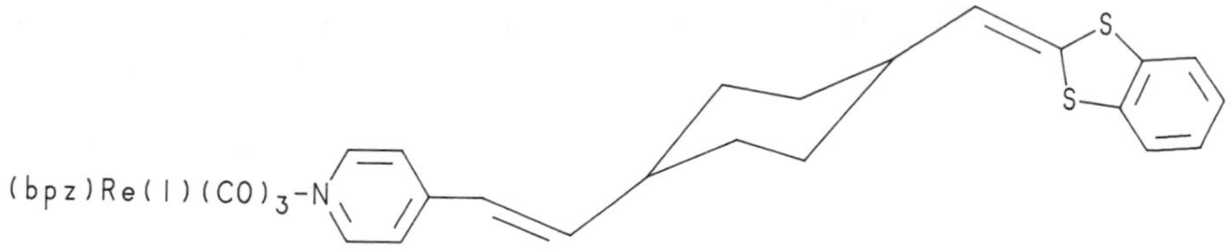

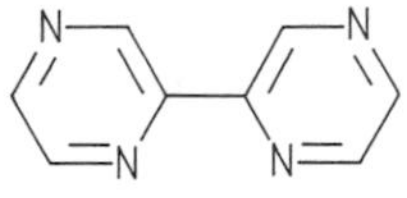

$(bpz)Re(I)(CO)_3$-N

bpz

The emission of the MLCT of the complex was found to be quenched relative to the compound $[(bpy)Re(I)(CO)_3]^+$, which has no electron donor. Assign a likely mechanism responsible for the quenching observed in $[(bpz)Re(I)(CO)_3$-ch-DTF$]^+$. (Perkins, T. A., Hauser, B. T., Eyler, J. R., and Schanze, K. S. *J. Phys. Chem.* **1990,** *94,* 8745.)

3. Construct an energy diagram for possible excitation and decay pathways accompanying optical transitions in **5.**
4. The relative quenching efficiencies of the fluorescence of the porphyrin–quinone molecules with $n = 2$, 4, and 6 were measured. These values were, in no particular order, 6.7×10^2, 1.0×10^4, and 3.9×10^4. Match these relative efficiencies with the compounds:

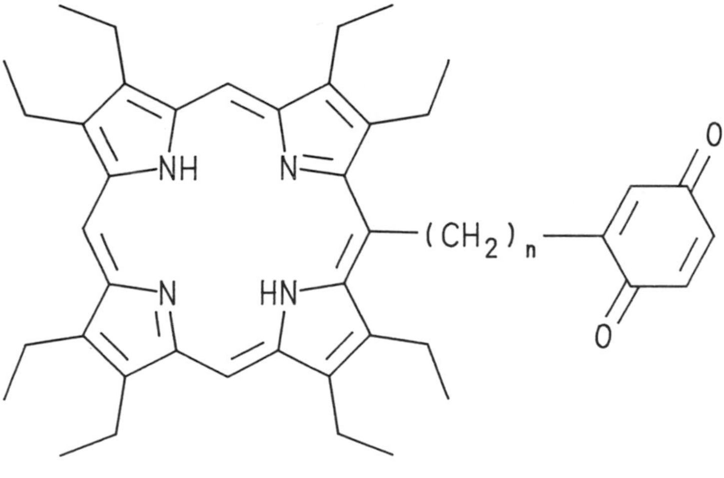

$(CH_2)_n$

$n = 2$

$n = 4$

$n = 6$

What is the likely quenching mechanism in cyclohexane? In proprionitrile? (Nishitani, S., Kurata, N., Sakata, Y., Misumi, S., Migita, M., Okada, T., and Mataga, N. *Tetrahedron Lett.* **1981,** 2099.)

5. Show how the energies in Table 4.2 were calculated assuming that the Coulombic term can be described by $-e^2/\varepsilon_s d_{cc}$. The measurements were performed in butyronitrile, which has a dielectric constant of $\varepsilon_s = 20$. Now calculate and compare the energies assuming the donor and acceptor are intermolecular pairs and can approach to encounter distances. What conclusion you can draw about the energetics in intramolecular and intermolecular photoinduced electron transfer.
6. Design a supramolecular system from various molecular components. Use components and connectors of your choice but that can function in photoinduced electron transfer to achieve efficient charge transfer.

CHAPTER

5

Photoinduced Electron Transfer in Organized Assemblies and the Solid State

We will now direct our attention to photoinduced electron transfer processes in organized and solid-state environments. This chapter will begin with a discussion of photoinduced electron transfer in *organized assemblies*. This will lead to an exploration of *vectorial* photoinduced electron transfer and then *artificial photosynthesis* and *artificial solar capture and storage*. We then turn to the role of semiconductors and examples of *semiconductor-mediated interfacial electron transfer*. The remaining sections will cover photoinduced electron-transfer processes in *imaging* and *xerography*. Since there are more examples in the literature than can be covered adequately in one chapter, the interested reader will want to consult some of the excellent review articles listed at the end of the chapter.

5.1. Photoinduced Electron Transfer in Organized Assemblies

As noted earlier in this book, rapid electron return in solution often reduces the lifetimes of radical ion pairs. The reason, as we have learned, is that these ions are formed in close proximity. Consequently, electron return is often rapid. To achieve the formation of long-lived ion pairs, a way must be found to maintain a large separation distance between the ionic partners. Only under conditions where the efficiencies for formation of these radical ions are high enough and their lifetimes sufficiently long will these intermediates be available to participate in potentially useful reactions. We have learned in Chapter 2 how polar media can assist in ion dissociation. Now we will show that well-separated, long-lived ion pairs can be generated in *organized assemblies*. As this story unfolds, it will become increasingly apparent that photoinduced electron transfer in these environments can be manipulated and used in a variety of interesting applications. One of the most promising areas is the capture, storage, and harvest of solar energy (Fig. 5.1). It is here where one can exploit photoinduced electron transfer to generate charge-separated species to perform useful chemical functions such as the cleavage of water into hydrogen and oxygen. Mother Nature has fine-tuned natural photosynthesis in green plants by making use of architecturally elaborate molecular networks. Artificial photosynthesis is an attempt to mimic the marvelous mechanism responsible for the conversion of solar energy into energy-rich molecules in plants.

Many of these systems can be regarded as supramolecular devices in the sense that they consist of several molecular subunits bound by weak forces such as van der Waals and electrostatic. These ensembles function by converting light into chemical energy. Figure 5.2 shows a typical configuration consisting of a sequential assemblage of molecular components. When exposed to light, an electron is transported via the relay to an acceptor well separated from the sensitizer. Similarly, an electron from a distant donor can return to the acceptor. This represents a simple one-dimensional assembly. More complex schemes can be imagined, but they must be capable of transmitting an electron(s) over large separation distances to achieve effective charge separation.

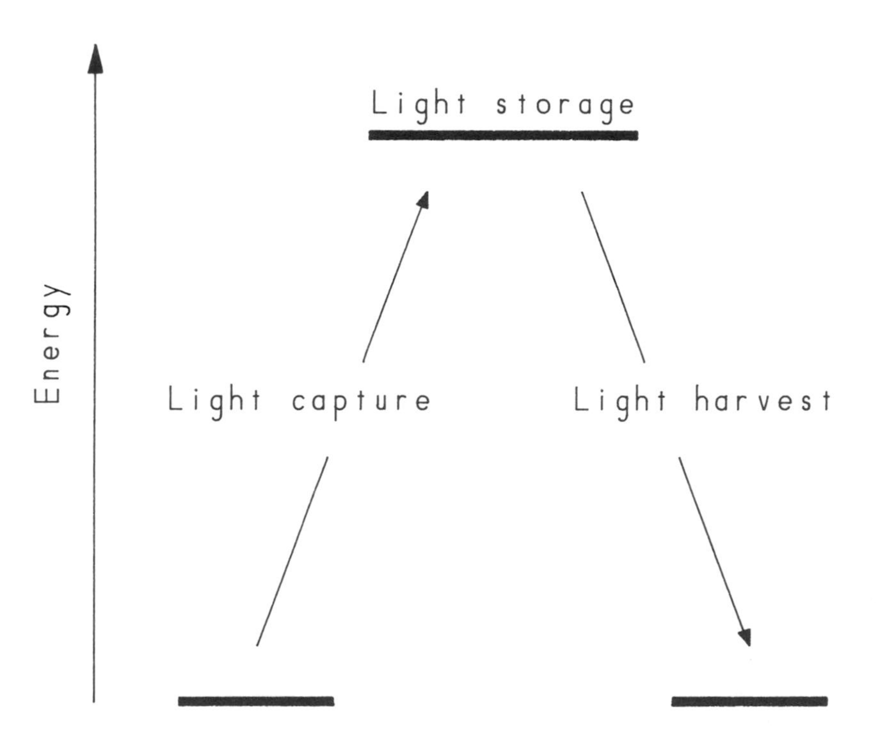

Figure 5.1 Photochemical energy capture, storage, and harvest.

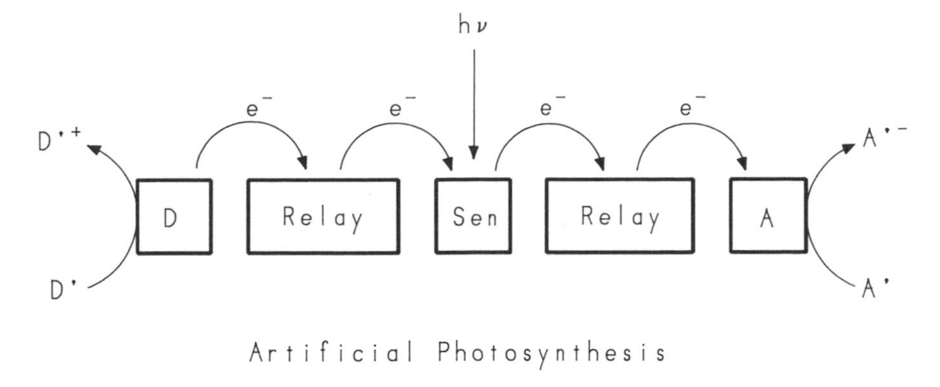

Figure 5.2 A simple artificial photosynthetic system demonstrating how effective charge separation can be achieved.

It is useful to first define the distinctive features of an organized environment and how it can be manipulated to assist in the formation of long-lived ion pairs.

For our purposes, an organized assembly is an ordered rearrangement of molecular aggregates. The molecular nature of the organized environment can have a profound effect on the distribution of the reactants and products. For example, an organized assembly often contains hydrophobic and hydrophilic components. Polar reactants and products are then attracted by electrostatic interactions to the hydrophilic regions. Nonpolar molecules are in turn attracted to the hydrophobic region. In this manner, an organized assembly can localize and separate molecules based on their molecular structures. The net result is efficient partitioning of both the reactants and products. Depending on the precise details of the molecular architecture, an organized assembly may assist in enhancing the forward flow of electrons and help in maintaining a fairly large separation distance between the charge-transfer intermediates after they are formed. This can effectively block electron return.

Perhaps the best way to see how effective charge separation can be achieved is to explore several types of organized assemblies. We start with *micelles* and *microemulsions*.

5.1.1. Photoinduced Electron Transfer in Micelles and Microemulsions

Micelles have proven to be valuable in attempts to generate long-lived ion pairs. A micelle is a spherical aggregate formed when amphiphilic molecules combine in an aqueous media (Fig. 5.3). Amphiphilic molecules consist of hydrophilic and hydrophobic functional groups. These molecular building blocks combine in such a fashion that the hydrophilic portions protrude into the aqueous layer and the hydrophobic portions extend into the interior of the sphere. Micelles are formed spontaneously when the concentration of the molecules exceeds the *critical micellar*

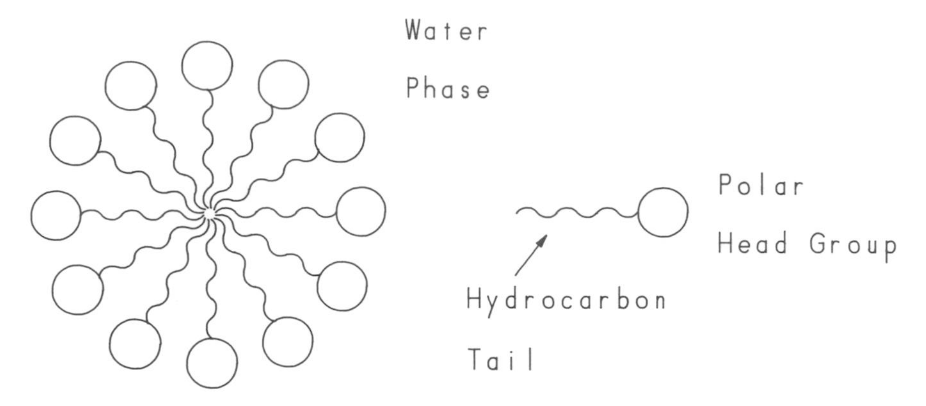

Figure 5.3 A micelle consists of a spherical aggregate of surfactant molecules in an aqueous medium. These long-chain molecules are arranged in such a fashion that the hydrocarbon alkyl trails extend toward the interior of the aggregate and the polar "heads" are aligned at the micellar surface.

concentration (*cmc*). The diameter of a typical spherical micellar aggregate may range from 15 to 60 Å.

Surfactant molecules are ideal candidates for building up micelles. Typically, these molecules consist of long organic chains connected to polar head groups (Fig. 5.4). The nature of the head group establishes whether the micelle is charged or neutral. For example, cetyl(hexadecyl)trimethylammonium chloride (CTAB) aggregates to form a cationic micelle whereas sodium dodecyl(lauryl) sulfate (SDS) forms an anionic micelle. *N*-Dodecyl-*N*,*N*′-dimethylbetaine is zwitterionic; polyoxyethylene(E9–10)t-octylphenol (Triton X-100) is neutral.

The contemporary picture of the micelle has evolved considerably since the introduction of an early model in 1936. The shape of the micelle—spherical or rod-like—depends on the structure and concentration of monomer. In general, at low concentrations of monomer, the micellar structure is spherical but becomes rod-like at higher concentrations. The core of the micelle is a liquid-like, hydrophobic environment made up of extended alkyl chains. The surface contains ionic head groups of monomer closely packed together. Surrounding the surface is a double layer comprised of a thin layer of water molecules and solvent gegenions (Fig. 5.5). The simplest picture of the double layer is based on a model proposed by Helmholtz (Fig. 5.6).[1] According to this model, the charge on the surface is neutralized by a single layer of charged ions in the electrolyte, and the potential varies linearly with distance. However, a more general picture is one in which the charges in solution attract opposite charges, or *gegenions*. The distribution of charges in solution is more diffused and the potential drop is nonlinear. This model is called the Gouy–Chapman double layer. In another model, proposed by Stern, the micelle is surrounded by a Stern and Gouy–Chapman layer. The charge of the Stern and Gouy–Chapman layers is measured by the zeta potential or mean surface potential,

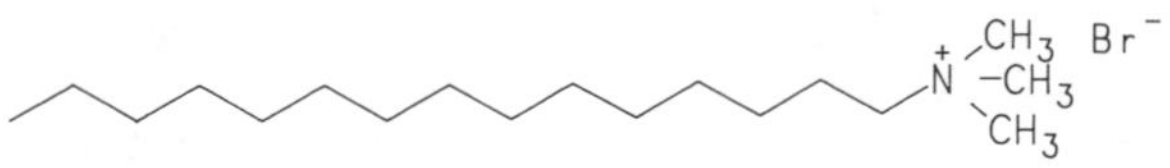

$\overset{+}{N}(CH_3)_3\ Br^-$

Cetyl(hexadecyl)trimethylammonium bromide (CTAB) cmc – 9.2×10^{-4} M

$OSO_3^-Na^+$

Sodium dodecyl(lauryl)sulfate (SDS) cmc – 8×10^{-3} M

Figure 5.4 Typical surfactant molecules used as components of micelles.

and acts as an electrostatic barrier to the passage of charged ions through and across the micellar surface. The value of the zeta potential usually ranges from 50 to 100 mV and drops with distance from the surface of the micelle.

When certain reactant molecules are introduced into a micellar solution, they may enter the micelle and arrange themselves depending on their own structures as well as the structure of the micelle. In micellar environments, the reactants may initially reside in the outer aqueous layer or within the micelle. The course of reaction taking place within the micelle is dictated by the charges on the reactants and products in the micelle.

The interactions of the reactants and products with the micellar environment are essentially electrostatic and hydrostatic. In contrast to a homogeneous solvent environment where ion formation usually takes place within a solvent cage, in micellar environments, the reactants can be maintained at certain distances. Nonpolar molecules may be attracted to the organic inner core of micelles, and ionic products may remain in the charged portion of the micelle. In cationic micelles, negative reactants and products associate with the micelle, whereas positive molecules may be repelled from the micelle. The double layer prevents the reentry of the positive species into the micelle. The reverse situation is encountered in anionic micelles.

A caveat must be added here. The representation of micellar environments to explain photoinduced charge-transfer phenomena is highly idealized and probably quite primitive. In many cases, the exact position of the reactants and products can only be conjectured. Even with this reservation, a few cases illustrating the unique properties of micelles will demonstrate the tremendous versatility of these organized assemblies. For an example, consider the photoinduced reactions of excited $Ru(bpy)_3^{2+}$ with viologen quenchers possessing long alkyl chains in micellar environments (Fig. 5.7).[2,3] It is known that the solubility of the oxidized forms of certain viologens, such as $C_{14}MV^{2+}$, is fairly high in aqueous solvents; the reduced form, $C_{14}MV^{\dot{+}}$, however, is relatively insoluble in water but associates more readily with cationic micelles. By taking advantage of this particular feature of the viologen,

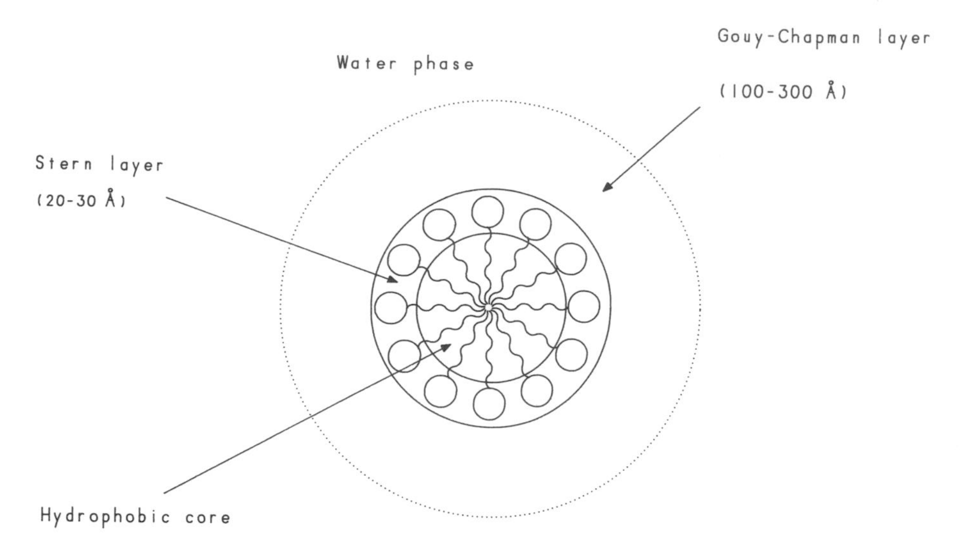

Figure 5.5 A simplified diagram of the micellar structure. A double layer exists at the surface of the micellar surface. The Gouy-Chapman layer is a diffused double layer of positive and negative ions.

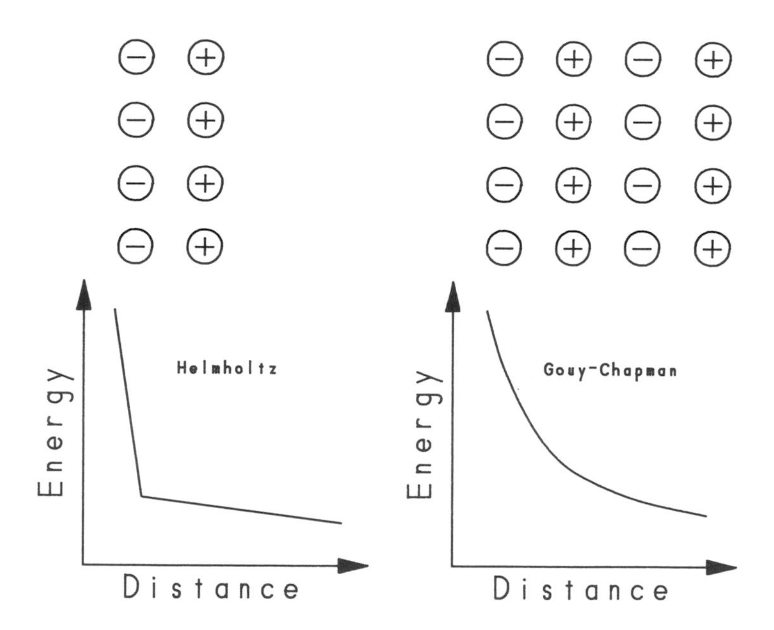

Figure 5.6 The Helmholtz and Gouy-Chapman layers.

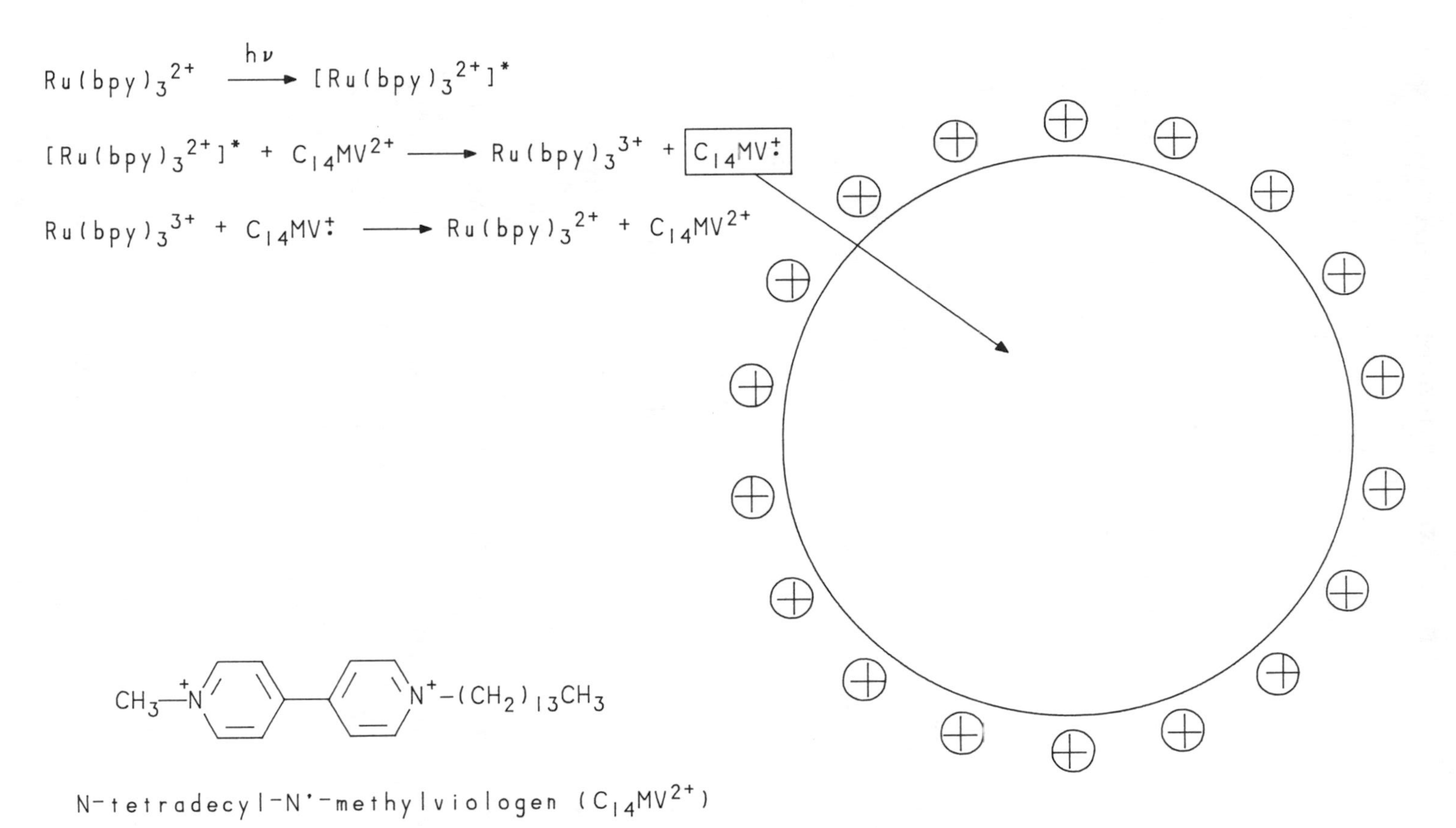

Figure 5.7 The sensitized photoreduction of a methylviologen in a solution containing a cationic micelle. $MV^{\dot{+}}$ becomes trapped by the micelle. However, the other charged intermediates are repelled by the micellar surface. This partitioning of the products inhibits electron return.

a long-lived ion pair can be generated by photoinduced electron transfer. The reduction of $C_{14}MV^{2+}$ by excited $Ru(bpy)_3{}^{2+}$ proceeds as shown below:

$$[Ru(bpy)_3{}^{2+}]^* + C_{14}MV^{2+} \xrightarrow{k_{el}} Ru(bpy)_3{}^{3+} + C_{14}MV^{\dot{+}} \tag{5.1}$$

$$Ru(bpy)_3{}^{3+} + C_{14}MV^{\dot{+}} \xrightarrow{k_{ret}} Ru(bpy)_3{}^{2+} + C_{14}MV^{2+} \tag{5.2}$$

In homogeneous solvents, the absorption peak at 602 nm due to $C_{14}MV^{\dot{+}}$ disappears rapidly, indicating a rapid electron return ($k_{ret} \sim 4 \times 10^9$ M^{-1} s^{-1}). However, in a cationic micelle, such as CTAC, the signal decays in two steps ($k_{ret} \sim 2 \times 10^7$ M^{-1} s^{-1}). These results can be understood by following the scheme in Fig. 5.7. Initially, $C_{14}MV^{2+}$ and $Ru(bpy)_3{}^{2+}$ are located in the aqueous phase. Association between the micelle and the reactants is poor because of electrostatic repulsion. When the solution is exposed to light, photoexcitation and electron transfer occur exclusively within the aqueous environment. Subsequent to its formation, $C_{14}MV^{\dot{+}}$ enters the cationic micelle. $Ru(bpy)_3{}^{3+}$ is prevented from entering the micelle because of the positive charge surrounding the micelle. Eventually, $C_{14}MV^{\dot{+}}$ and $Ru(bpy)_3{}^{3+}$ become separated, and electron return is effectively reduced, in this case by a factor over 400.

The quenching of the porphyrin, $ZnTMPyP^{4+}$, by $C_{14}MV^{2+}$ is another example[4]:

$$^1[ZnTMPyP^{4+}]^* + C_{14}MV^{2+} \xrightarrow{k_{el}} ZnTMPyP^{5+} + C_{14}MV^{\dot{+}} \tag{5.3}$$

$$[ZnTMPyP^{5+} + C_{14}MV^{\dot{+}} \xrightarrow{k_{ret}} ZnTMPyP^{4+} + C_{14}MV^{2+} \tag{5.4}$$

This system is of particular interest since the quenching of singlet porphyrins in homogeneous media is notoriously inefficient because of rapid electron return. In CTAC, $k_{ret} \sim 10^7$ s^{-1}, as compared to $k_{ret} \sim 5 \times 10^9$ s^{-1} in solution.

Effective charge separation was noted when chlorophyll *a* and duroquinone were injected into anionic sodium lauryl sulfate (NaLS) micelles.[5] These neutral reactants are initially arranged *within* the micelle. Flash photolysis studies revealed a long-lived transient species with an absorption of duroquinone anion at 685 nm. This observation suggested partitioning of the radical ion pair initially formed within the micelle (Fig. 5.8). The anionic micelle was thought to eject the duroquinone anion because of electrostatic repulsions between the negatively charged species. The anion is prevented from reentry by the electrostatic barrier of the anionic micellar surface.

5.1.2. Photoinduced Electron Transfer in Reversed Micelles

In nonpolar, nonaqueous media, some surfactants can aggregate to form *reversed micelles*. In forming reversed micelles, sometimes referred to as *inverted micelles*, the surfactant molecules aggregate so that their charged head groups form the core of the micelle and their nonpolar, hydrophobic tails extend into the exterior (Fig. 5.9). Because of its hydrophilic nature, the core attracts water molecules and charged

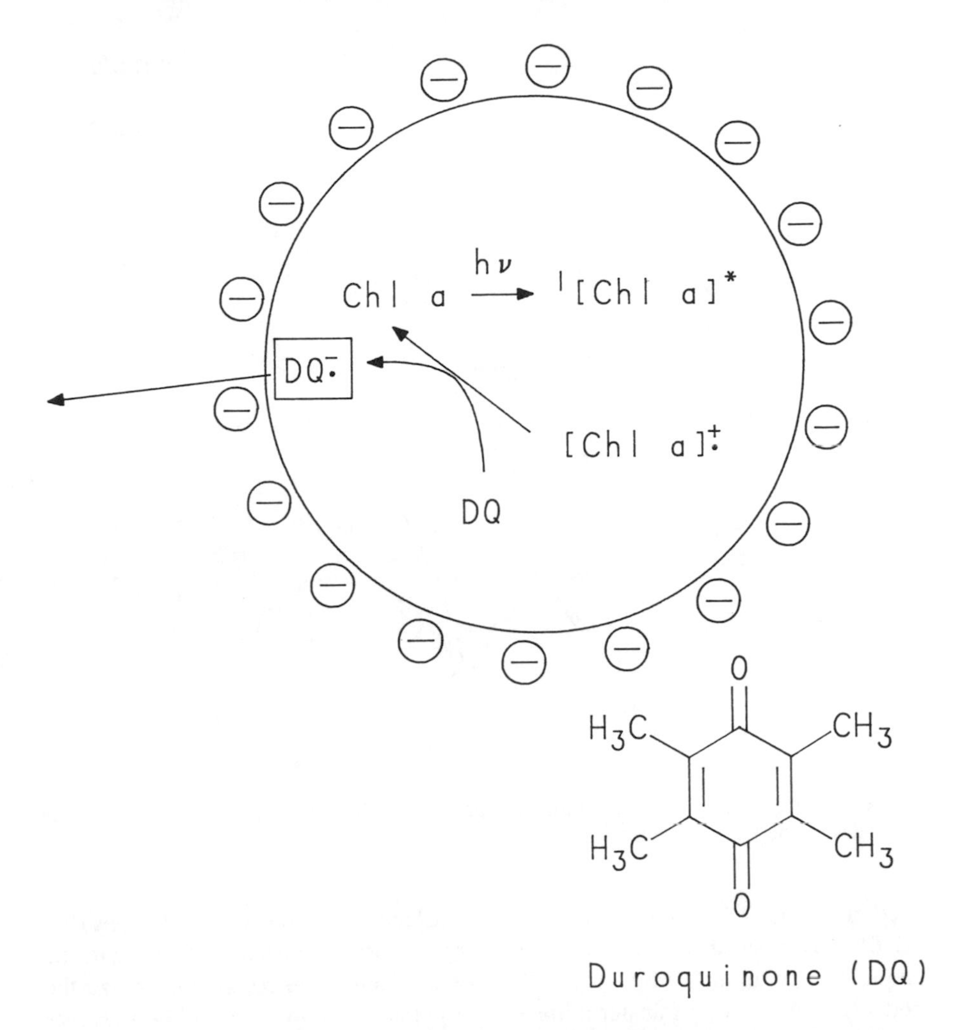

Figure 5.8 Duroquinone is reduced within this anionic micelle by chlorophyll *a* (Chl a). In this case, $DQ^{\cdot-}$ is easily ejected from the micelle, preventing electron return.

molecules, whereas nonpolar molecules establish themselves easily in the hydrocarbon phase. An example of photoinduced electron transfer in reversed micelles is shown in Fig. 5.10.[6] This reaction was carried out in water–toluene reversed micelles in the presence of EDTA and benzylnicotinamide ($BNA^{\cdot+}$). Photoexcitation of $Ru(bpy)_3^{2+}$ takes place within the interface between the aqueous core and surrounding hydrophobic shell. Subsequently, $[Ru(bpy)_3^{2+}]^*$ reduces $BNA^{\cdot+}$ to its neutral form, BNA, which then migrates into the organic phase. When the reaction is carried out in the presence of a dye such as 4-dimethylaminoazobenzene dissolved in the toluene layer, the absorption of the dye at $\lambda = 402$ nm vanishes, suggesting reduction of the dye by BNA. In the meantime, $Ru(bpy)_3^{2+}$ is regenerated by

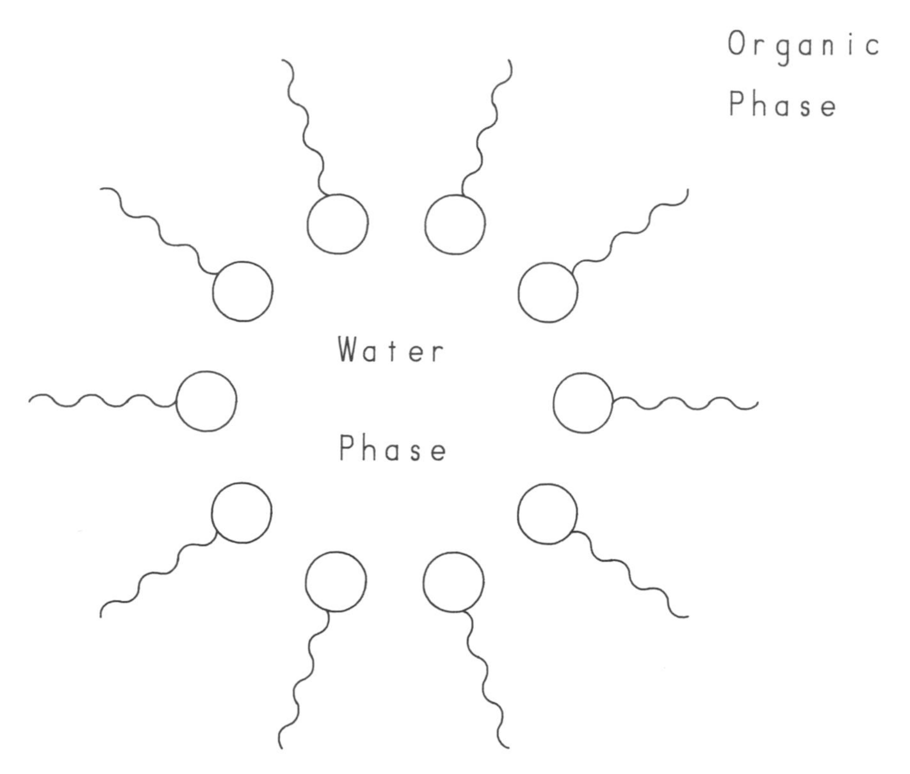

Figure 5.9 Reversed micelles are formed in organic media by the aggregation of surfactant molecules as shown in the figure.

reaction with EDTA within the aqueous core. Electron return between $(Ru(bpy)_3{}^{3+}$ and BNA is blocked because of a fairly large separation distance. Thus, with its unique configuration, a reverse micelle, like a micelle, can compartmentalize the products of an electron transfer initiated by light and thereby enhance the lifetimes of the charge-separated ionic species.[7]

5.1.3. Magnetic Effects in Micellar Cages

The subject of magnetic effects on photoinduced electron transfer in homogeneous solution was introduced in earlier chapters of this book. Micelles may also exert a dynamic magnetic effect on photoinduced electron transfer. Consider the oxidation of diphenylamine by triplet 1-acetonaphthone in SDS micelles (Fig. 5.11).[8,9] Quenching of triplet 1-acetonaphthone can occur within the micellar cage with the generation of a triplet radical pair. This step is quickly followed by a proton transfer. Immediate electron return to ground state reactants within the micelle is not favored because of the triplet alignment of the radical pair. Instead, ejection from the cage occurs. If, however, the triplet pair undergoes intersystem crossing to a singlet ion pair, this will be followed by recombination of singlet radical pairs. In

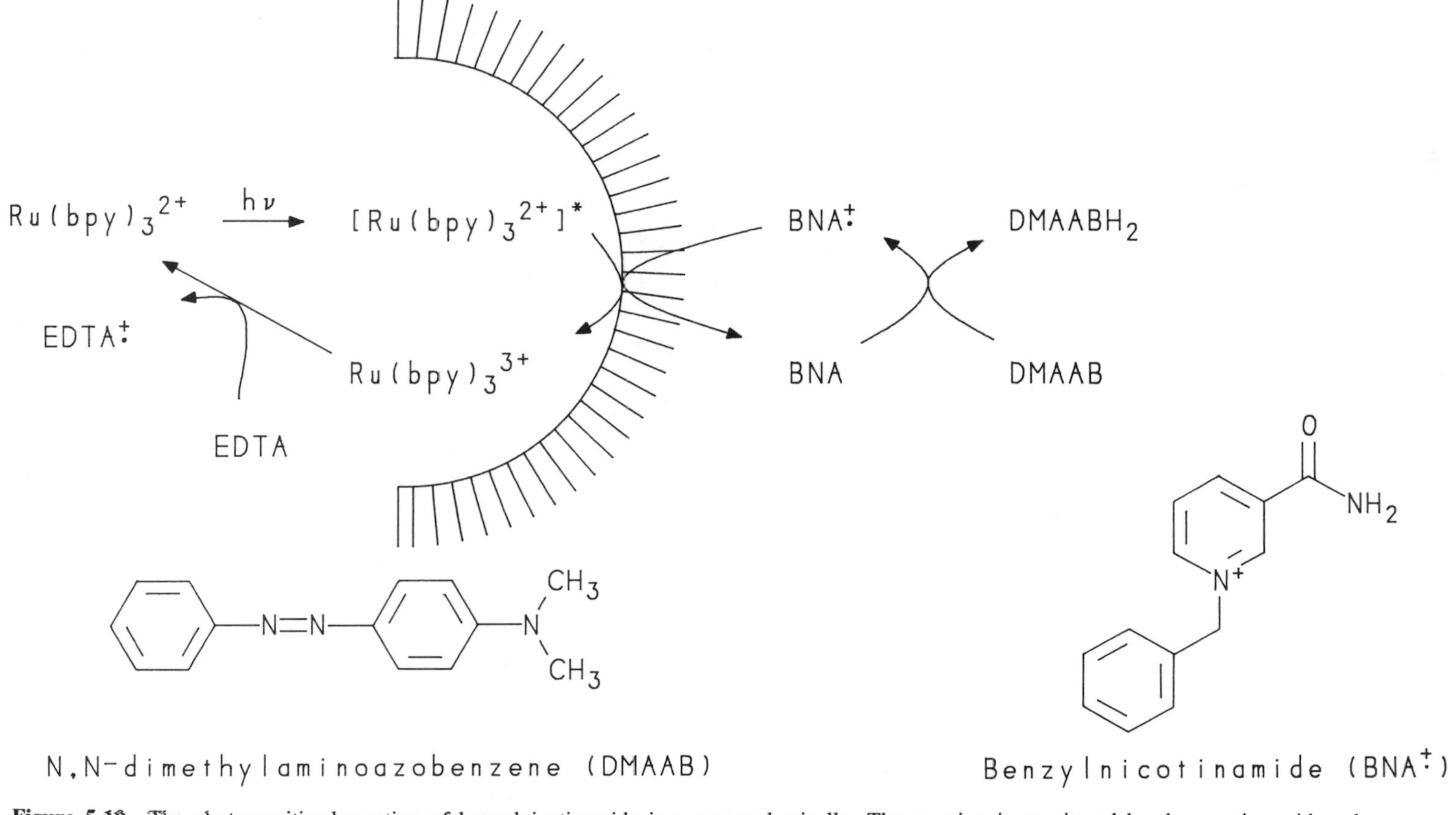

Figure 5.10 The photosensitized reaction of benzylnicotinamide in a reversed micelle. The reaction is monitored by the reaction with a dye (DMAAB).

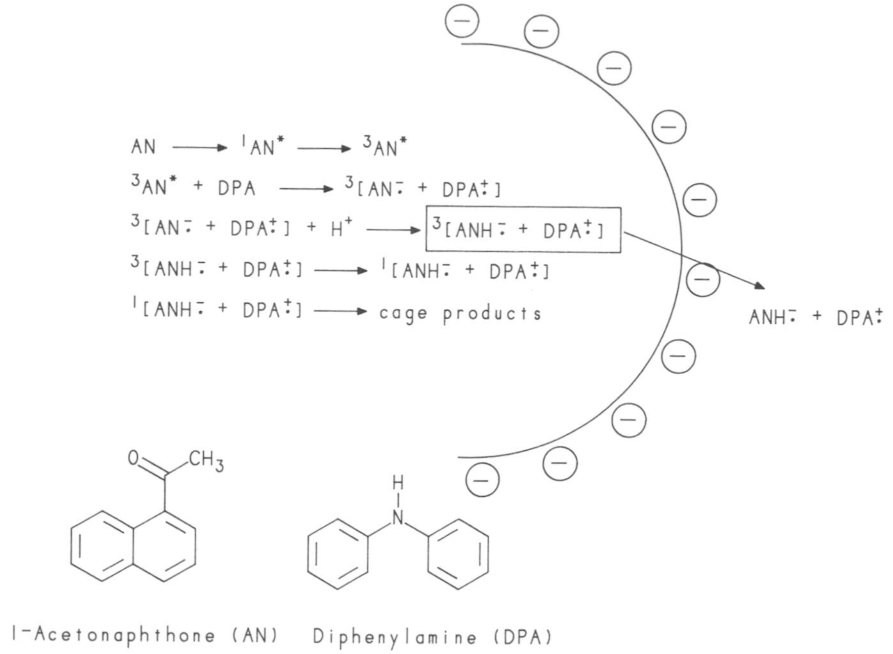

$AN \longrightarrow {}^1AN^* \longrightarrow {}^3AN^*$

${}^3AN^* + DPA \longrightarrow {}^3[AN^{\overline{\cdot}} + DPA^{+\cdot}]$

${}^3[AN^{\overline{\cdot}} + DPA^{+\cdot}] + H^+ \longrightarrow {}^3[ANH^{\overline{\cdot}} + DPA^{+\cdot}]$

${}^3[ANH^{\overline{\cdot}} + DPA^{+\cdot}] \longrightarrow {}^1[ANH^{\overline{\cdot}} + DPA^{+\cdot}]$

${}^1[ANH^{\overline{\cdot}} + DPA^{+\cdot}] \longrightarrow$ cage products

$ANH^{\overline{\cdot}} + DPA^{+\cdot}$

1-Acetonaphthone (AN) Diphenylamine (DPA)

Figure 5.11 An external magnetic field can exert an effect on photoinduced electron transfers in micelles such as the example in this figure. When triplet ion pairs are formed within a micelle, their triplet alignment inhibits electron return. After the ions separate, they are ejected from the micelle into the solution phase. This effect is more likely in the presence of a strong magnetic field, since in this case intersystem crossing of the triplet ion pair to the singlet state becomes increasingly improbable.

fact, in the presence of an external magnetic field, the degeneracy that exists between the three triplets and singlet (which allows triplet to singlet conversion) is lifted because of the Zeeman interaction (Section 2.7). Electron return becomes less likely since fewer singlet ion pairs are produced. As the magnetic field increases, the yield of long-lived radicals also increases because of enhanced charge separation of triplet ions.

5.1.4. Photoinduced Electron Transfer across Langmuir–Blodgett Films

Earlier in this century, Irving Langmuir and Katherine Blodgett developed a technique to spread thin films of active substances on the surface of water.[10] These films have earned the designation of *Langmuir and Blodgett* or *LB films*. Actually, the idea of floating organic layers was known since ancient times. In Roman times, divers floated an oil layer on water to reduce stray light and make the bottom of the sea more visible. Benjamin Franklin calmed a pond in England by pouring a small amount of oil on the surface of the water. The water spread quickly and produced a smooth, glassy surface.

LB films consist of monomolecular carboxylic acids with long alkyl chains. Since the carboxylic acid portion or its salt is hydrophilic and the hydrocarbon chain hydrophobic, these materials will be oriented in water with the alkyl chains extending away from the water. To make a "floating" LB film, the material is dissolved in an organic solvent and added slowly and carefully to the water surface. As the solvent evaporates, a monomolecular film is formed on the water surface. LB films can also be deposited on substrates such as aluminum oxide. Typically, as the substrate is passed vertically through a "floating" LB monolayer, the carboxylic groups become attached to the aluminum oxide surface and the chains orient themselves away from the surface of the substrate. Bilayers can be formed by a second passage. In fact, multilayers consisting of hundreds of monolayers can be built up by sequential depositions.

LB films have been used in a variety of interesting applications including microlithography as thin photoresists, surface acoustic wave (SAW) technology, and

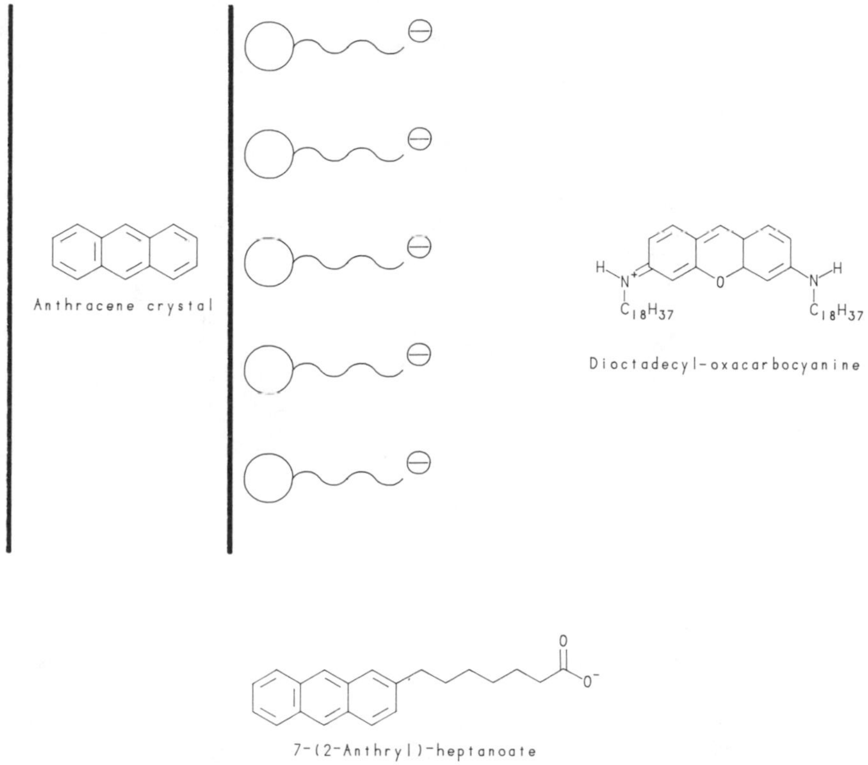

Anthracene crystal

Dioctadecyl-oxacarbocyanine

7-(2-Anthryl)-heptanoate

Figure 5.12 Photosensitized hole injection. An anthracene crystal is separated from the sensitizer by an acid monolayer. The alkyl chain of the acid probably enhances the tunneling of the electron from cyanine to anthracene.

nonlinear optics. In this section, we will touch on the use of LB films as organized assemblies for the study of photoinduced electron transfer.

Photoinduced electron transfer across thin films consisting of successive monolayers has attracted the interest of chemists interested in devising ways to maximize the efficiency of charge separation. Monolayers are excellent environments for photoinduced charge transfer since reactants can be readily immobilized and distributed at known distances. Accordingly, photoinduced electron transfer in Langmuir–Blodgett films has received much attention.[11–17]

As an example, monolayers have been used in *photosensitized hole injection*.[18] When an organic crystal such as anthracene or perylene is covered with a dye, illumination of the dye leads to hole injection into the bulk of the crystal in competition with electron return. Although efficient hole injection normally requires a large electric field, it was noted that when a monolayer of 7-(2-anthryl)-heptanoic acid (2A7) is deposited between an aromatic crystal and the dye (Fig. 5.12), the process occurs at lower applied fields. When the monolayer is deposited on the crystal, a hole is injected into the 2A7 monolayer from the photoexcited dye. This process is followed by injection of the hole by the monolayer into the crystal.

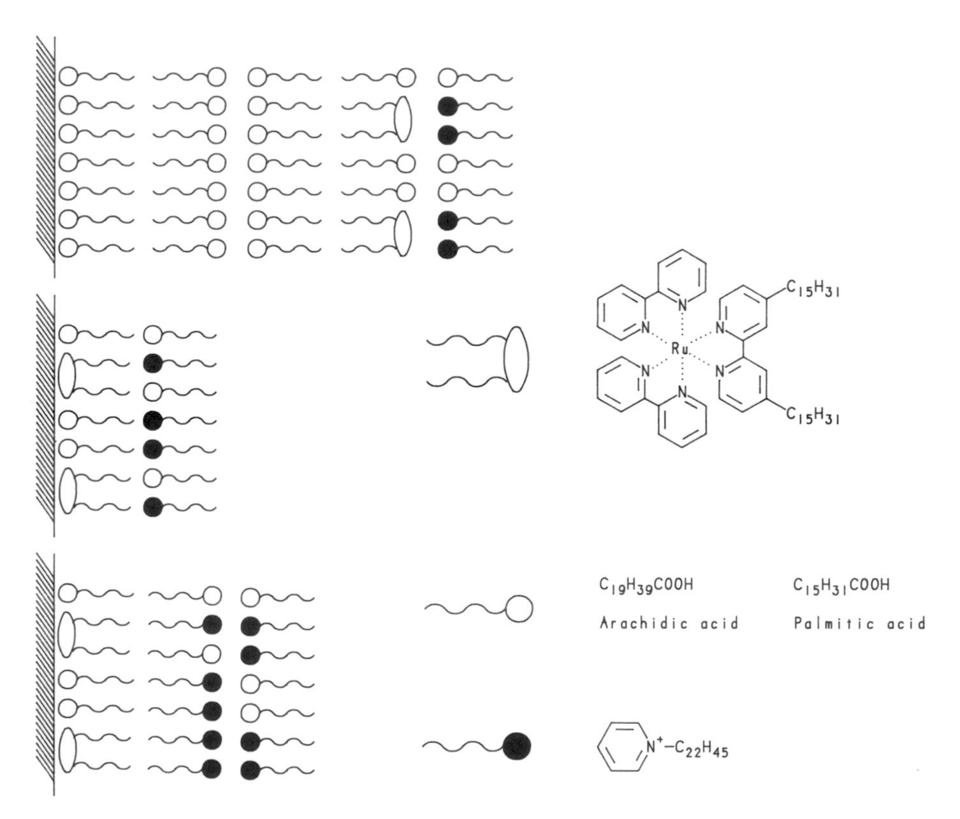

Figure 5.13 Photoinduced electron transfer in a multilayer assemblage of long chain molecules. The luminescence of the ruthenium sensitizer is quenched by an electron acceptor in a manner dependent on the mutual separation distance between the reactants.

Multilayered Langmuir–Blodgett layers containing $Ru(bpy)_3^{2+}$ and amphiphilic electron acceptors have been studied by nanosecond laser photolysis.[19] The conclusions from these studies are interesting. For example, mixed layers consisting of arachidic acid, palmitic acid, and methyl arachidate in various ratios were deposited on hydrophilic quartz plates. The amphiphiles were then spread onto this phase. The $Ru(bpy)_3^{2+}$ sensitizer was covalently attached to two long alkyl chains and incorporated within the layers. Three types of layers were prepared in this fashion, as shown in Fig. 5.13. In the first case, the ruthenium sensitizer was placed in contact with the electron acceptor, and then the sensitizer and quencher were separated by one and two layers, respectively. On photoexcitation, each system displayed luminescence. The decay of this emission was found to be related to the separation distance. Apparently, the sensitizer and acceptor in contact decay the fastest, whereas the slowest decay occurs when the sensitizer and acceptor are separated by two alkyl chains.

5.1.5. Photoinduced Electron Transfer in Lipid Bilayers

Figure 5.14 shows another type of organized assembly—a spherical vesicle consisting of amphiphilic phospholipid molecules. Phospholipids, one of the major

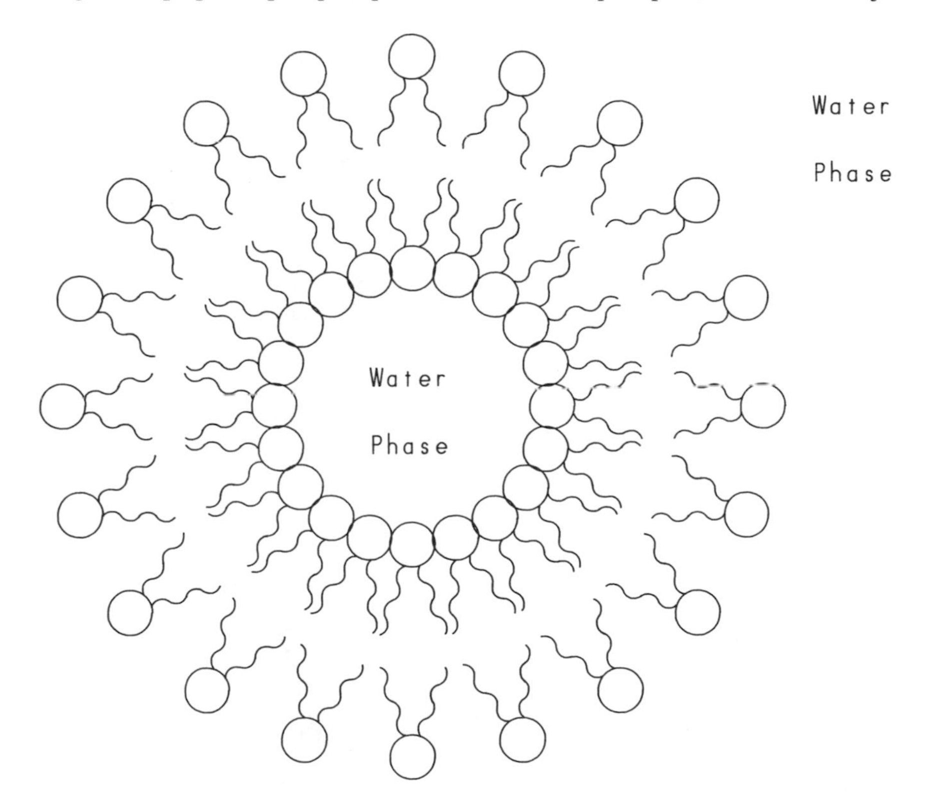

Figure 5.14 An organized assemblage of amphiphilic molecules arranged as a spherical bilayer.

components of biological membranes, consist of fatty acid chains attached to glycerol (Fig. 5.15). Although these amphiphilic molecules are abundant in nature, naturally occurring phospholipids and synthetic analogues can also be prepared in the laboratory.

In lipid vesicles, the polar groups of the phospholipid molecules may be distributed on the inner and outer surfaces of each vesicle. The diameters of single-bilayer vesicles typically fall within the range of a few hundred angstroms. The layer thickness may be as large as ~50 Å. In contrast to micellar formation, vesicles consisting of lipid surfactants can be formed at lower cmcs. Surfactant vesicles can also be polymerized within their bilayers and across their surfaces. By coating the exterior and interior portions of the vesicles with polymeric units, one enhances the stability of the vesicle. Figure 5.16 shows a schematic representation of a vesicle in which adjacent tails are chemically linked.

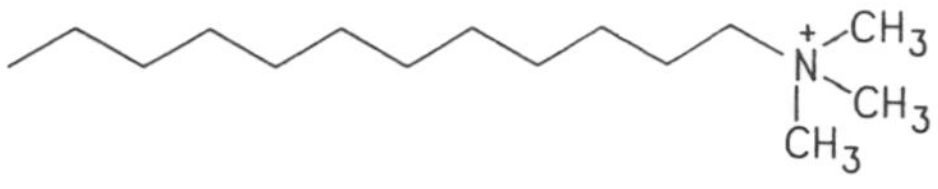

Dodecyldimethylammonium (DODAC)

Dihexadecyl phosphate (DHP)

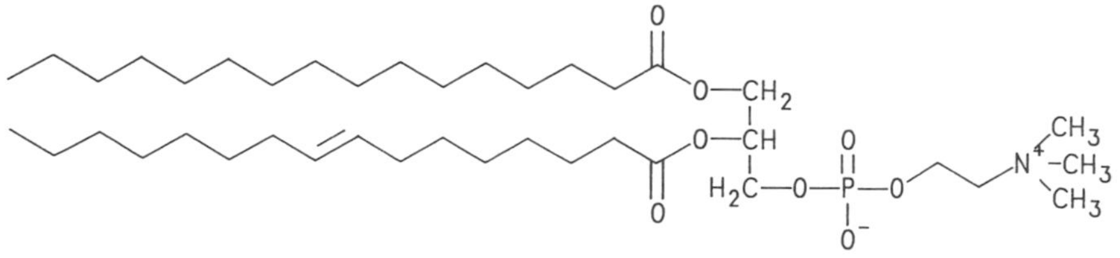

1-Palmitoyl-2-oleoylphosphatidyl choline (PC)

Figure 5.15 Phospholipids—both naturally-occurring and synthetic—are examples of molecular building blocks of vesicles.

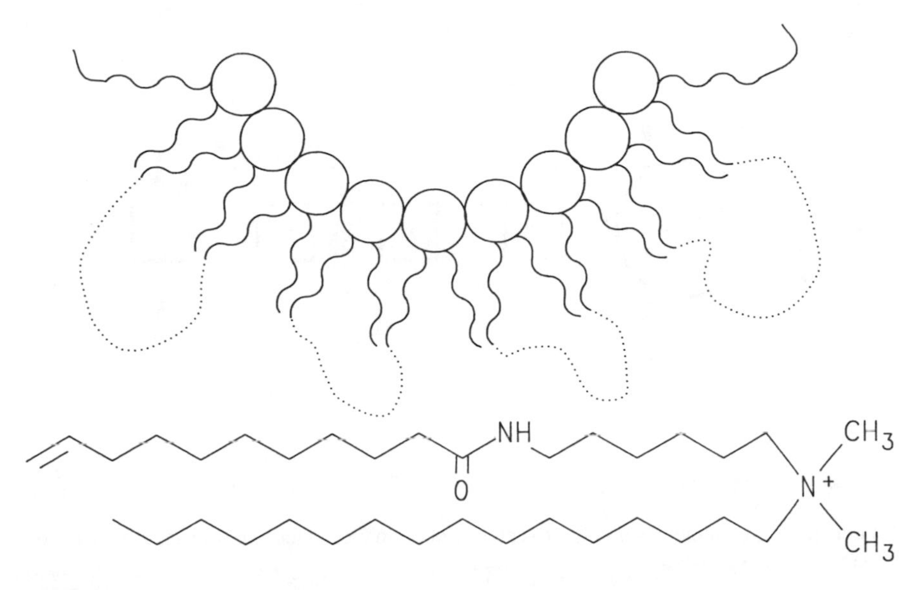

Figure 5.16 Surfactant vesicles can be polymerized across their surfaces, thereby enhancing the stabilities of these bilayer structures. The dotted lines represent linear chains connecting the vesicles. In this case, linkage can be achieved by reactions between the ethylenic moieties.

When donor and acceptor molecules are incorporated in lipid bilayers, nonpolar molecules can be trapped within the core of hydrophobic alkyl tails whereas polar molecules are distributed in close proximity to the hydrophilic head groups. These effects are strongly affected by electrostatic interactions and hydrogen bonding.[20,21] The partitioning of the "microenvironment" of the lipid vesicle into hydrophilic and hydrophobic domains (as in the micelle) is useful in maintaining maximum separation between charged species. The phospholipid bilayer can be regarded as a self-assembled supramolecular system containing the basic elements of the energy conversion machinary of living organisms.

An interesting feature of phospholipid bilayers is that they can be formed by the spontaneous assembling of component phospholipid molecules. Once formed in this manner, phospholipids can easily be manipulated in the laboratory to form models of energy conversion processes.

Vectorial photoinduced electron transfer across phospholipid membranes over distances from ~40 to ~50 Å consists of a series of short hops (Fig. 5.17). Donor and acceptor molecules are bound to the surface of the membrane or within its interior, which is usually hydrophobic.[22] These systems are vastly simplified versions of the photosynthetic reaction center in green plants.

An early example of photoinduced electron transfer in lipid vesicles was reported for a phospholipid dispersion containing $Ru(bpy)_3^{2+}$ attached to two 1-hexadecyl chains (Fig. 5.18).[23] If EDTA and MV^{2+} are added to aqueous dispersions of the phospholipid encapsulated sensitizer, EDTA can be trapped within the interior aqueous core of the vesicle, and the remaining EDTA can be removed by gel

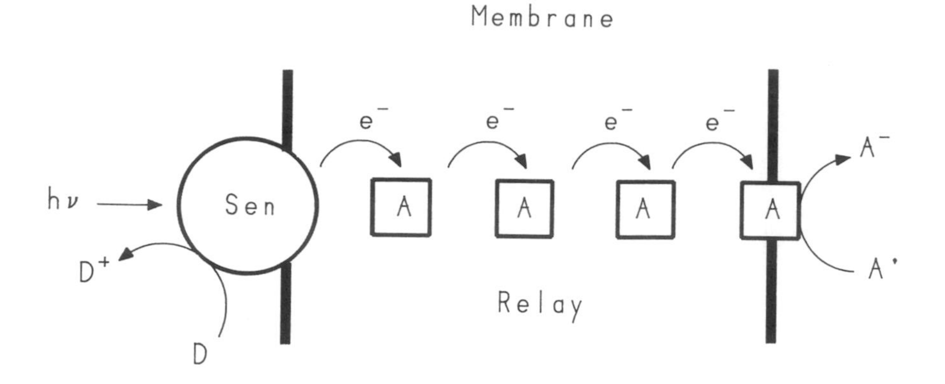

Figure 5.17 A simple schematic of vectorial photoinduced electron transfer across a membrane.

filtration. These dispersions can then be added to aqueous solutions of the MV^{2+}, which remain in the outer aqueous layer. When the vesicles are illuminated with visible light, $MV^{\dot{+}}$ can be detected by its absorption at 603 nm. The following sequence was proposed: after reduction of the aqueous MV^{2+}, the oxidized sensitizer is reduced EDTA. The structure of the vesicle prevents close contact between $Ru(bpy)_3{}^{3+}$ and $MV^{\dot{+}}$, preventing electron return. Thus, like micelles and reversed micelles, lipid bilayer membranes are useful environments for achieving effective charge separation.

In lipid bilayers, electron transfer can occur sequentially by a series of electron shuttles between several electron donors or electron acceptors arranged adjacent to one another. The concept of *transmembrane electron transport* (i.e., electron transfer across a membrane) has received experimental and theoretical support.[24,25] For example, chemists have designed novel vesicular systems to evaluate the role of *microenvironmental organization*. An example of this organizational control of transmembrane electron transfer is photoinduced electron transfer in vesicles synthesized from dihexadecylphosphate.[26] Zinc and manganese porphyrins were anchored symmetrically on both the inner and outer side of a vesicle, as shown in Fig. 5.19. A propyl viologen was then attached to the inner and outer aqueous phases. Light excitation to generate a triplet Zn porphyrin was followed by reduction of the "inner" dipropyl viologen (PVS). By time-resolved techniques, transmembrane electron transfer from reduced propyl viologen to the Mn porphyrin was shown to take place in competition with electron return. The thickness of this vesicle was estimated between 40 and 50 Å, illustrating dramatically that transmembrane electron transport can occur over fairly large distances.

In a zinc porphyrin–viologen molecule linked by a flexible chain of four meth-

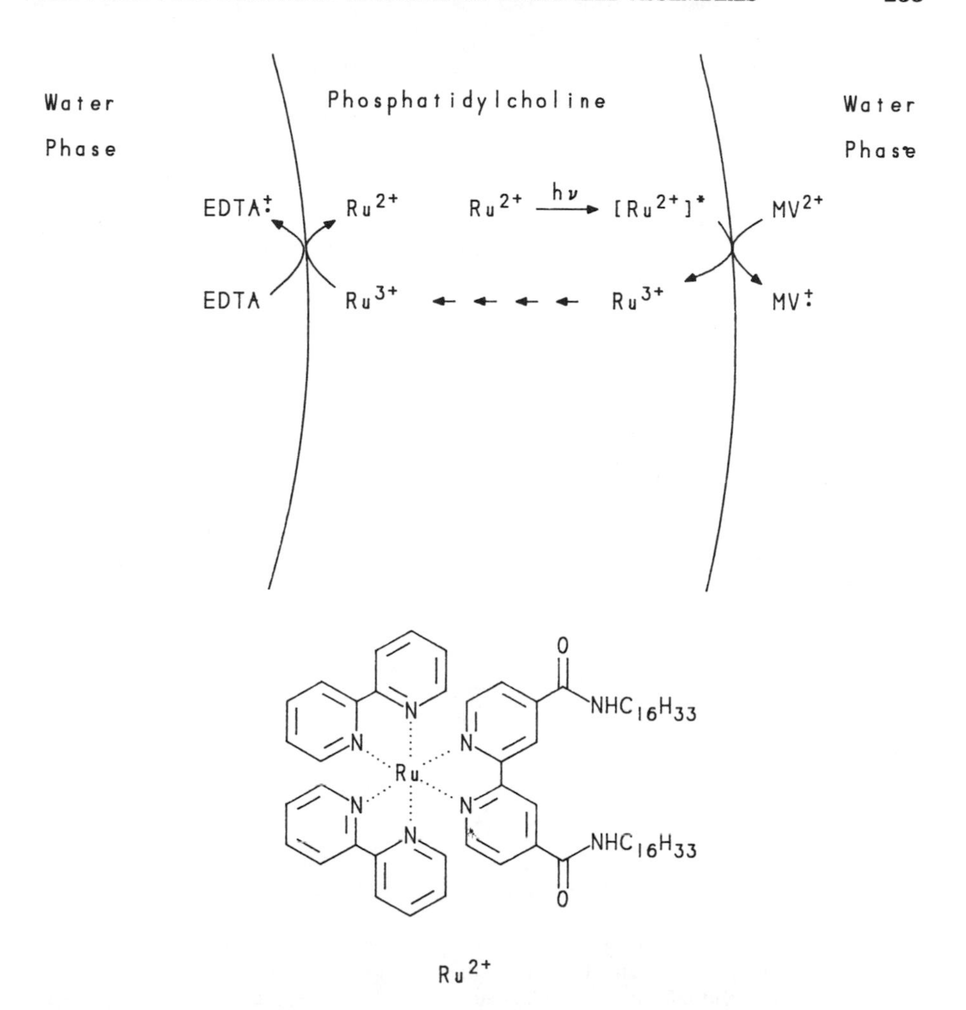

Figure 5.18 Photoinduced electron transfer in a phospholipid dispersion. Charge-separation is attainable because of the compartmentalization of the reactants by the polar and nonpolar phases of the phospholipid dispersion in an aqueous medium. In this example, EDTA is used as a sacrificial donor.

ylene groups, ion-pair formation was found to be enhanced by 35% in dihexadecyldimethylammonium chloride bilayer dispersions than in aqueous acetonitrile.[27] Apparently, the structure of the bilayer permits the porphyrin and viologen to be held in a favorable orientation for maximizing the efficiency of charge separation.

The examples presented in this section are simple models of the biological machinary in green plants responsible for photosynthesis. A discussion of natural photosynthesis is reserved for Chapter 6 since we will need some background theory to appreciate the subtle features by which plants can capture and utilize solar energy. Without going into the details of photosynthesis, it suffices to say that

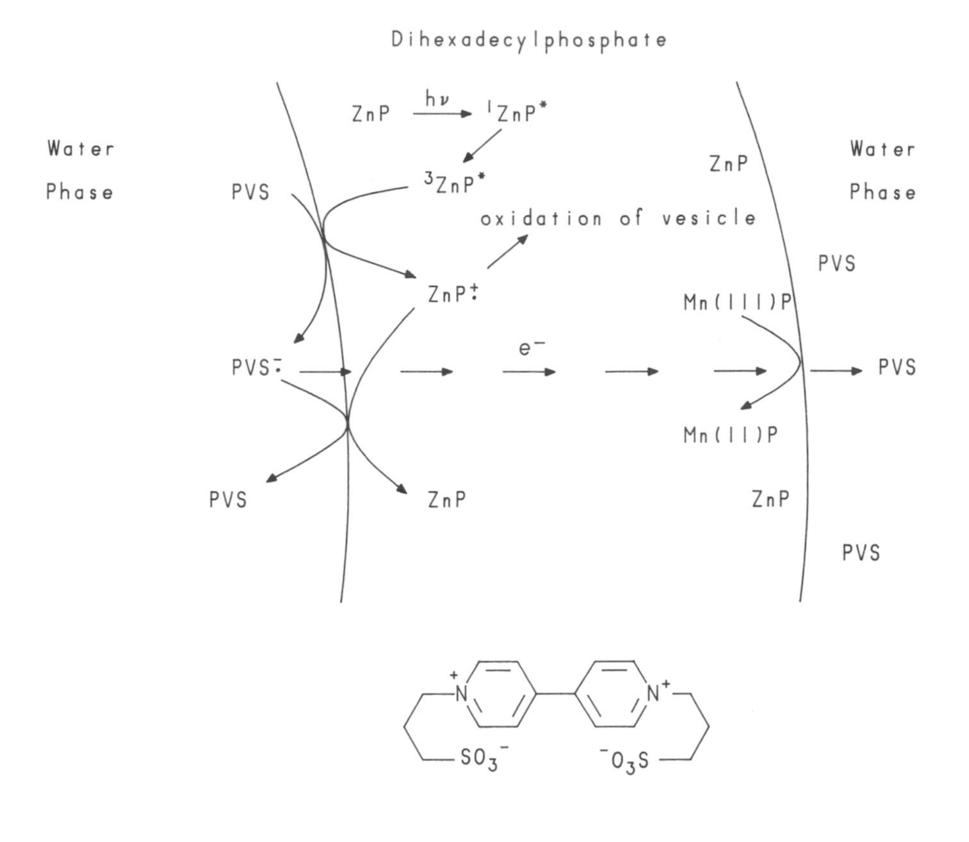

Figure 5.19 Another example of the partitioning of reactants in phospholipid vesicles. The reactants are arranged symmetrically within this vesicle. However, photoinduced electron transfer proceeds vectorially across the phospholipid membrane, similar to the transmembrane electron transfer in photosynthesis.

an electronically excited state of chlorophyll, which belongs to the porphyrin family of molecules, donates an electron to various electron acceptors separated from the chlorophyll by a hydrophobic barrier. Electrons are actually transported or "shuttled" across the barrier in a *vectorial* transport. Effective charge separation is ultimately achieved so that the subsequent processes necessary for metabolism and sustainance can take place. The examples of photoinduced electron transfer using organized assemblies such as lipid bilayers represent attempts by chemists to imitate the photochemical conversion processes used by green plants. These so-called *biomimetic* systems are actually supramolecular assemblies formed by the spontaneous aggregation of small molecular components. Supramolecular assemblies such as the lipid bilayers discussed in this section are relatively easy to produce in the laboratory and are able to mimic the fundamental features of the photosynthetic apparatus in nature.

5.2. Photoinduced Electron Transfer on Semiconductor Surfaces

In an electrically conductive material, the orbitals of the constituent atoms or molecules are normally closely packed together. The filled bonding orbitals are known as the *valence band* (VB), and the vacant antibonding orbitals comprise the *conduction band* (CB). These bands are continuous and separated by an energy spacing known as the *band gap* (Fig. 5.20). In good electrical conductors such as metals, there is a continuum of energy states. Electron mobility in metals is enhanced because the electrons can easily flow into partially vacant orbitals and flow through the material on application of an external voltage. In a poor conductor—an insulator—a wide energy gap between the VB and CB orbitals prevents electron flow.

Semiconductors are materials characterized by nonoverlapping bands. These materials are intermediate between conductors and insulators (Fig. 5.20). Semiconductors are nonconductive when the VB orbitals are populated and the CB orbitals are vacant. However, there may be a slight excess of electrons in the conduction band or slight excess of positive holes in the valence band of semiconductors. Mobile electrons and holes, both of which are designated as *charge carriers,* can be created by thermally or photochemically exciting the electrons in the

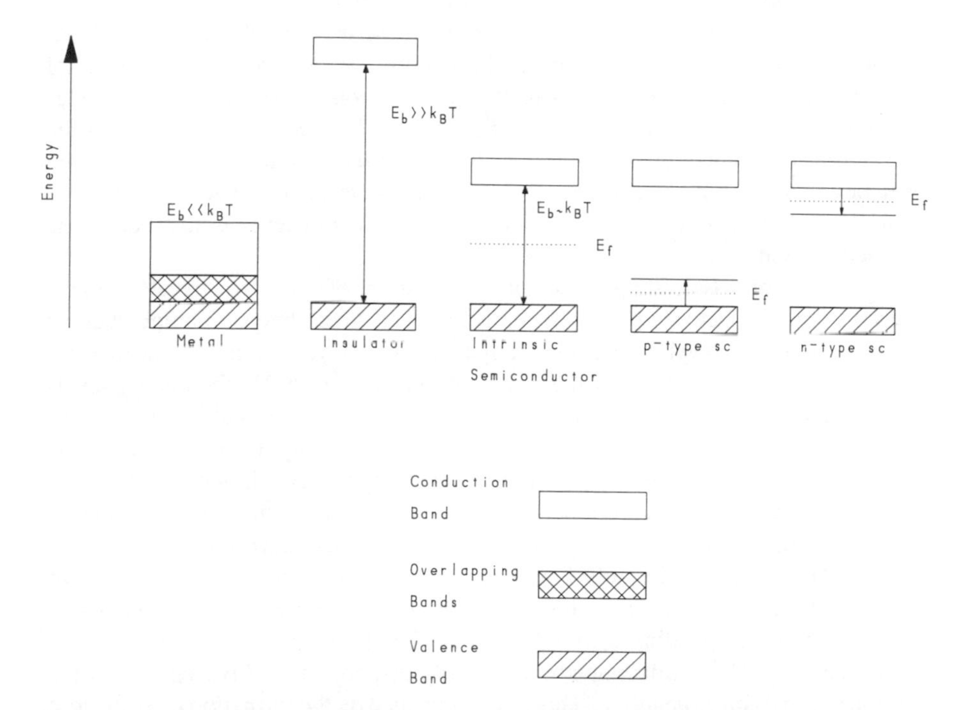

Figure 5.20 Conductivity is influenced by the energy gap between the conduction and valence bands of solid materials. *p*-type semiconductors are doped with electron acceptors to create positive charge carriers. Similarly, n-type semiconductors are doped with electron donors to form negative charge carriers. E_f is the energy of the Fermi level (see text).

valence band to orbitals in the conduction band. The thermal or photoexcitation energy must be sufficient to overcome the band gap energy. Mobile charge carriers can also be created by the introduction of doping material into the semiconductor. In doping, a few atoms in a semiconductor are replaced by atoms of an impurity. For example, if a semiconductor made of silicon or germanium—two Group IV elements—is doped with boron or aluminum, each atom of the doping material contributes one less electron to the lattice and create a hole in the valence band. This semiconductor is termed a *p*-type semiconductor since the material bears an excess of positive charge (positive doping). The holes in a *p*-type semiconductor are the major charge carriers, and the electrons in the conduction band are the minor charge carriers. An *n*-type semiconductor can be created when an electron donor impurity is added to the semiconductor (negative doping). Arsenic (Group V), for example, can be used to dope a silicon or germanium semiconductor since it introduces one electron into the conduction band. In the *n*-type semiconductor, the electrons (major charge carriers) in the conduction band outnumber the holes (minor charge carriers) in the valence band.

Now let us note an important distinction between a metal and semiconductor. In a metal, the orbitals are normally so closely packed that when an electron is promoted from a low-lying to higher-lying orbital, the electron almost instantaneously decays back to the ground state. In a semiconductor, the electron does not "relax" so rapidly since it must pass over a wider band gap.

The *Fermi level* is a term commonly used to describe the behavior of semiconductors. We define the Fermi level as the energy level where the probability of electron occupancy is exactly 1/2. In an undoped semiconductor (which is sometimes called an *intrinsic* semiconductor), the Fermi level lies exactly midway between the valence and conduction bands (Fig. 5.20). In a *p*-type semiconductor, the Fermi level lies slightly above the valence band, which indicates that vacant holes exist in the Fermi level; in an *n*-type semiconductor the Fermi level is just below the conduction band.

In a *p*-type or *n*-type semiconductor, the position of the Fermi level can shift when the semiconductor is placed in contact with an electrolyte containing charged species. The Fermi level of the bulk semiconductor moves in an attempt to equilibrate to the redox potential of the redox couple. In Fig. 5.21, the energy levels in the semiconductor and the solution are arranged on a vertical scale with the positive direction at the bottom: electrons flow spontaneously downhill (toward the positive direction) and holes travel spontaneously uphill. As shown in Fig. 5.21, when an *n*-type semiconductor is immersed in a solution containing a redox couple, negative charge carriers begin to flow from the semiconductor to acceptors in solution. The energy of the bulk Fermi level deep in the semiconductor moves to equilibrate to the potential of the redox couple. The positions of the band edges do not move, however, leading to *band bending*. The bands are said to be "pinned" at the interface. The semiconductor eventually acquires a positive charge as the solution becomes more negative. This is accompanied by the formation of an electric field distributed within the region of the semiconductor–electrolyte interface. This region is the space charge layer. Within this layer, for an *n*-type semiconductor a

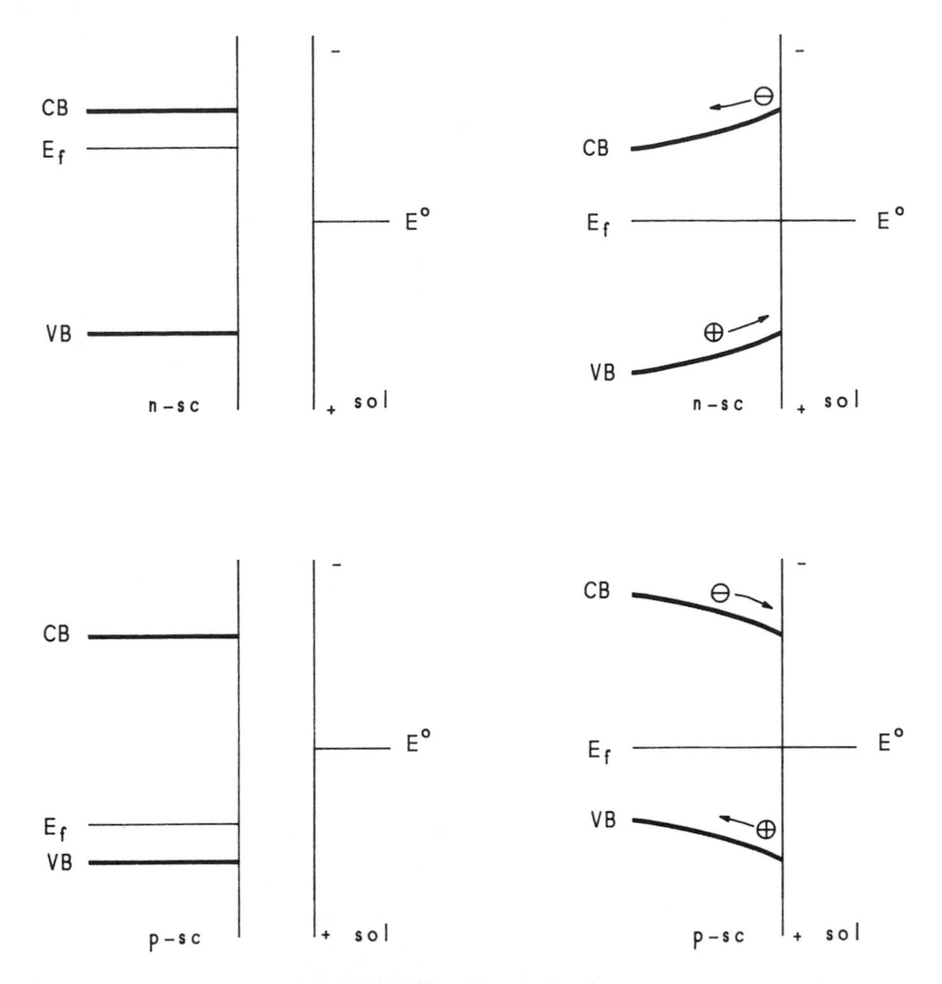

Figure 5.21 The semiconductors are separated from the electrolytic solution—their energies are at the flat band potentials shown on the left. When semiconductor and solution are brought within contact, the resulting electric field or space charge leads to band bending.

mobile electron is favored to move from the CB into the bulk of the semiconductor; a hole moves from the VB toward the interface. In a *p*-type semiconductor (Fig. 5.21), band bending occurs in an opposite direction.

Now let us examine the events accompanying illumination of a semiconductor immersed in the solution with the redox couple. Figure 5.22 shows the creation of an *electron-hole pair* when the semiconductor is illuminated with light. An important point is that once an electron-hole pair is formed by photoexcitation, band bending prevents electron return within electron-hole pairs since the electron and hole begin to move in opposite directions. Relaxation of the electron in the conduction band to the valence band is prevented by the energy gap between these states. In an illuminated *n*-type semiconductor, excess holes will move spontaneously from the

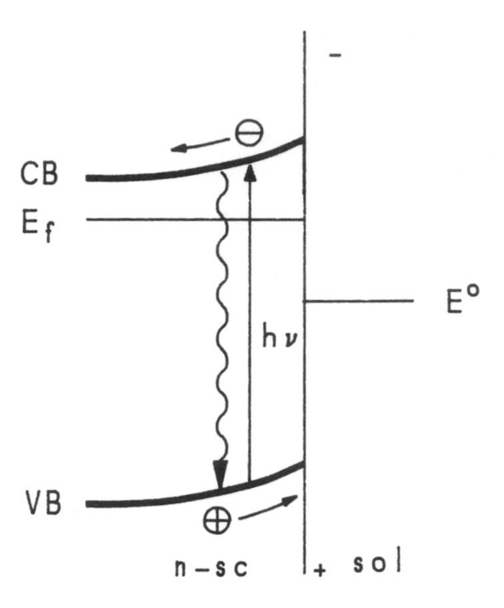

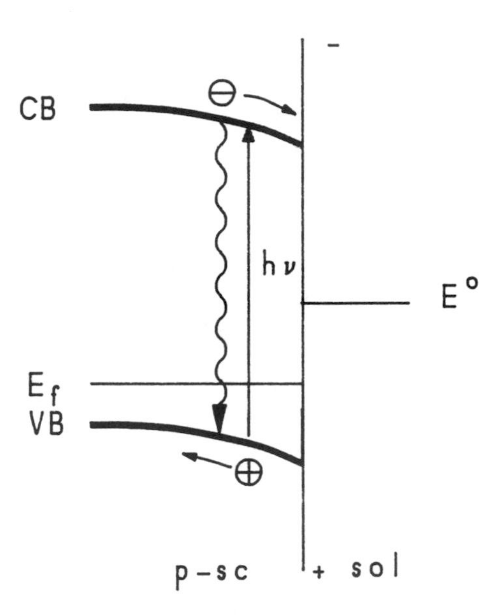

Figure 5.22 The electron-hole pairs which are created by photoexcitation are stabilized by the movement of the charge carriers in opposite directions.

VB upward to the solution to oxidize any donor with a redox potential lying above the VB. For this reason, a photoirradiated *n*-type semiconductor can induce oxidations. In an analogous manner, an illuminated *p*-type semiconductor can function as a reductant. Thus, if the redox potential of the acceptor is *positive* with respect to the conduction band edge, excess electrons in the conduction band can move spontaneously to electron acceptors in the electrolyte.

Figure 5.23 compares the energies of the band gap of several *n*- and *p*-type semiconductors with the redox potentials of selected half-reactions. The purpose of this figure is to show that the ability of a semiconductor to reduce or oxidize any substrate depends on the energy gap between the conduction and valence bands, as well as the actual relationship of these bands with the redox potentials of the half-reactions. If the substrate is to be reduced, its redox potential should be positioned below the CB of the semiconductor. Likewise, for oxidation of a substrate, its redox potential should lie above the VB. For example, band excitation of SiC and GaAs semiconductors produces excited states capable of carrying electrons and holes. Their CBs can serve as electron donors to Fe^{3+} and $Ru(bpy)_3{}^{3+}$. The excited states of ZnO and TiO_2 function primarily as electron acceptors since their VBs lie below the redox potentials of the solution substrates.

We shall now describe briefly several photoelectrochemical cells in which semiconductors can be employed as electrodes. Figures 5.24 and 5.25 illustrate three configurations. Each consists of a semiconductor electrode and in inert counter electrode (e.g., platinum) immersed in an electrolyte. In a *photovoltaic cell,* only one reducible or oxidizable species is present in solution (Fig. 5.24). When light excites the semiconductor to generate an electron-hole pair, an oxidation or reduction may take place on the surface of the electrode. Since the reverse reaction occurs on the inert electrode, an electric current between the electrodes is established. The overall chemical composition in the solution does not change.

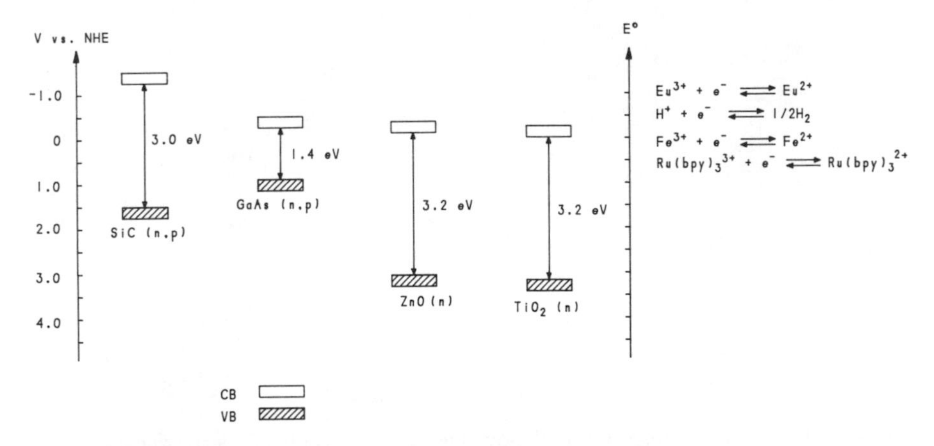

Figure 5.23 The relative ordering of the band energies of the photoexcited semiconductor and the redox potentials determines the direction of electron flow.

In a *photosynthetic cell* (Fig. 5.25), illumination of the semiconductor surface can drive an endothermic electron transfer. The absorbed light is ultimately stored as a chemical potential. In a *photocatalytic cell,* light is used to drive an exothermic but kinetically "slow" electron transfer. A photocatalytic cell thus offers a faster route. These designs require that the energy of the photon be capable of exciting the band gap. The energy of the photon is subsequently stored in the electron-hole pair. From the previous discussion, the electron in the conduction band and hole in the valence band may recombine (thermally or radiatively) in the same sense that electron return takes place between a radical cation and anion. Opposing this tendency for recombination, the space charge assists in preserving the electron-hole pairs.

In *dye sensitization,* electron transfer at the junction of semiconductors and electrolytes can be accomplished by irradiation of a dye adsorbed onto the semiconductor surface (Fig. 5.26). If, for example, a donor molecule is immobilized on the semiconductor and then photoexcited, it can inject an electron into the conduction band of an *n*-type semiconductor. Reduction of the electron acceptor takes place at the counter electrode.

In the previous cases, the semiconductors are employed as solid electrodes. However, semiconductors can also be used as suspended particles. If suspended nonuniformly in the medium, the particles can be microheterogeneously dispersed as colloids or microemulsions. Colloidal semiconductors, for instance, are distinguished by a relatively small size with typical diameters ranging from less than ~100 Å to greater than ~300 Å. The diameters of macroparticles can exceed ~1000 Å. Colloids usually produce clear solutions but macroparticles yield turbid solutions. Titanium dioxide suspensions are typical of the colloids used as particulate semiconductors.[28]

If the semiconductor particles are separated from the reactants, they are said to be heterogeneously dispersed (e.g., powders, solid supports, and surfaces).

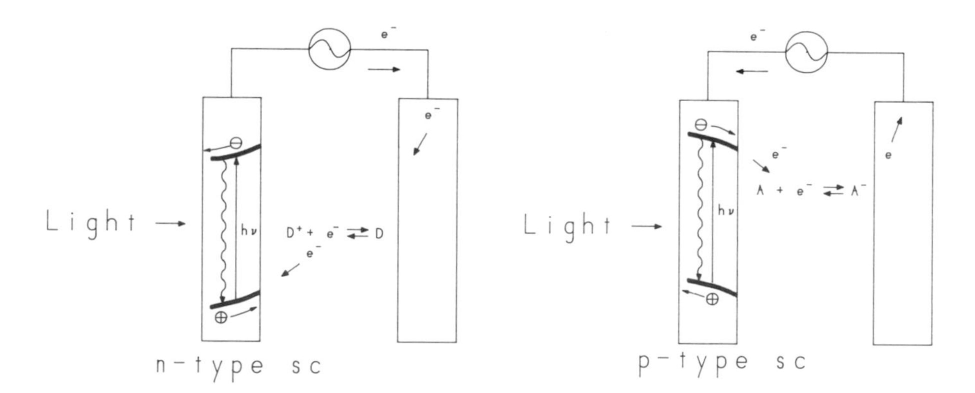

Figure 5.24 Two representations of photovoltaic cells consisting of a semiconductor electrode and inert platinum electrode. Illumination of the semiconductor results in an electron-hole pair which drives a redox reaction in solution. An electric current is generated in the cell.

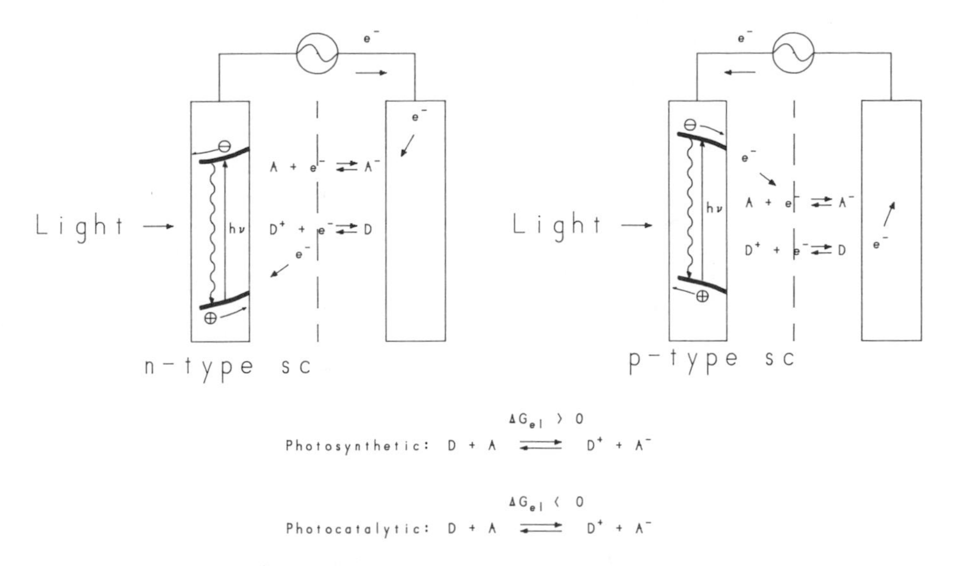

Figure 5.25 Representation of photosynthetic and photocatalytic cells. A membrane separates the products formed at each electrode.

Semiconductor particles offer some advantages over solid semiconductor electrodes. For example, when TiO_2 suspensions are illuminated in the presence of so-called "hole scavengers," each particle becomes electron rich. A TiO_2 particle of roughly 120 Å diameter can be populated by up to 300 electrons. In aqueous solution there is a double layer at the TiO_2–H_2O interface so that the particle functions like a capacitor. The charge of these particles when immersed in water can be sustained for several weeks! This is a dramatic improvement over larger and traditional semiconductor configurations.

To improve the adsorptive characteristics of the semiconductor electrode or particle, donor or acceptor sensitizers can be chemically attached to its surface. The coating can be done by surface painting or chemical derivatization.[29–32] These techniques enable capture of a wide portion of the visible spectrum.[33] Also, surface modification of the semiconductor may overcome some of the inefficiency due to diffusion of free sensitizer molecules. Certain bifunctional molecules can be attached to semiconductors having hydroxyl groups at the surfaces.[34,35] In Fig. 5.27, two examples of coated semiconductors are shown.[36] Both cases differ by the distance between the sensitizer and the semiconductor.

A sensitizer can be incorporated into a polymer, which is then coated as a thin film onto the surface of the semiconductor. "Metal-loaded" polymers such as nickel:tetrakis (dialkylphosphino)benzene have been described.[35] Electrodes can also be modified by electropolymerization.[37] This technique may involve the reduction or oxidation of an organic substrate on the electrode to initiate electropolymerization. Coating helps to protect the semiconductor from degeneration due to

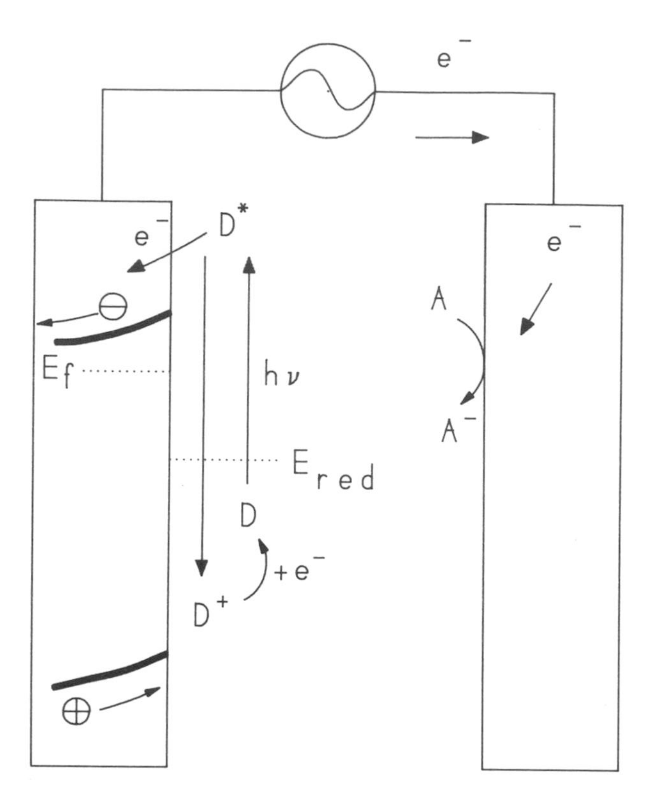

Figure 5.26 In dye-sensitization, a dye may be covalently linked onto the surface of the semiconductor. The dye captures visible light energy.

the oxidation of the material as holes migrate to the surface. Photocorrosion may seriously degrade the performance of a semiconductor.

Semiconductors can be used in connection with bilayer lipid membranes. Such semiconductive-lipid configurations may involve deposition of an inorganic semiconductive material onto the surface of a phospholipid or surfactant polymer. These films may be stable for days. Semiconductors can also be used in the form of polycrystalline thin films. These thin-film devices are particularly advantageous in that the charge carriers can migrate to opposite sides of the film, enhancing the efficiency of charge separation.

5.2.1. Photocatalysis on Semiconductor Surfaces

A growing number of studies are concerned with exploiting semiconductors to enhance the yield of certain photochemical reactions involving organic substrates. Photocatalysis on semiconductor surfaces has, for example, attracted considerable interest because a donor and acceptor can be adsorbed onto the surface of an oxidative and/or reductive electrode. Phototransformations of organic substrates on

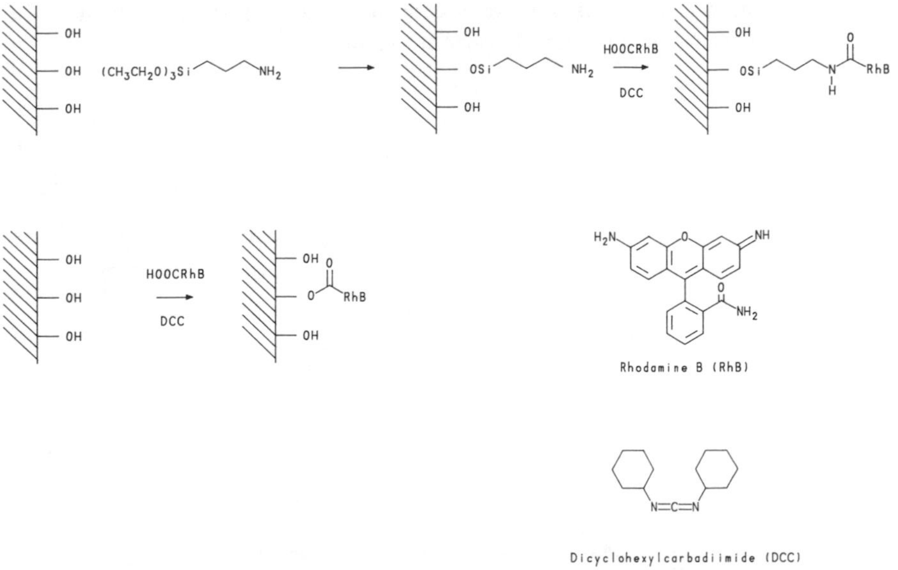

Figure 5.27 Various methods can be used to attach sensitizers to semiconductor surfaces. In this example, the dye, rhodamine B, is attached to the surface of a TiO_2 semiconductor. The two methods shown here lead to different sensitizer-semiconductor separation distances.

semiconductors include rearrangements, isomerizations, bond cleavages, oxidations, and reductions.

The principle of photoinduced electron transfer on semiconductor surfaces involving organic substrates has been described in the previous section. Briefly, the surface of the semiconductor is dipped into a solution containing the organic substrate and exposed to light from a conventional source or from a laser. The substrate must become complexed to the semiconductor surface and preferably become strongly bound before photoexcitation. Once photon absorption has occurred and an electron-hole pair is created, an electron can be captured by a hole at the surface of the illuminated semiconductor or an electron can be removed from the CB by the substrate, thus initiating oxidation or reduction of the substrate. This is followed by the production of radical intermediates and eventually by the formation of stable products.

Electron transfer at the interface of the semiconductor is often referred to as *interfacial electron transfer*. The requirements of photocatalysis on semiconductors (choice of semiconductor, effect of pH, surface effects, etc.) are critical factors in achieving interfacial electron transfer. Energetics (i.e., the relative ordering of the redox potentials of the donors and acceptors with respect to the bands of the semiconductor) also play a critical role. Mechanistic interpretations of photoinduced catalysis on semiconductor surfaces are often difficult to make since homogeneous reactions can also take place.

An interesting example of interfacial electron transfer is the photoinduced dimerization of phenyl vinyl ether (**1**) on ZnO surfaces[38]:

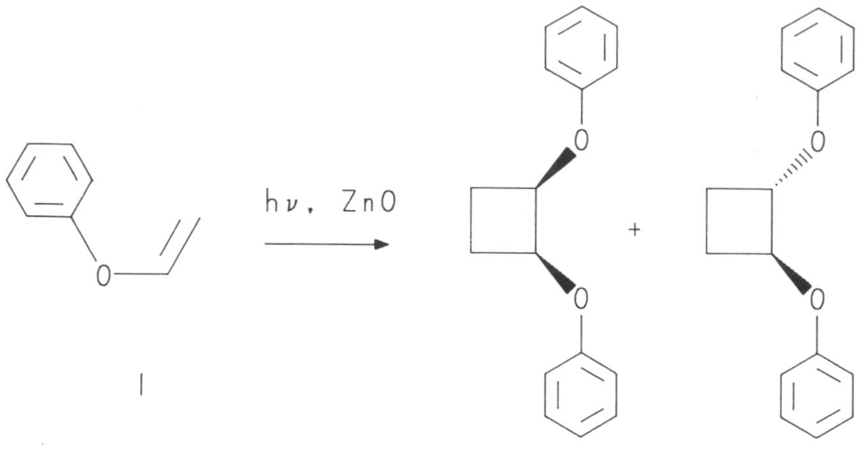

$$1 \xrightarrow{h\nu,\ ZnO} 2 + 3 \qquad (5.5)$$

In homogeneous solution, this dimerization can be photosensitized by 9,10-dicyanoanthracene (Section 3.2.3). On the semiconductor surface, photoinduced generation of an electron-hole pair is followed by oxidation of phenyl vinyl ether by a hole on the semiconductor surface. The redox potential of **1** is 1.25 V, which is above the band-edge position of VB of ZnO (Fig. 5.23) so oxidation of **1** is thermodynamically feasible. Oxidation is followed by a rearrangement of the phenyl vinyl ether radical cation:

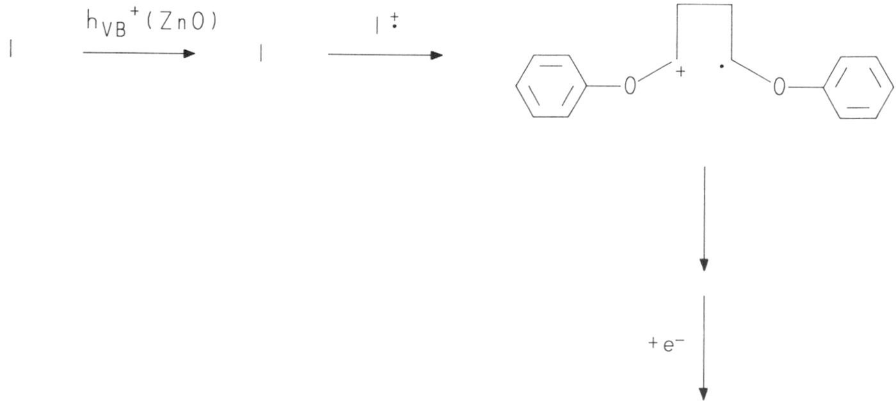

$$1 \xrightarrow{h_{VB}^{+}(ZnO)} 1 \xrightarrow{1^{\ddagger}} \cdots \xrightarrow{+e^-} 2 + 3 \qquad (5.6)$$

The subsequent pathways involve a complex series of surface and solution phase reactions. Although dimeric products can be formed in solution and on the surface of the semiconductor, the formation of dimer in the adsorbed state dominates. The *trans–cis* ratio is ~0.4 on the ZnO surface, which suggests that the surface plays a role in controlling stereochemistry.

The valence isomerization of quadricyclane (**4**) to norbornadiene (**5**) on a surface-illuminated semiconductor is shown below[39]:

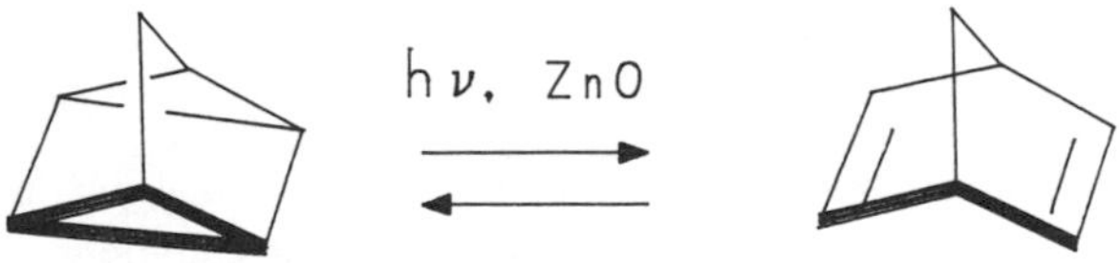

$$\mathbf{4} \underset{}{\overset{h\nu,\ ZnO}{\rightleftarrows}} \mathbf{5} \qquad (5.7)$$

The redox potentials of **4** and **5** are 0.91 and 1.48 V vs. SCE, so both of these substrates can be oxidized on illuminated ZnO (Fig. 5.23). Evidence for formation of the intermediacy of a radical cation was obtained from quenching experiments using 1,2,4,5-tetramethoxybenzene [$E^0(D^{+\cdot}/D) = 0.81$ V]. Quenching can compete by electron donation to the hole of the semiconductor.

In suspensions of semiconductor colloids, certain amines and carbonyl compounds were noted to undergo photocatalytic conversions. For example, in ZnS suspensions, acetaldehyde (**6**) can be photoreduced and photooxidized to ethanol (**7**) and acetic acid (**8**), **9**, and small quantities of hydrogen[40]:

$$\underset{6}{CH_3CHO} \xrightarrow[H_2O]{h\nu,\ ZnS} \underset{7}{CH_3CH_2OH} + \underset{8}{CH_3COOH} + \underset{9}{CH_3COCH(OH)CH_3} + H_2 \qquad (5.8)$$

The following scheme has been proposed:

$$ZnS \xrightarrow{h\nu} e_{CB}^-(ZnS) + h_{VB}^+(ZnS) \qquad (5.9)$$

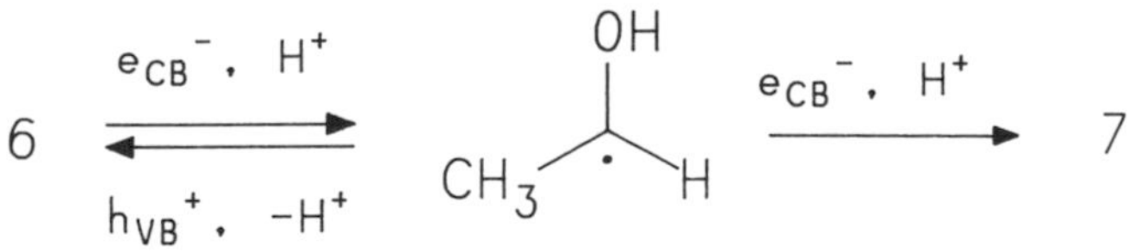

$$6 \underset{h_{VB}^+,\ -H^+}{\overset{e_{CB}^-,\ H^+}{\rightleftarrows}} CH_3\dot{C}(OH)H \xrightarrow{e_{CB}^-,\ H^+} 7 \qquad (5.10)$$

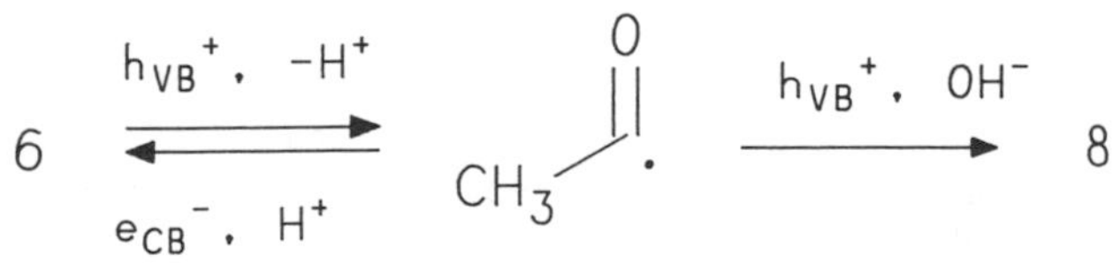

$$6 \underset{e_{CB}^-,\ H^+}{\overset{h_{VB}^+,\ -H^+}{\rightleftarrows}} CH_3\dot{C}{=}O \xrightarrow{h_{VB}^+,\ OH^-} 8 \qquad (5.11)$$

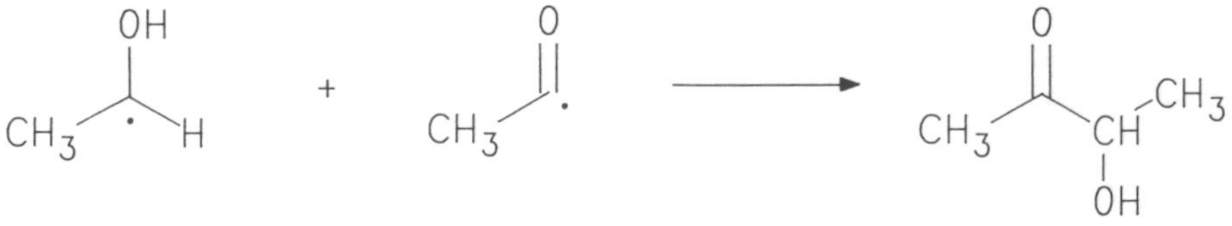

9 (5.12)

One of the more promising applications of photocatalysis on semiconductor surfaces is *oxidative degradation* of organic compounds. In photocatalytic oxidation, oxygen participates in the reaction resulting in the formation of oxidized products by reacting directly with the semiconductor or with radical species formed during the course of the reactions.

For instance, lactams can be photocatalytically oxidized to imides on oxygenated suspensions of TiO_2[41]:

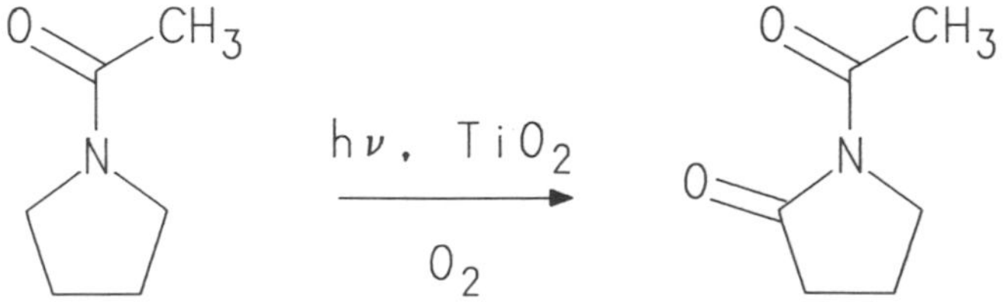

$h\nu$, TiO_2 / O_2

10 11 (5.13)

Amines such as *N*-methyl-4-phenylbutylamine (**12**) have been reported to be oxidized on platinized TiO_2[42]:

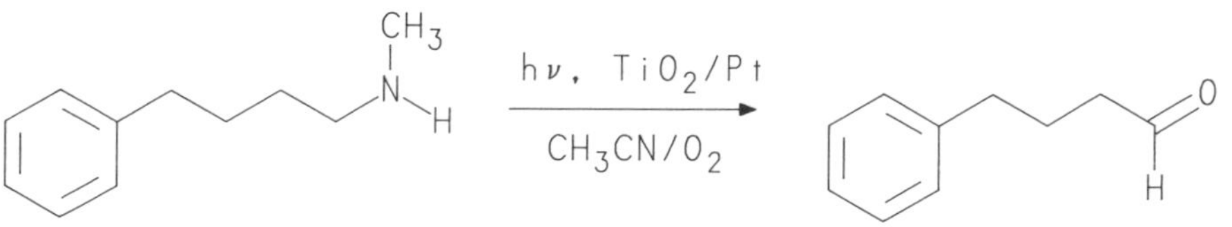

$h\nu$, TiO_2/Pt / CH_3CN/O_2

12 13 (5.14)

In reactions 5.13 and 5.14, irradiation of the band in TiO_2 excites an electron into the CB. The newly created hole lies between 3 and 3.5 V, so any surface-adsorbed molecule with a less positive redox potential will suffer oxidation. In 5.14, the oxidized amine converts rapidly to an α-amino radical, which can react with oxygen and eventually form the aldehyde (**13**):

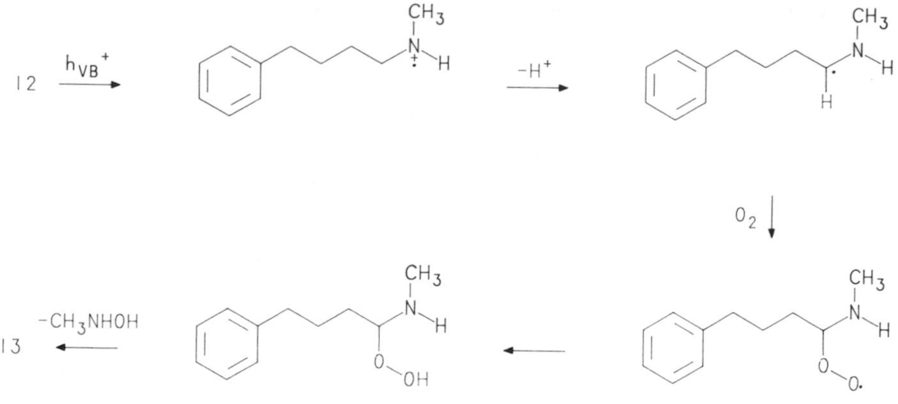

(5.15)

An alternative mechanism is also possible. Superoxide can be formed directly after generation of the electron-hole pair in TiO_2:

$$O_2 \xrightarrow{e_{CB}^-} O_2^{\cdot -} \tag{5.16}$$

Reduction of oxygen to superoxide by this route is thermodynamically feasible since the redox potential of oxygen, $E^0(A/A^-) = -0.33$ V vs. SCE, lies very near the energy of CB of TiO_2. The α-amino alkyl radical can lose an electron, followed by nucleophilic attack by superoxide:

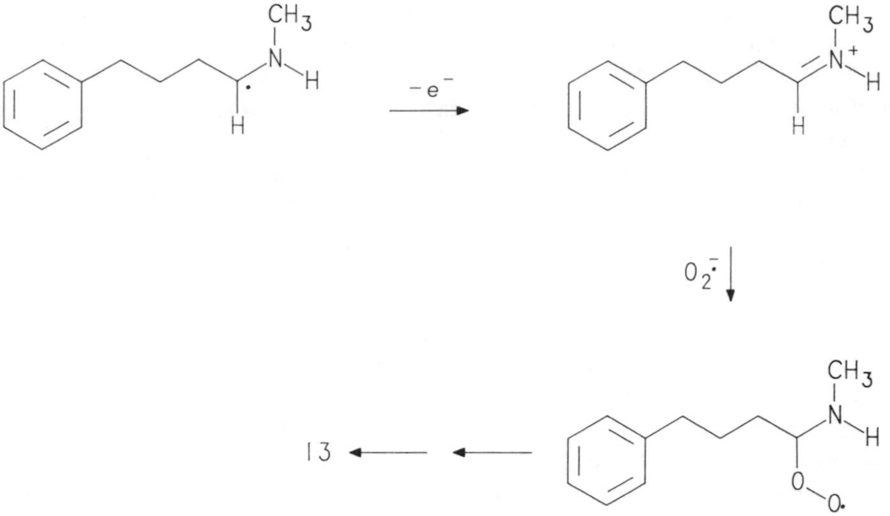

(5.17)

Olefins have also been reported to undergo oxidation, e.g.,[43]

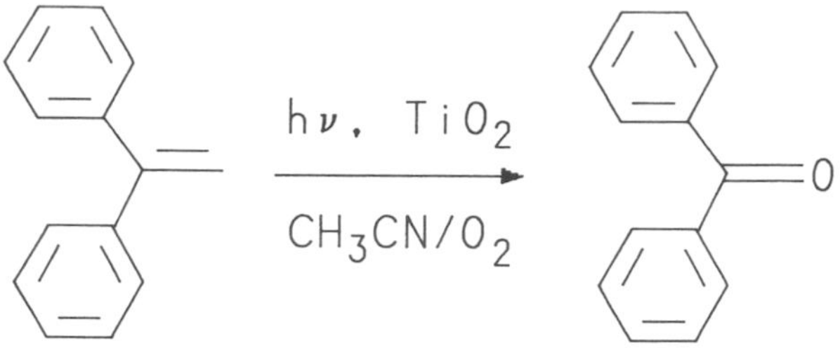

$$\mathbf{14} \xrightarrow[CH_3CN/O_2]{h\nu,\ TiO_2} \mathbf{15} \qquad (5.18)$$

Reaction 5.18 should be compared with the photosensitized dimerization of diphenylethylene in solution in the presence of electron acceptors (reaction 3.57).

5.3. The Photocleavage of Water

A reaction that has attracted considerable attention because of its potential application as an energy source is the cleavage of water into hydrogen and oxygen. The production of hydrogen in this manner holds promise as a fuel for energy purposes. In terms of storage capacity, hydrogen is about three times superior to oil and about four times better than coal. The combustion of hydrogen is "clean" and does not result in harmful emissions into the environment, in contrast to the burning of fossilized fuels. One serious drawback to hydrogen as a fuel is that mixtures of hydrogen and oxygen can recombine explosively. Nonetheless, much effort has been invested into seeking efficient and inexpensive ways to harness solar energy to decompose water into hydrogen.

The decomposition of water into hydrogen and oxygen requires the input of at least ~55 kcal mol^{-1} [28]:

$$H_2O \rightarrow H_2 + \tfrac{1}{2}O_2 \qquad (5.19)$$

Reaction 5.19 is endergonic. However, if energy, say in the form of light, can be added to cleave water into hydrogen and oxygen, and the hydrogen formed in this manner stored, the energy obtained is relatively inexpensive. The burning of hydrogen with oxygen can then be used as a source of electrical energy. This was done successfully on the first Apollo space mission to the moon.

In capturing sunlight to cleave water into hydrogen or oxygen or both, the photochemical system must contain a species capable of absorbing a substantial fraction of the visible spectrum. Many metal complexes have been used for this purpose because of their ability to absorb visible solar light. In Fig. 5.28, a high

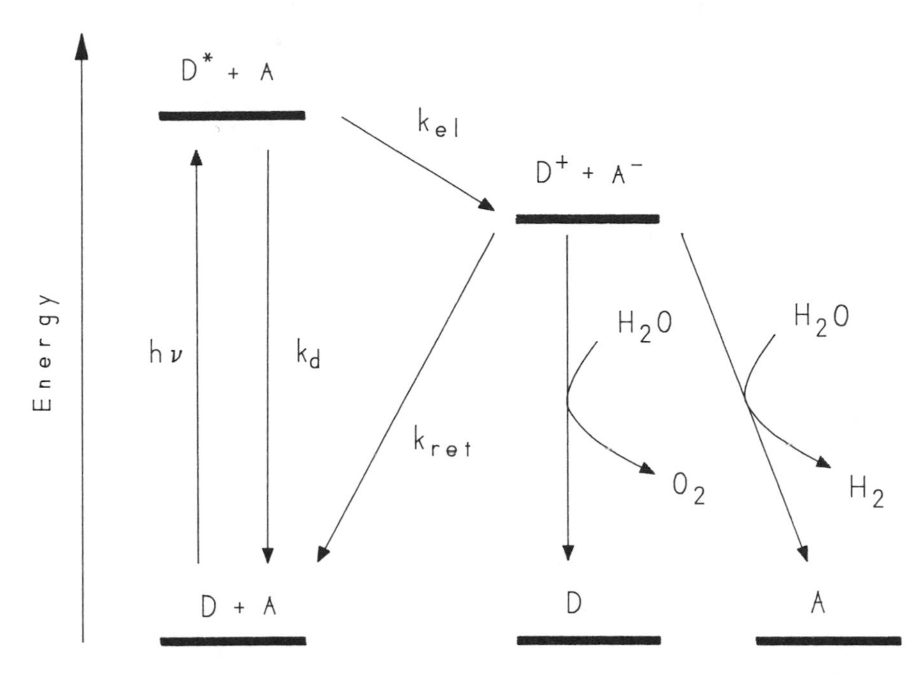

Figure 5.28 The cleavage of water can be integrated into a scheme involving photoinduced electron transfer.

energy ion pair is produced via electron transfer. The subsequent cleavage of water into hydrogen or oxygen competes with electron return.

The half-reactions for the splitting of water into hydrogen and oxygen are

$$2e^- + 2H_2O \rightarrow 2OH^- + H_2 \tag{5.20}$$

$$2H_2O \rightarrow 4H^+ + O_2 + 4e^- \tag{5.21}$$

Both of these processes are multielectron transfers: hydrogen production is a two-electron and oxygen production is a four-electron process. So both involve the generation of transient radical intermediates and accordingly proceed with high activation energy barriers (Fig. 5.29). The function of a catalyst is to stabilize these transients and also to deliver the required number of electrons. To satisfy this kinetic requirement, the catalyst must be integrated properly into the photochemical cell. Colloidal suspensions of platinum, rhodium-ethylenediamine, and heteropolyanion systems, among others, have been employed successfully for this purpose.[44,45]

To exploit solar energy for hydrogen production, the choice of reactants, medium, nature and size of catalysts, pH, and temperature demands scrupulous attention.[46–49] Although there have been many novel approaches to the problem of water cleavage into hydrogen and oxygen, these attempts have resulted in only modest success. The values in Table 5.1 demonstrate that the production of hydrogen is generally quite small. Still, these systems deserve examination, if only because

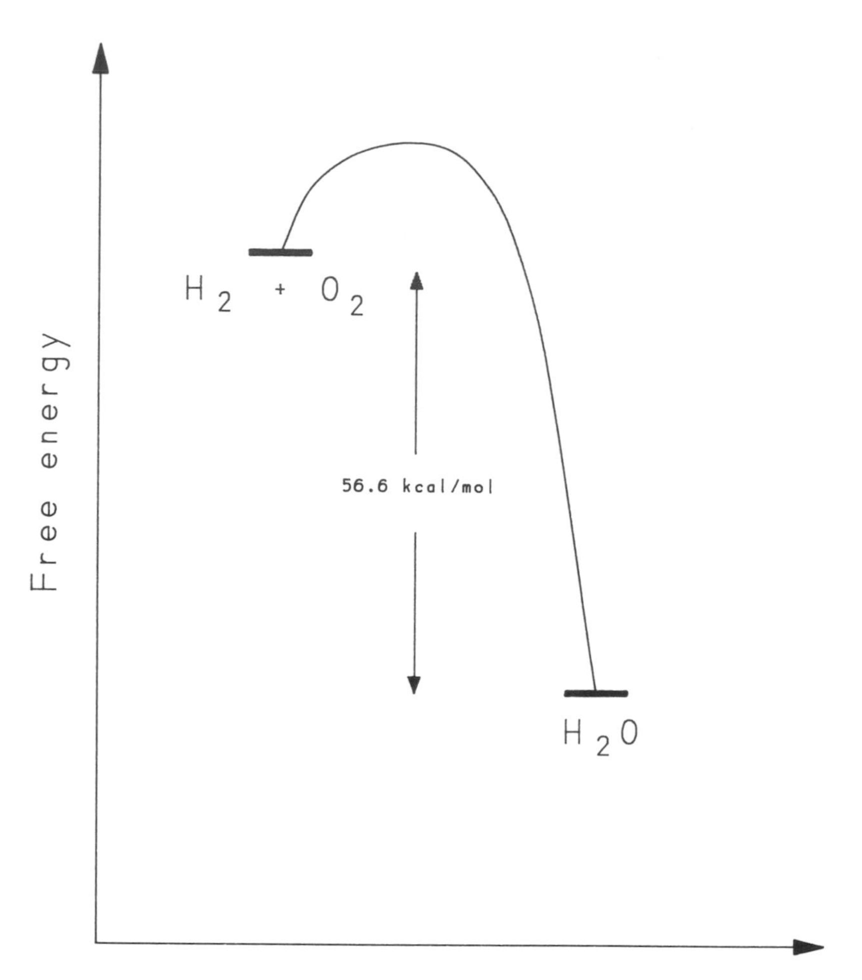

Figure 5.29 The cleavage of H_2O into hydrogen using an electron donor involves a large activation barrier. For this reason, a catalyst is normally employed in photochemical schemes to generate hydrogen from water.

they demonstrate interesting chemistry. One of the early attempts was the well-known reaction[4,50]:

$$2MV^{\dot{+}} + 2H_2O \xrightarrow{Pt} H_2 + 2OH^- + 2MV^{2+} \tag{5.22}$$

To exploit this reaction, $MV^{\dot{+}}$ must be continuously replenished. For example, $MV^{\dot{+}}$ can be generated by reduction of MV^{2+} by $[Ru(bpy)_3{}^{2+}]^*$:

$$[Ru(bpy)_3{}^{2+}]^* + MV^{2+} \rightarrow Ru(bpy)_3{}^{3+} + MV^{\dot{+}} \tag{5.23}$$

However, this reaction is quickly followed by electron return:

$$Ru(bpy)_3{}^{3+} + MV^{\dot{+}} \rightarrow Ru(bpy)_3{}^{2+} + MV^{2+} \tag{5.24}$$

Table 5.1 HYDROGEN YIELDS IN THE PHOTOCLEAVAGE OF WATER

Reaction System	ml H_2/hour[a]	Reference
$Ru(bpy)_3^{2+}/Rh^{3+}$/TEA/Pt	2	62
$Ru(bpy)_3^{2+}/MV^{2+}$/Pt	2.6	50
$Cr[4,7\text{-}(CH_3)_2phen]_3^{3+}$/EDTA/Pt	2.1	63
Anthracene sulfonate/MV^{2+}/TEA/Pt	2	64
$Ru(bpy)_3^{2+}/MV^{2+}/Pt/TiO_2/RuO_2$	1.5	52
SiO_2/PVS/Pt	0.5	56
$SiO_2/Ru(bpy)_3^{2+}$/PVS/Pt	0.15	56
CdS/Rh/DHP/PhSH	0.1	58,59

[a] The volumes of H_2 are only approximate and have not been normalized with respect to reactant concentrations, conditions, and incident light energy. See references for complete details.

But in the presence of an amine scavenger, such as EDTA, electron return can be blocked:

$$Ru(bpy)_3^{3+} + EDTA \rightarrow Ru(bpy)_3^{2+} + EDTA^{\dot{+}} \tag{5.25}$$

By preventing electron return with a scavenger, a sizable concentration of $MV^{\dot{+}}$ accumulates and becomes available for generating hydrogen in combination with a platinum catalyst. Small particles of Pt (~100 Å) result in the largest yields, reportedly about 12 liters a day per liter of solution.[50] A similar system employing excited polypyridine Cr^{3+} complexes has been tried.[51] But both of these strategies are nonregenerative in the sense that sufficient levels of the sacrificial donor, EDTA, cannot be renewed. Eventually, more quantitities of EDTA must be added to replenish the amount required to scavenge the oxidized sensitizer.

In the cyclic decomposition of water into hydrogen and oxygen, the strategy is to renew the initial reactants by electron return and to couple these reactions with water splitting on catalysts placed on the semiconductor surface. An early approach was based on a simple relay system illustrated in Fig. 5.30.[52] The reaction was carried out in aqueous solution containing $Ru(bpy)_3^{2+}$, an electron acceptor, and a TiO_2 semiconductor coated with platinum and RuO_2 particles. In the scheme shown in Fig. 5.30, photoexcitation of $Ru(bpy)_3^{2+}$ is immediately followed by reduction of an electron acceptor. $MV^{\dot{+}}$ subsequently injects an electron into the conduction band of TiO_2. The electron traverses the particle and becomes available to reduce water on the platinum surface. The oxidation of water takes place on the RuO_2 surface:

$$Ru(bpy)_3^{3+} + H_2O \xrightarrow{RuO_2} Ru(bpy)_3^{2+} + 2H^+ + \tfrac{1}{4}O_2 + e^- \tag{5.26}$$

In attempting to optimize the yield of hydrogen and oxygen from water, much attention has been devoted to modifying the semiconductor surfaces. Novel spinels and perovskite semiconductors as well as polymer-coated platinum particles were among the materials tested to enhance the efficiency of the photocleavage of water as well as to stabilize the semiconductor surface.[53] In an elegant study, Wrighton

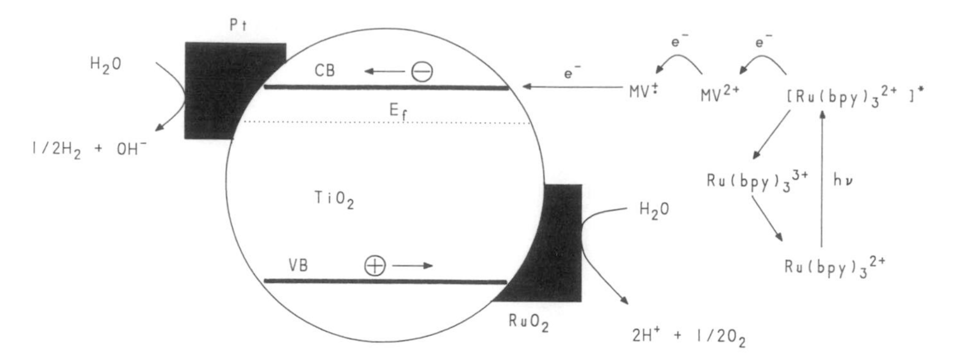

Figure 5.30 The photochemical cleavage of water to give hydrogen and oxygen using a semiconductor particle of TiO_2. A platinum coating catalyzes the generation of hydrogen. RuO_2 is used to catalyze the oxidation of water.

and co-workers described the photogeneration of hydrogen using *p*-type silicon.[54,55] This system consisted of a chemically modified silicone semiconductor coated with a silicon oxide layer and a modified silane polymer layer containing SiV^{2+}, in addition to small amounts of platinum metal clusters (Fig. 5.31). The modified layer not only protects the semiconductor from the effects of photocorrosion but also increases the lifetime of the conduction band electrons by reaction with SiV^{2+}. By incorporating platinum within the polymer layer, Wrighton was able to achieve hydrogen generation.

A novel strategy to generate hydrogen is based on electrostatic effects. This approach uses colloidal SiO_2 particles on which TiO_2 particles are immobilized.[56] The success of the procedure depends on the fact that microemulsions of SiO_2 are negatively charged. Thus, when a negatively charged electron relay is formed near the surface of the colloid, it is ejected electrostatistically into the bulk of the solution. Electron return is thereby suppressed. When photoexcited directly with ultraviolet light, TiO_2 reduces an electron relay (in this case, *N,N'*-bis(3-sulfonatopropyl)-4,4'-bipyridinium) (Fig. 5.32). The negative charge on the surfaces forces $PVS^{\overline{\cdot}}$ away from the semiconductor. $PVS^{\overline{\cdot}}$ is then free to react with water on the surface of the colloidal platinum to generate hydrogen. Alternatively, a sensitizer bearing a positive charge [such as $Ru(bpy)_3{}^{2+}$] can be bound to the surface of negative SiO_2 (Fig. 5.33). The oxidized sensitizer remains of the surface of the microemulsion to prevent electron return.

The microenvironment of the semiconductor particle is quite important in enhancing hydrogen evolution from water. For example, some studies have noted that β-cyclodextrin molecules provide favorable environments for dye sensitization of semiconductor surfaces.[57] In closely related work, semiconductor dispersions within lipid vesicles have been evaluated. This approach was pursued because of the difficulty of maintaining stabilized colloidal suspensions in solution. Colloidal CdS was coated with rhodium and dispersed within dihexadecylphosphate (DHP) surfactant vesicles (Fig. 5.34).[58,59] Efficient hydrogen production was accomplished on illumination of the semiconductor in the presence of various electron acceptors.

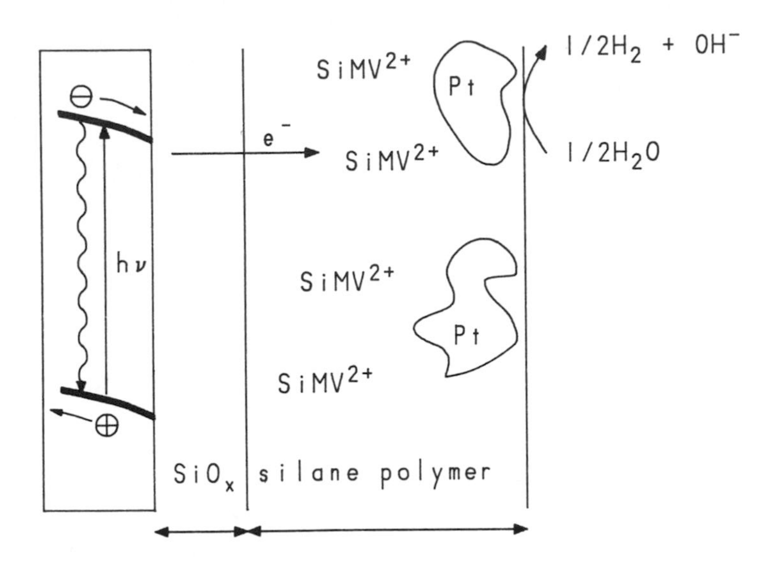

$(CH_3O)_3Si$ $Si(OCH_3)_3$

$SiMV^{2+}$

Figure 5.31 Photoinduced electron transfer on a silanized semiconductor.

CdS particles mixed with RuO_2-mixed TiO_2 have also proven to be excellent substrates for the photoproduction of hydrogen

$$CdS \xrightarrow{h\nu} e_{CB}\text{-}(CdS) + h(CdS) \tag{5.27}$$

$$e_{CB}\text{-}(CdS) + h(CdS) \rightarrow CdS \tag{5.28}$$

$$e_{CB}\text{-}(CdS) + TiO_2/RuO_2 \rightarrow CdS + e_{CB}\text{-}(TiO_2/RuO_2) \tag{5.29}$$

$$e_{CB}\text{-}(TiO_2/RuO_2) + H_2O \rightarrow \tfrac{1}{2}H_2 + OH^- \tag{5.30}$$

In one interesting attempt to exploit semiconductors, Tien demonstrated the generation of hydrogen from artificial seawater in a photovoltaic cell.[60,61] In this

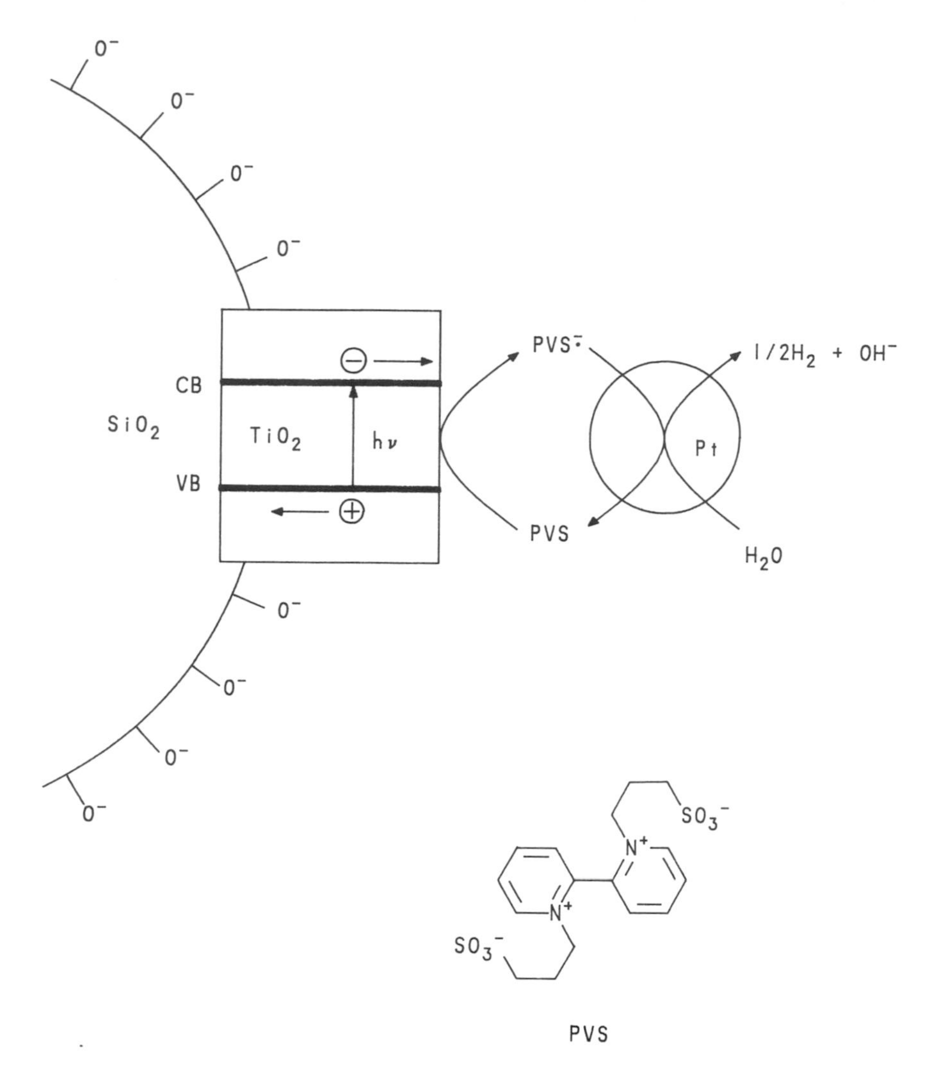

Figure 5.32 Direct photoexcitation of a semiconductor located in the outer region of a colloidal particle. Electron return is prevented by electrostatic effects between the charged surface and the charged products.

design, semiconductive polycrystalline CdSe was painted on an Ni foil. The coated foil was glued between two transparent chambers and then incorporated into an electrochemical cell as shown in Fig. 5.35. When the semiconductor is irradiated with visible light, electron-hole pairs are generated, and electrons travel through the conduction bond to the metal to reduce water at the surface with the subsequent release of hydrogen. The holes in the valence band oxidize Fe^{2+} to Fe^{3+}. The Fe^{2+} is regenerated by reduction on the surface of the left electrode, whereas the oxidation of Pb to Pb^{2+} on the right electrode replenishes the electron through the

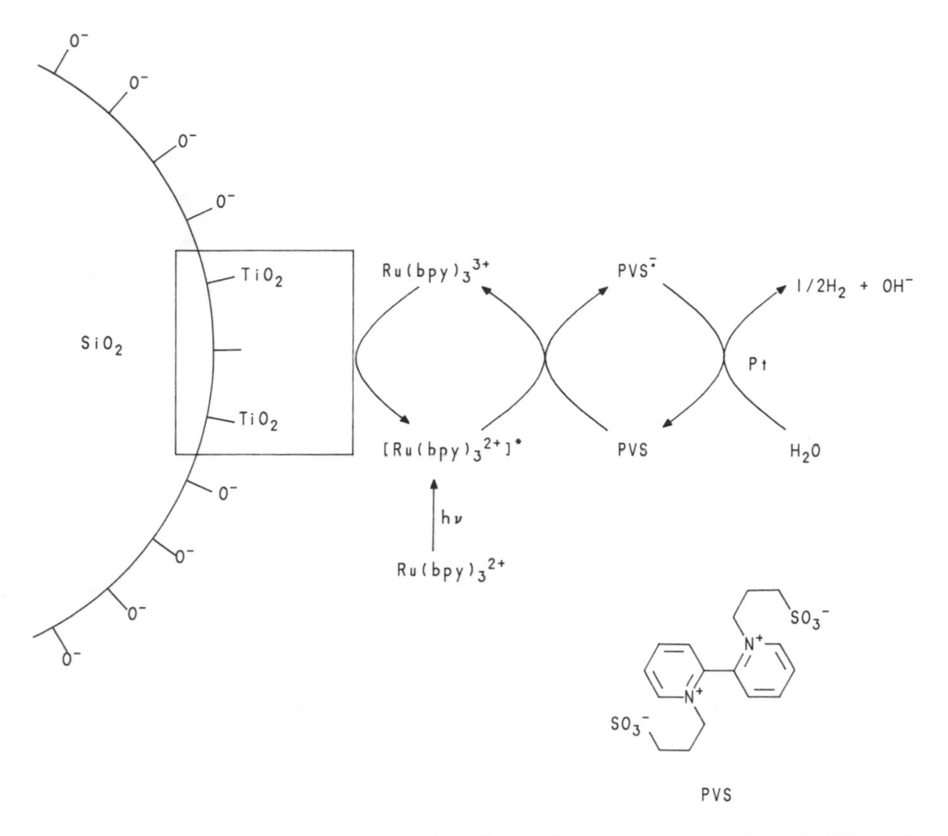

Figure 5.33 In this example, a dye-sensitized reduction occurs on the surface of a SiO_2/TiO_2 particle.

external circuit. Thus, this configuration represents an ingenious attempt to convert solar energy into electricity and to generate hydrogen simultaneously from seawater!

It remains to be determined if any of these processes can be made sufficiently efficient to warrant large-scale production. In most cases, the yields of hydrogen (or oxygen) gas are quite modest even after extensive "fine-tuning." Moreover, these techniques involve the use of expensive semiconductor materials and catalysts. These problems must be resolved before hydrogen- or oxygen-generating systems can be used as a cheap energy source.

5.4. Photoinduced Electron Transfer in Imaging Applications

There is almost no one who does not make daily use of the remarkable technology called *photoimaging*. Photoimaging includes important applications such as

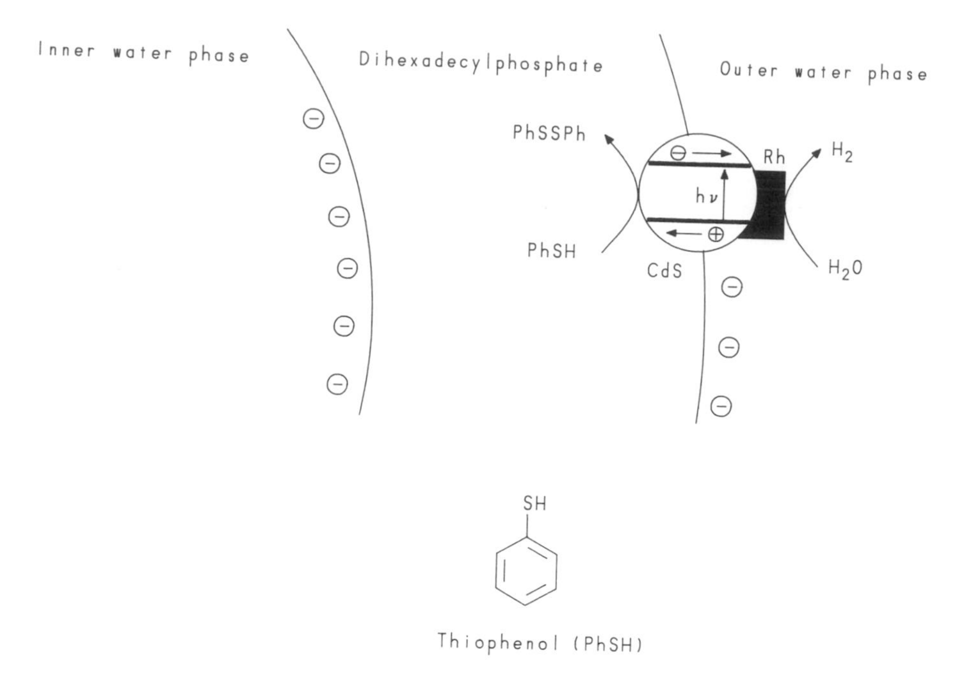

Figure 5.34 Photoinduced electron transfer can take place on the surfaces of semiconductor particles dispersed in lipid vesicles. H_2 generation is catalyzed by rhodium catalyst impregnated onto the semiconductor surface.

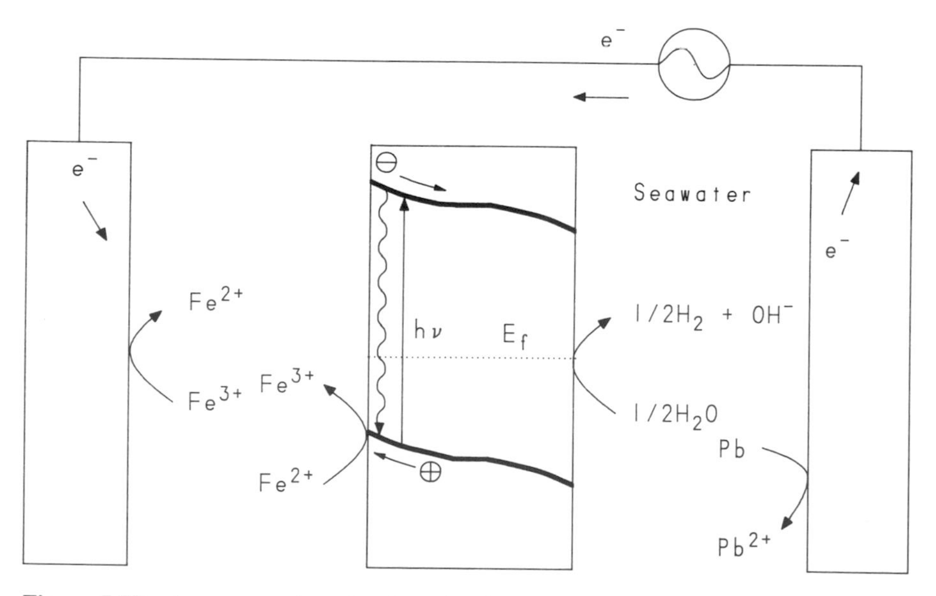

Figure 5.35 A representation of a photochemical cell used to generate hydrogen gas from seawater. (Adapted with permission from Tien, H. T. and Chen, J.-W. *Photochem. Photobiol.* **1989,** *49*, 527. Copyright 1989 Pergamon Press.)

photography, office copying, compact-disc technology, the manufacturing of printed circuit boards and integrated circuits, and the use of electronic media for storing and outputting images and information. These applications have become rooted in our culture and society. It is desirable, therefore, to know something about the way photoimaging works and how the materials used in photoimaging and related technologies are processed and used. However, this topic would cover several volumes. Our objective here will be more modest. We will simply attempt to show that photoinduced electron transfer lies at the heart of "conventional" and "not-so-conventional" photoimaging technologies. For detailed information of all of the ramifications of photoimaging, the reader will be encouraged to consult one of the excellent reviews listed at the end of the chapter. This section draws from these papers.

5.4.1. What Is Photoimaging?

Photoimaging is the art of capturing, rendering, and presenting information using light. It is often said that photoimaging consists of three stages: capture, rendition, and readout. The most important requirement is that a way must be found to capture the photons of the image. This is the *image capture step,* which must involve the use of a light-sensitive medium. Following absorption of light by the medium, thermal reactions take place. These processes constitute rendition. The region of the medium exposed to light must ultimately be clearly resolved from unexposed portions for the technique to be of any value. This stage is usually accomplished in the development step to provide visual contrast between the image and the background. Changes in optical properties (e.g., opacity, scattering, refractive index) provide the means for discerning the original image.

To give the reader a taste of how photoinduced electron transfer is connected with photoimaging, we shall give examples involving conventional photographic processes.

5.4.2. Conventional Silver Halide Imaging

In conventional black and white or silver halide photography, an opaque silver image is developed on unexposed photographic paper. In practice, the silver halide materials used in photography consist of microcrystalline grains embedded in a gelatin, which in turn is layered onto a solid support such as a glass plate or paper. Silver halide crystals are actually wideband gap semiconductors (Fig. 5.36). These crystals are photoconductive in the sense that electrons can be promoted from the

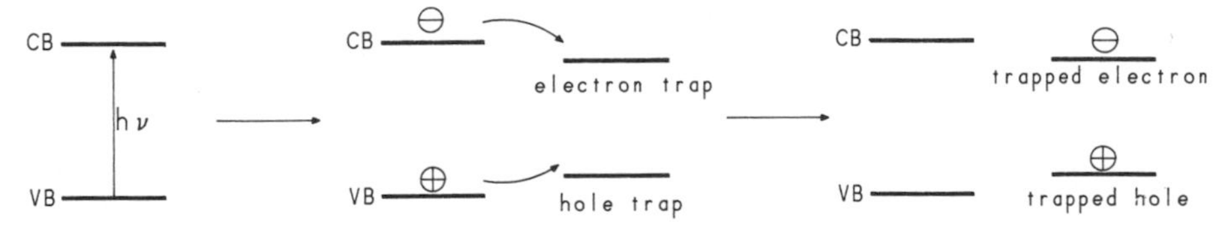

Figure 5.36 Photoexcitation of the band gap in silver halide crystals leads to formation of trapped electron-hole pairs.

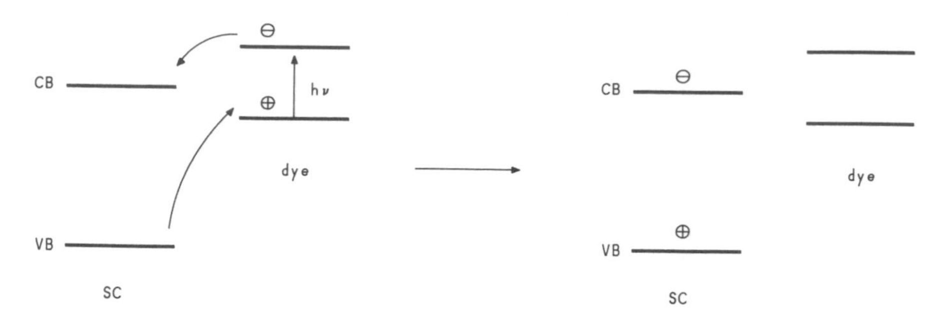

Figure 5.37 In dye sensitization, a sensitizer attached to the semiconductor surface is photoexcited.

VB into the CB on light irradiation. Thus, when the silver halide crystal is exposed to light, an electron-hole in the semiconductor is created. The electron-hole eventually suffers recombination or may become trapped, say, at a defect site in the crystal. Development involves the supply of electrons to the trapped hole to transform the exposed region into metallic silver. The tiny specks of silver actually grow at the same location as the original atom exposed to light. The collective growth of silver atoms results visually in the dark area.

Physical imperfections in silver halide crystals play a crucial role in the photographic process since electrons and holes are trapped at these sites.

One of the features of silver halide photography concerns the amplification of the image following the initial exposure to light. How does this happen? Typically, a silver grain has about 10^9 silver ions. Since each microcrystal absorbs only a few photons and the entire grain is reduced to metallic silver during development, the "signal" is "amplified" by over a million!

In dye sensitization, a dye is attached to the silver halide semiconductor. The dye is chosen for its ability to absorb at longer wavelengths. This particular feature of the dye enables increased sensitivity to wavelengths that are not absorbed by silver halide. Figure 5.37 shows the process by which dye sensitization works. The primary step is absorption of a photon by the dye. This is immediately followed by the transfer of an electron from the LU molecular orbital of the dye into the CB of the silver halide. Note that the energy levels must be favorably disposed for this step (i.e., the LU molecular orbital of the dye must lie at a higher level than the CB). An electron in the VB of the silver halide is eventually removed by the hole in the dye. Dye sensitization creates an electron-hole pair in the silver halide in much the same way as direct excitation.

5.4.3. Xerography

In *xerography* or *electrophotography*, an electric field is used to create an image. Figure 5.38 shows an overview of what is involved in this technology. A *charge-transport* and a *charge-generation* medium are applied onto a grounded substrate material. An electric field is applied with a corona discharge (Fig. 5.38). The corona

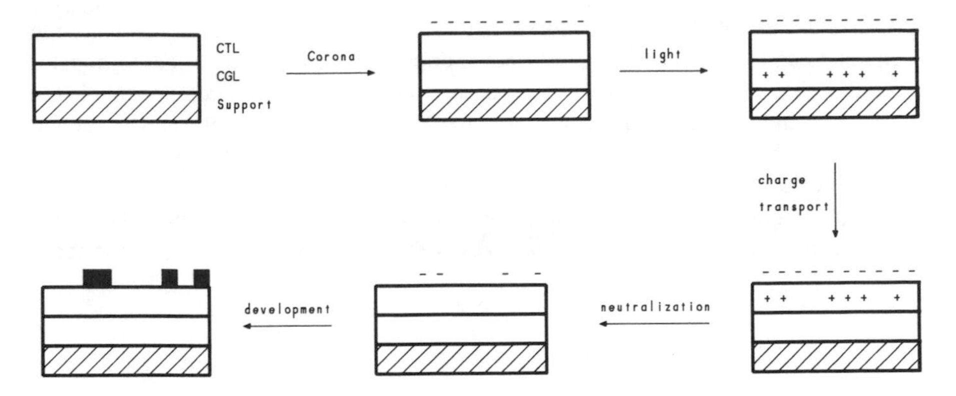

Figure 5.38 Processes taking place in electrophotography. (Adapted with permission from Eaton, D. F. *Top. Curr. Chem.* **1990,** *156,* 199. Copyright 1990 Springer-Verlag.)

is supplied by small wires connected to a power supply. In air, the corona can consist of an "umbrella" of positive or negative ions generated at the tip of a wire.

After the application of negative corona, for example, the medium is exposed to light and electron-hole pairs are subsequently formed within the charge-generation layer. The charge generation may consist of an inorganic semiconducting or organic conductive material. Once the electron-hole pair is formed, the hole migrates into the charge-transport layer toward the charged surface where a portion of the exposed surface is neutralized. A charged pigment is applied to the surface to form an image. This step is referred to as *toning*. The pigment is finally transferred to paper, which is subsequently heated to melt the pigment to form the permanent image.

A variety of charge-generation and charge-transport materials have been used. Zinc oxide is one example. This material also serves as the charge-transport material so that this system is essentially one-layer. The electric field created by a negative corona separates the electron-hole pairs in the zinc oxide, which are formed by illumination with ultraviolet light. The migrating holes eventually neutralize the negative ions on the surface.

An interesting class of organic materials used in xerography is *organic photoconductors*. One of the early examples was a donor–acceptor pair consisting of polyvinylcarbazole (**16**) and 2,4,7-trinitrofluorenone (**17**) shown below:

N

16

NO_2

O_2N NO_2

O

17

As in the case of zinc oxide, these materials perform the dual role of charge generators and transporters. The polymer is doped with **16** forming a charge-transfer complex that absorbs at longer wavelengths than each separate component. Light absorption induces electron transfer to generate a radical cation of **16.** The radical cation is mobile in the polymer chain and migrates to the surface. Application of a positive corona can cause the radical anion of **17** to migrate.

Two-layer systems using phthalocyanines and *N,N*-dimethylperylene-3,4,9-tetracarboxylic diimide (**18**) or squarylium dyes (**19**), for example, have been described:

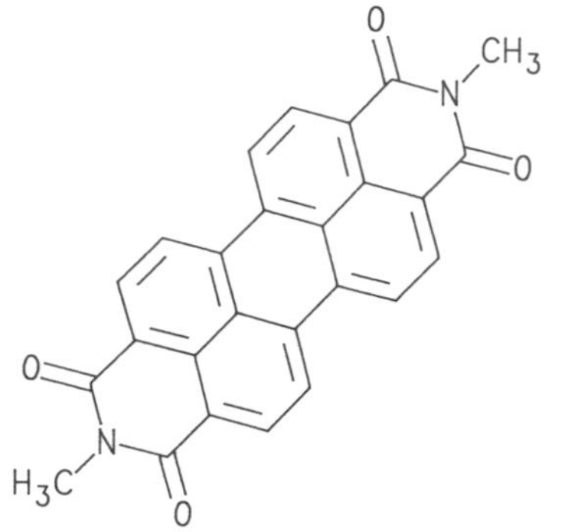

18 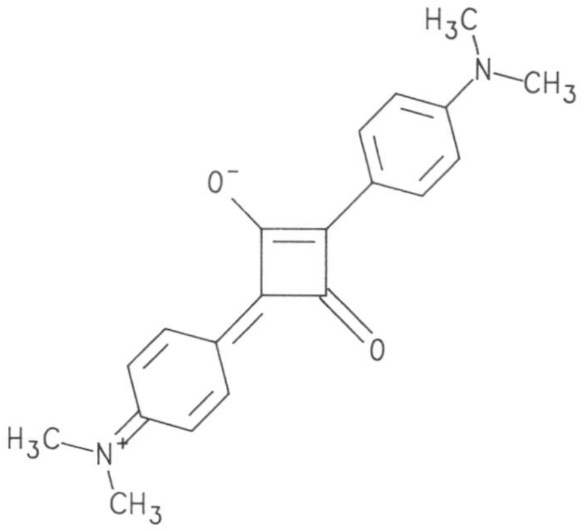 19

These materials are used for the charge-generation layer. As highly pigmented materials, they are characterized by high light absorption.

For charge-transport properties, the material must possess good oxidation/reduction properties. For example, nitroaromatics and aromatic ketones such as fluorenone are good electron-transport materials. These materials can be used in conjunction with a positive corona. Materials capable of transporting positive ions include aromatic amines or compounds like hydrazones (**20**):

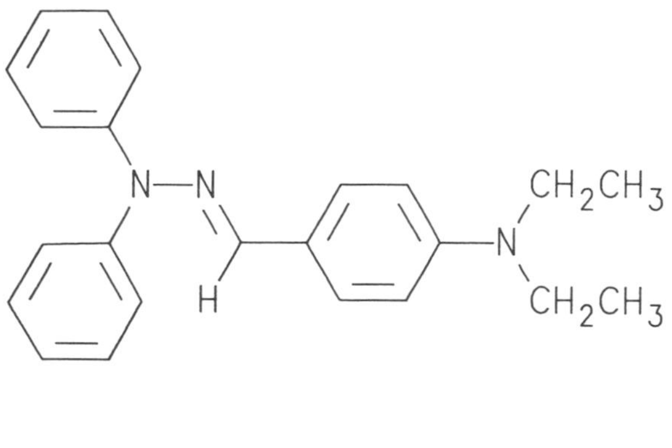

20

Photoimaging technology will continue to grow as long as there is a pressing demand for improved and inexpensive technology. Two significant applications making use of photoimaging are computerized axial tomography (CAT) and magnetic resonance imaging (MRI). Both of these procedures are used in diagnostic medicine to create digitized views of body sections. The information within a CAT or MRI image may consist of several megabytes, which gives some indication of the tremendous amount of detail that can be processed and used to make diagnoses.

Suggestions for Further Reading

Photoinduced Electron Transfer in Organized Assemblies

Meyer, T. J. *Acc. Chem. Res.* **1989,** *22,* 163. Artificial photosynthetic systems are reviewed

Photoinduced Electron Transfer in Micelles and Microemulsions

Kalyanasundaram, K. *Photochemistry in Microheterogeneous Systems.* Academic Press, New York, 1987, Chapter 1. An excellent treatise on photochemistry taking place within a variety of solid-state and microheterogeneous environments.

Thomas, J. K. *Acc. Chem. Res.* **1977,** *10,* 133. A review on micellar structures and how they may be used to study homogeneous reactions.

Magnetic Effects in Micellar Cages

Steiner, U. E., and Ulrich, T. *Chem. Rev.* **1989,** *89,* 51. A comprehensive review on magnetic effects in photochemistry including photoinduced electron transfer in organized environments.

Photoinduced Electron Transfer Across Langmuir–Blodgett Films

Richardson, T. *Chem. Br.* **1989,** 1218. A brief survey of L-B films.

Photoinduced Electron Transfer in Lipid Bilayers

Fendler, J. H. *Science* **1984,** *223,* 886.

Fuhrhop, J.-H., and Mathieu, J. *Angew. Chem. Int. Ed. Engl.* **1984,** *23,* 100. A survey of synthetic vesicles and their role in photoinduced electron transfer.

Photoinduced Electron Transfer on Semiconductor Surfaces

Bard A. J. *Science* **1980,** *207,* 139.

Wrighton, M. S. *Acc. Chem Res.* **1979,** *12,* 303.

Bard, A. J., and Faulkner, L. R. *Electrochemical Methods: Fundamentals and Applications*. Wiley, New York, 1980, p. 636. Photoeffects on semiconductors.

Baral, S., and Fendler, J. H. In *Photoinduced Electron Transfer. Part B. Experimental Techniques and Medium Effects,* Fox, M. A., and

Chanon, M., eds. Elsevier, Amsterdam, 1988, Chapter 3.2. Photoinduced electron transfer in vesicles and bilayer lipid membranes.

Fendler, J. H., and Baral, S. In *Photochemical Energy Conversion,* Norris, J. R., Jr., and Meisel, D., eds. Elsevier, New York, 1989, p. 148.

Fox, M. A. In *Photocatalysis and Environment,* Schiavello, M., ed. Kluwer Academic Publishers, New York, 1988, p. 445. A review paper on the photooxidation of organic compounds on semiconductor surfaces.

Gerischer, H. In *Solar Energy Conversion. Solid-State Physics Aspects,* Seraphin, B. O., ed. Springer-Verlag, Berlin, 1979, p. 115. A review on solar energy conversion using semiconductors.

Henglein, A., and Weller, H. In *Photochemical Energy Conversion,* Norris, J. R., Jr., and Meisel, D., eds. Elsevier, New York, 1989, p. 161.

Holden, A. *The Nature of Solids*. Columbia University Press, New York, 1965, Chapter 14.

Kalyanasundaram, K. In *Energy Resources through Photochemistry and Catalysis,* Grätzel, M., ed. Academic Press, New York, 1983, Chapter 7. A review of heterogeneous photocatalysis.

Pichat, P., and Fox, M. A. In *Photoinduced Electron Transfer. Part D. Photoinduced Electron Transfer Reactions: Inorganic Substrates and Applications,* Fox, M. A., and Chanon, M., eds. Elsevier, Amsterdam, 1988, Chapter 6.1. A review of photocatalysis.

Sakata, T., and Kawai, T. In *Energy Resources through Photochemistry and Catalysis,* Grätzel, M., ed. Academic Press, New York, 1983, Chapter 10. A review of photocatalysis using semiconductor powders.

Sariciftci, N. S., Smilowitz, L., Heeger, A. J., and Wudl, F. *Science* **1992,** *258,* 1474. A report on photoinduced electron transfer from a semiconducting polymer to buckminsterfullerene (C_{60}).

Serpone, N., Pichat, P., Herrmann, J.-M., and Pelizzetti, E. In *Supramolecular Photochemistry,* Balzani, V., ed. Reidel, Dordrecht, 1987, p. 415. Discusses a novel approach based on coupled semiconductor particles.

Tuller, H. In *Ceramic Materials for Electronics: Processing, Properties, and Applications,* Buchanan, R. C., ed. Marcel Dekker, New York, 1991, p. 379. A discussion of conduction in oxide materials. Good background information on doping and defects.

The Photocleavage of Water

Bolton, J. R. *Science* **1978,** *202,* 705. Discusses important aspects of the capture of solar energy.

Gray, H. B., and Maverick, A. W. *Science* **1981,** *214,* 1201. The role of metal complexes in solar energy capture and conversion is explored.

Greenbaum, E. In *Photochemical Energy Conversion,* Norris, J. R., Jr., and Meisel, D., eds. Elsevier. New York, 1989, p. 184.

Heller, A. In *Energy Resources through Photochemistry and Catalysis,* Grätzel, M., ed. Academic Press, New York, 1983, Chapter 12. The application of semiconductors layered with platinum-group metal catalysts to hydrogen-generating solar cells is discussed.

Kalyanasundaram, K., and Grätzel, M. *Photochem. Photobiol.* **1984,** *40,* 807. A 1984 review on developments in solar energy.

Kiwi, J., Kalyanasundaram, K., and Grätzel, M. *Struc. Bonding* **1984,** *49,* 39.

Parmon, V. N., and Zamaraev, K. I. In *Photochemical Energy Conversion,* Norris, J. R., Jr., and Meisel, D., eds. Elsevier, New York, 1989, p. 316.

Photoinduced Electron Transfer in Imaging Applications

Alfimov, M. V., and Sazhnikov, V. A. In *Photoinduced Electron Transfer. Part D. Photoinduced Electron Transfer Reactions: Inorganic Substrates and Applications,* Fox, M. A., and Chanon, M., eds. Elsevier, Amsterdam, 1988, Chapter 6.5.

Eaton, D. F. *Top. Curr. Chem.* **1990,** *156,* 199.

References

1. Potter, E. C. *Electrochemistry. Principles and Applications.* Cleaver-Hume Press, London, 1961, Chapter 7.
2. Brugger, P.-A., and Grätzel, M. *J. Am. Chem. Soc.* **1980,** *102,* 2461.
3. El Torki, F. M., Schmehl, R. H., and Reed, W. F. *J. Chem. Soc., Faraday Trans. 1* **1989,** *85,* 349.
4. Grätzel, M. *Ber. Bunsenges. Phys. Chem.* **1980** *84,* 981.
5. Wolff, C., and Grätzel, M. *Chem. Phys. Lett.* **1977,** *52,* 542.
6. Willner, I., Ford, W. E., Otvos, J. W., and Calvin, M. *Nature (London)* **1979,** *280,* 823.
7. Pileni has proposed an alternate model to explain the drastic reduction of k_{ret} in reverse micelles. For details see Brochett, P., Zemb, T., Mathis, P., and Pileni, M.-P. *J. Phys. Chem.* **1987,** *91,* 1444.
8. Tanimoto, Y., Takayama, M., Itoh, M., Nakagaki, R., and Nagakura, S. *Chem. Phys. Lett.* **1986,** *129,* 414.
9. Ulrich, T., and Steiner, U. E. *Tetrahedron* **1986,** *42,* 6131.
10. Blodgett, K. B., and Langmuir, I. *Phys. Rev.* **1935,** *51,* 964.
11. Tien, H. T. *Nature (London)* **1970,** *227,* 1232.
12. Quina, F. H., and Whitten, D. G. *J. Am. Chem. Soc.* **1977,** *99,* 877.
13. Renschler, C. L., and Faulkner, L. R. *Faraday Discuss. Chem. Soc.* **1980,** *70,* 311.
14. Möbious, D. *Ber. Bunsenges. Phys. Chem.* **1978,** *82,* 848.
15. Ilani, A., Woodle, M., and Mauzerall, D. *Photochem. Photobiol.* **1989,** *49,* 673.
16. Palazzotto, M. C., Sahyun, M. R. V., Serpone, N., and Sharma, D. K. *J. Chem. Phys.* **1989,** *90,* 3373.
17. Kuhn, H. *Pure Appl. Chem.* **1979,** *51,* 341,.

18. De Schryver, F. C., Van der Auweraer, M, Verschuere, B., and Willig, F. In *Supramolecular Photochemistry,* Balzani, V., ed. Reidel, Dordrecht, 1987, p. 385.

19. Fujihara, M., and Aoki, K. *Proc Symp. Photoelectrochemistry and Electrosynthesis on Semiconducting Materials,* Vol. 14, Ginley, D. S., Nozik, A., Armstrong, N., Honda, K., Fujishima, A., Sakata, T., and Kawai, T., eds. Electrochemical Society, Pennington, NJ, 1988, p. 280.

20. Rong, S., Brown, R. K., and Tollin, G. *Photochem. Photobiol.* **1989,** *49,* 107.

21. Grimaldi, J. J., Boileau, S, and Lehn, J. M. *Nature (London)* **1977,** *265,* 229.

22. Armitage, B., and O'Brien, D. F. *J. Am. Chem Soc.* **1992,** *114,* 7396.

23. Ford, W. E., Otvos, J. W., and Calvin, M. *Nature (London)* **1978,** *274,* 507.

24. Ford, W. E., Otvos, J. W., and Calvin, M. *Proc. Natl. Acad. Sci. U.S.A.* **1979,** *76,* 3590.

25. Ford, W. E., and Tollin, G. *Photochem. Photobiol.* **1982,** *35,* 809.

26. Rafaeloff, R., Maliyackel, A. C., Grant, J. L., Otvos, J. W., and Calvin, M. *Nouv. J. Chim.* **1986,** *10,* 613.

27. Nakamura, H., Motonaga, A., Ogata, T., Nakao, S., Nagamura, T., and Matsuo, T. *Chem. Lett.* **1986,** 1615.

28. Grätzel, M. In *Photoinduced Electron Transfer. Part D. Photoinduced Electron Transfer Reactions: Inorganic Substrates and Applications,* Fox, M. A., and Chanon, M., eds. Elsevier, New York, 1988, Chapter 6.3.

29. Dare-Edwards, M. P., Goodenough, J. B., Hamnett, A., Seddon, K. R., and Wright, R. D. *Faraday Discuss. Chem. Soc.* **1980,** *285,* 285.

30. Wrighton, M. S. *J. Chem. Ed.* **1983,** *60,* 335.

31. Levy-Clement, C. In *Photochemical Energy Conversion,* Norris, J. R., Jr., and Meisel, D., eds. Elsevier, New York, 1989, p. 267.

32. Wrighton, M. S. *Science* **1986,** *231,* 32.

33. Fox, M. A. *Pure Appl. Chem.* **1988,** *60,* 1013.

34. Bard, A. *J. Chem. Ed.* **1983,** *60,* 302.

35. Fox, M.A., and Chandler, A. In *Supramolecular Photochemistry,* Balzani, V., ed. Reidel, Dordrecht, 1987, p. 405.

36. Fujihara, M., Ohishi, N., and Osa, T. *Nature (London)* **1977,** *268,* 226.

37. Coche, L., and Moutet, J.-C. *Electrochim. Acta* **1985,** *30,* 1063.

38. Draper, A. M., Ilyas, M., de Mayo, P., and Ramamurthy, V. *J. Am. Chem. Soc.* **1984,** *106,* 6222.

39. Draper, A. M., and de Mayo, P. *Tetrahedron Lett.* **1986,** *27,* 6157.

40. Yanagida, S., Ishimaru, Y., Miyake, Y., Shiragami, T., Pac, C., Hashimoto, K., and Sakata, T. *J. Phys. Chem.* **1989,** *93,* 2576.

41. Pavlik, J. W., and Tantayanon, S. *J. Am. Chem. Soc.* **1981,** *103,* 6755.

42. Fox, M. A., and Chen, M.-J. *J. Am. Chem. Soc.* **1983,** *105,* 4497.

43. Fox, M. A., and Chen, M.-J. *J. Am. Chem. Soc.* **1981,** *103,* 6757.

44. Zamaraev, K. I., and Parmon, V. N. In *Energy Resources Through Photochemistry and Catalysis,* Grätzel, M., ed. Academic Press, New York, 1983, Chapter 5.

45. Shafirovich, V. Ya., and Shilov, A. E. In *Photochemical Energy Conversion,* Norris, J. R., Jr., and Meisel, D., eds. Elsevier, New York, 1989, pp. 173–183.

46. Kiwi, J., and Grätzel, M. *J. Am. Chem. Soc.* **1979,** *101,* 7214.

47. DeLaive, P. L., Sullivan, B. P., Meyer, T. J., and Whitten, D. G. *J. Am. Chem. Soc.* **1979,** *101,* 4007.

48. Miller, D., and McLendon, G., *Inorg. Chem.* **1981,** *20,* 950.

49. Visser, A. J. W. G., and Fendler, J. H. *J. Phys. Chem.* **1982,** *86,* 2406.

50. Kiwi, J., and Grätzel, M. *Nature (London)* **1979,** *281,* 657.

51. Ballardini, R., Juris, A., Varani, G., and Balzani, V. *Nouv. J. Chim.* **1980,** *4,* 563.

52. Borgarello, E., Kiwi, J., Pelizzetti, E., Visca, M., and Grätzel, M. *Nature (London)* **1981,** *289,* 158.

53. Kiwi, J. In *Energy Resources through Photochemistry and Catalysis,* Grätzel, M., ed. Academic Press, New York, 1983, Chapter 9.

54. Bookbinder, D. C., Bruce, J. A., Dominey, R. N., Lewis, N. S., and Wrighton, M. S. *Proc. Natl. Acad. Sci. U.S.A.* **1980,** *77,* 6280.

55. Simon, R. A., Mallouk, T. E., Daube, K. A., and Wrighton, M. S. *Inorg. Chem.* **1985,** *24,* 3119.

56. Frank A. J., Willner, I., Goren, Z., and Degani, Y. *J. Am. Chem. Soc.* **1987,** *109,* 3568.

57. Willner, I., Eichen, Y., and Frank, A. J. *J. Am. Chem. Soc.* **1989,** *111,* 1884.

58. Tricot, Y.-M., and Fendler, J. H. *J. Am. Chem. Soc.* **1984,** *106,* 2475.

59. Tricot, Y.-M., and Fendler, J. H. *J. Am. Chem. Soc.* **1984,** *106,* 7359.

60. Tien, H. T., and Chen, J.-W. *Photochem. Photobiol.* **1989,** *49,* 527.

61. Tien, H. T. *Adv. Mater.* **1990,** *2,* 263.

62. Lehn, J.-M., and Sauvage, J.-P. *Nouv. J. Chim.* **1977,** *1,* 449.

63. Ballardini, R., Juris, A., Varani, G., and Balzani, V. *Nouv. J. Chim.* **1980,** *4,* 563.

64. Rehak, V. *Collect. Czech. Chem. Commun.* **1987,** *52,* 1666.

Problems

1. Sato et al. observed an enhancement in electron transfer between $[Ru(bpy)_3{}^{2+}]^*$ and *N,N'*-dimethyl-4,4'-bipyridinium in the presence of sodium dodecyl (lauryl) sulfate (SDS) micelle. Using a schematic, suggest an explanation for the rate enhancement. (Sato, H., Kawasaki, M., Kasatani, K., and Ban, T. *Chem. Lett.* **1982,** 1139.)
2. The singlet of oxonine (shown below) can be quenched by *N,N*-tetramethyl-*p*-phenylenediamine in water–oil microemulsions. It is found that ion pairs recombine more readily when an external magnetic field is applied than in zero field. Why? Radical ion pair decay curves display a rapid as well as a slower

rate of recombination. What is a possible cause of the biphasic nature of the decay curves?

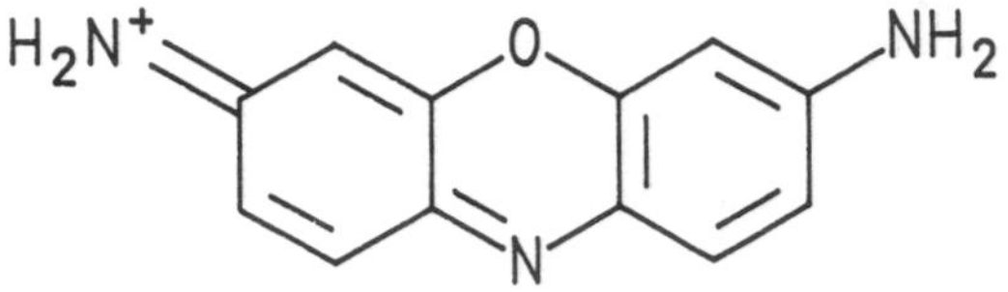

(Baumann, D., Ulrich, T., and Steiner, U. E. *Chem. Phys. Lett* **1987,** *137,* 113.)

3. The valence isomerization of Dewar benzene on semiconductor surfaces has been reported on illuminated *n*-type semiconductor surfaces. Suggest a mechanism. Hint: refer to reaction 3.22. (Al-Ekabi, H., and de Mayo, P. *J. Phys. Chem.* **1986,** *90,* 4075.)
4. Suggest mechanisms for reactions 5.13 and 5.18.

CHAPTER

6

Theories of Photoinduced Electron Transfer

To this point, we have presented a variety of examples illustrating the scope of photoinduced electron transfer. Beyond these examples, however, there remains much to be discussed. The reader perhaps may have wondered about what factors control the rates of photoinduced electron-transfer reactions. If indeed there are differences in rates, and in fact there are, what are the factors influencing these differences? The origin of these differences can be traced to the nature of the molecular structures of the reactants and of the medium. In Chapter 1, we were primarily concerned with the role of energetics in photoinduced electron-transfer reactions. Master equations were derived to aid the reader in estimating the free-energy changes in these processes. However, in some instances thermodynamic calculations may suggest a favorable electron transfer, yet the reaction may in fact proceed quite slowly. This reaction may be prevented from occurring because of kinetic barriers. Figure 6.1 illustrates the effect of a kinetic barrier preventing an electron-transfer reaction from occurring even though the process may be quite exothermic. This situation is somewhat analogous to a tank filled with water. If the container develops a large leak, water will flow out rapidly. If the leak is tiny, water will trickle out. The small leak acts as a barrier preventing a rapid loss of water. In the same sense, a barrier can slow down electron transfer. Why should the quenching of $[Ru(bpy)_3{}^{2+}]^*$ by methyl viologen (MV^{2+}) to give $Ru(bpy)_3{}^{3+}$ and $MV^{\overset{+}{\cdot}}$ be faster by 10^4 than quenching by Eu^{3+}, even though ΔG_{el} for both processes is about -9 kcal mol^{-1}?[1,2] The reason that quenching by Eu^{3+} is so slow is due to kinetic barriers. These barriers act to prevent the electron transfer. To successfully manipulate excited-state electron transfer, it is desirable to understand what the kinetic barriers are and how they can be controlled. In this chapter, we

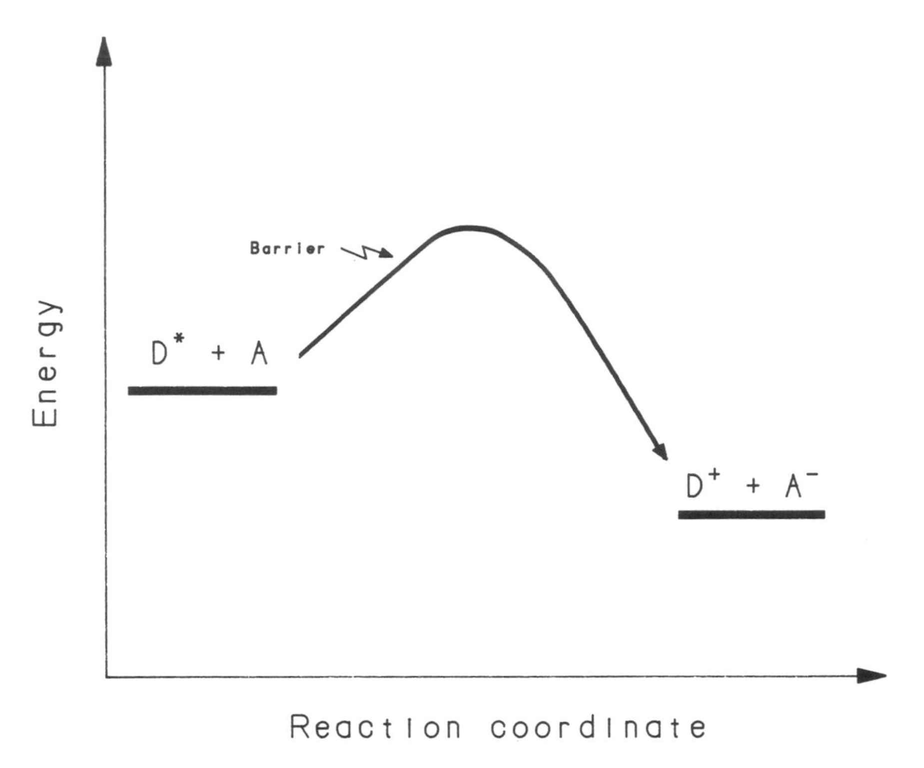

Figure 6.1 The kinetic barrier prevents a reaction from occurring even though the process may be quite exothermic.

will consider the kinetic barriers that influence the rate of electron transfer between an excited state and ground state.

6.1. Diffusion-Controlled and Electron-Transfer Rate Constants

Scheme 6.1 is a good starting point to begin our analysis (we assume that the excited state is an electron donor; a similar analysis applies to an excited electron acceptor):

$$D^* + A \underset{k_{-\mathrm{dif}}}{\overset{k_{\mathrm{dif}}}{\rightleftharpoons}} \underset{\mathrm{EC}}{D^* \cdots A} \underset{k_{-\mathrm{el}}}{\overset{k_{\mathrm{el}}}{\rightleftharpoons}} [D^+ \cdots A^-] \xrightarrow{k_{\mathrm{ret}}} D + A \qquad (6.1)$$

k_{dif} is the rate constant representing the rate of diffusive encounters between reactants. $k_{-\mathrm{dif}}$ denotes the rate of separation of the reactants after collision. k_{el} is the first-order rate constant of electron transfer. The reverse step is designated by the rate constant $k_{-\mathrm{el}}$.

We refer to a collision between two reactant molecules as an *encounter,* and the complex resulting from the encounter a *collision* or *encounter complex.* In 6.1, this is designated by EC. In small molecules, an enconter complex may have a lifetime as small as $\sim 10^{-9}$ to 10^{-10} s. During this time, two molecules may typically undergo numerous collisions before they react or separate from the encounter complex into the bulk of solvent molecules. In excited-state electron transfer, an encounter complex can be visualized as an ensemble consisting of a sensitizer and quencher surrounded by several layers or shells of solvent molecules. The sensitizer and quencher are said to be contained within a *solvent cage.* Within the cage, the spherical reactant molecules may be separated by a certain center-to-center distance usually taken as $d_{cc} \sim 7$ Å. This distance implies that spherical molecules with typical radii of ~3–4 Å are within a contact distance. Nonspherical molecules, such as ellipsoids, may be able to approach to closer distances. It is useful to think of molecules engaging in collisions or in sets of collisions by reencounters. After these repeated encounters, the sensitizer and quencher may undergo reaction or eventually escape from the cage. If reaction does take place within the solvent cage, the products that are formed may revert back to starting reactants, undergo irreversible chemical reaction, or leave the solvent cage by ion dissociation.

A steady-state treatment of Scheme 6.1 leads to the following kinetic expression:

$$k_q = \frac{k_{dif}}{1 + (k_{-dif}/k_{el})\,[1 + (k_{-el}/k_{ret})]} \tag{6.2}$$

Equation 6.2 can be simplified with the reasonable assumption that $k_{ret} \gg k_{-el}$. The latter assumption follows from the fact that reverse electron transfer (k_{-el}) is usually a thermodynamically uphill process while electron return (k_{ret}) is thermodynamically downhill. Thus, we obtain

$$k_q = \frac{k_{dif}}{1 + (k_{-dif}/k_{el})} \tag{6.3}$$

or

$$k_q = \frac{k_{dif}k_{el}}{k_{el} + k_{-dif}} = k_{dif}\,\alpha \tag{6.4}$$

where α is defined as

$$\alpha = \frac{k_{el}}{k_{el} + k_{-dif}} \tag{6.5}$$

If $k_{el} \ll k_{-dif}$, the quenching rate constant becomes

$$k_q = K_{eq}k_{el} \tag{6.6}$$

where $K_{eq} = k_{dif}/k_{-dif}$ is the equilibrium rate constant for formation of the encounter complex. This is the case where the electron transfer is slow as compared to k_{-dif}.

The rate of diffusion of two spherical reactants to an encounter distance (assuming that d_{cc} equals the sum of the reactant molecules' radii) can be calculated from Eq. 6.7:

$$k_{dif} = 4\pi D_{DA} d_{cc} \tag{6.7}$$

where D_{DA} is the combined diffusion constant for approach of D and A. Equation 6.7 is the simplest form of the Smoluchowski equation. It suggests that the observed rate is a function of the combined diffusion constants of the two reactants and the separation distance. For small molecules in aqueous solution, $D_{DA} \sim 1.4 \times 10^{-5}$ cm^2 s^{-1}. Assuming that $d_{cc} \sim 6$ Å, we estimate from Eq. 6.7 that $k_{dif} \simeq 4 \times 10^9$ M^{-1} s^{-1}. This is a typical value for k_{dif} in nonviscous fluids. In the solid state where diffusion is extremely slow, the diffusion constant is $D \leqq 10^{-12}$ cm^2 s^{-1} for small molecules. To calculate the diffusion constant for one reactant, we use the Stokes–Einstein equation:

$$D = \frac{k_B T}{6\pi\eta r} \tag{6.8}$$

where k_B is the Boltzmann constant, η is the viscosity of the solution, and r is the radius of the reactant.

Table 6.1 contains a tabulation of diffusion constants for a range of viscosities and two radii. As expected, the diffusion constant increases with a decrease in the solvent viscosity or decrease in r. The diffusion constants are also predicted to increase as the temperature of the solution rises.

Table 6.1 DIFFUSION CONSTANTS CALCULATED FROM STOKES–EINSTEIN EQUATION[a]

		D(cm^2 s^{-1})	
		r	
Solvent	η(cP)[b]	4 Å	7 Å
Diethyl ether	0.242	2.2×10^{-5}	1.3×10^{-5}
Acetonitrile	0.345	1.6×10^{-5}	8.9×10^{-6}
Benzene	0.649	8.3×10^{-6}	4.7×10^{-6}
Water	1.002	5.4×10^{-6}	3.1×10^{-6}
Ethanol	1.078	5.0×10^{-6}	2.8×10^{-6}
Dodecane	1.508	3.6×10^{-6}	2.0×10^{-6}
Ethylene glycol	26.09	2.1×10^{-7}	1.2×10^{-7}
Glycerin	1490	3.6×10^{-9}	2.1×10^{-9}

[a] Diffusion constants calculated using Eq. 6.8. A temperature of 293 K is assumed.
[b] cP, centipoise; 1 cP = 0.01 poise.

Table 6.2 DIFFUSION RATE CONSTANTS AT VARYING VISCOSITIES[a]

Solvent	η(cP)	k_{dif}(M^{-1} s^{-1})
Diethyl ether	0.242	2.7×10^{10}
Acetonitrile	0.345	1.9×10^{10}
Benzene	0.649	1.0×10^{10}
Water	1.002	6.5×10^{9}
Ethanol	1.078	6.0×10^{9}
Dodecane	1.508	4.3×10^{9}
Ethylene glycol	26.09	2.5×10^{8}
Glycerin	1490	4.4×10^{6}

[a] Diffusion rate constants calculated with Eq. 6.10 at $T = 293$ K.

Substitution of Eq. 6.8 into Eq. 6.7 leads to

$$k_{dif} = \frac{8k_BT}{3\eta} \tag{6.9}$$

Finally, inserting molar units in Eq. 6.9 gives a well-known version of the Smoluchowski equation:

$$k_{dif} = \frac{8RT}{3000\eta} \tag{6.10}$$

where N is Avogadro's number and R is the gas constant. For an instantaneously fast electron transfer, reaction will take place at every collision after the reactants have diffused to the encounter distance. Thus, $k_{el} \gg k_{-dif}$ and $k_q \sim k_{dif}$ (where k_q is the rate of quenching). The reaction is *diffusion controlled* (Section 1.7). According to Eq. 6.10, a fast quenching process should depend on the temperature and the viscosity of the solvent. In fact, the rate constants of diffusion-controlled reactions can readily be calculated simply from viscosity values obtained at given temperatures with the aid of Eq. 6.10. Diffusion-controlled rate constants for common solvents are listed in Table 6.2.

We now define k_a, the bimolecular rate constant for the activated rate of electron transfer:

$$k_a = K_{eq}k_{el} \tag{6.11}$$

Now, combining Eq. 6.11 with Eq. 6.3 gives

$$\frac{1}{k_q} = \frac{1}{k_{dif}} + \frac{1}{k_a} \tag{6.12}$$

k_a can easily be determined from Eq. 6.12, since k_q is an experimental quantity measureable by Stern–Volmer techniques, and k_{dif} can be calculated from Eq. 6.10. With knowledge of k_a, we can deduce a value for k_{el} from Eq. 6.11.

In the solid state or in other media where diffusion is impeded by the nature of the environment (so-called *diffusionless electron transfer*). Thus, for

$$D^* \cdots A \underset{k_{-el}}{\overset{k_{el}}{\rightleftharpoons}} [D^+ \cdots A^-] \tag{6.13}$$

the following holds: $k_q \simeq k_a \simeq k_{el}$.

The key to understanding reactivity in electron transfer depends on classifying and separating the rate-determining factors into *nuclear* and *electronic*. The rate-determining factors in electron transfer—those that affect k_{el}—consist of electronic interactions and nuclear motions. Thus, the first-order rate constant of electron transfer can be written as

$$k_{el} = \nu_n \kappa_n \kappa_{el} \tag{6.14}$$

ν_n is the *nuclear frequency* and may range from $\sim 10^{12}$ to $\sim 10^{14}\ s^{-1}$. The significance of ν_n is discussed later in this chapter. κ_{el} is defined as the *electronic factor,* and κ_n is the *nuclear factor*. The latter two quantities are dimensionless and range between 0 and 1. An equation for the nuclear factor is given by Eq. 6.15:

$$\kappa_n = \exp(-\Delta G_{el}{}^{\neq}/RT) \tag{6.15}$$

where $\Delta G_{el}{}^{\neq}$ is the free-energy of activation for the electron transfer. Accordingly, the complete expression for k_{el} can be written as

$$k_{el} = \nu_n \kappa_{el} \exp(-\Delta G_{el}{}^{\neq}/RT) \tag{6.16}$$

and for k_a as

$$k_a = K_{eq} \nu_n \kappa_{el} \exp(-\Delta G_{el}{}^{\neq}/RT) \tag{6.17}$$

$K_{eq}\nu_n\kappa_{el}$ may be regarded as the "preexponential" term whereas the activation energy is an "exponential" term.

Equations 6.16 and 6.17 "contain" the electronic and nuclear factors that affect reactivity in electron transfer. Two general approaches will be used to learn about these rate-determining factors. The first approach is based on the classical theory of Marcus, which was originally developed to explain the variation of the rates of electron transfer in solution. Marcus theory, together with various modifications, has the great advantage of allowing one to predict rates of electron transfer from experimental parameters. When applying the classical theory of Marcus to reactions in solution, electronic barriers are usually neglected, so that $\kappa_{el} \sim 1$ (although there are exceptions). With a few more assumptions, one can extend the classical ideas of Marcus to photoinduced electron transfer and subsequently make predictions on the electron-transfer reactivity of excited-states.

The second approach is *quantum mechanical* or *nonclassical*. It is based on the overlap of nuclear and electron wavefunctions between initial reactant and final product states. Like Marcus theory, there is a separation of nuclear and electronic factors; however, in the nonclassical, quantum mechanical theory, more emphasis is placed on electronic barriers ($\kappa_{el} < 1$). Using both approaches, we shall explore the factors that determine the rates in excited-state electron transfer. We start with the classical theory of Marcus and then lead on to the nonclassical theories.

6.2. Classical Theories of Electron Transfer

6.2.1. Electron and Nuclear Motion

In an electron exchange, an electron is "exchanged" or transferred between two ions of the same species:

$$M^n + M^{n+1} \rightleftharpoons M^{n+1} + M^n \tag{6.18}$$

An excited-state electron exchange can be written as

$$[M^n]^* + M^{n+1} \rightleftharpoons M^{n+1} + [M^n]^* \tag{6.19}$$

for the case where the excited state is an electron donor. A similar reaction can be written for an excited electron acceptor. In these processes, there is no net chemical change. The rates of electron exchanges span roughly the age of the universe in seconds!

In 1952, Libby suggested that electron-exchange reactions can be accommodated by the Franck–Condon principle (i.e., that one could take into account the relative velocities of electron and nuclear motion to develop a reasonable explanation for the "slowness" of certain electron exchanges).[3,4] He suggested that the Franck–Condon principle used to explain the vertical transitions during photoexcitation could also help explain the wide range of rate constants observed in electron transfer. According to Libby, the slow electron-exchange reaction $Co(NH_3)_6^{2+} + Co(NH_3)_6^{3+} \rightleftharpoons Co(NH_3)_6^{3+} + Co(NH_3)_6^{2+}$ is due to the large change in the Co—N bond length accompanying the movement of the electron between the two ions. Nuclear motion, Libby argued, proceeds much slower than "instantaneous" electron motion because the mass of an electron is lighter than the mass of a nucleus. During the transfer of an electron, the nuclei are "frozen." For a simple electron exchange, the sequence of events is electron transfer followed by nuclear adjustments or *nuclear reorganization,* which refers to the motion of the nuclei in the reacting molecules and surrounding solvent molecules during the electron transfer. After an electron is transferred into a new nuclear configuration, the nuclei respond to the force exerted by the electron by undergoing vibrational motions. These vibrations are so-called *vibronic transitions,* or electronic transitions coupled with nuclear vibrations. The energy required for nuclear reorganization is the energy of the transition state, or the *activation energy*.

The connection between photoexcitation and electron transfer is revealing. Numerous observations in photochemistry and electron transfer can be examined on the basis of the relative motions of electrons and nuclei. Because electronic and nuclear motion proceed on much different time scales, it is necessary to examine the electronic properties on the basis of the assumption that the nuclei remain in fixed positions. This assumption, the *Born–Oppenheimer approximation,* was

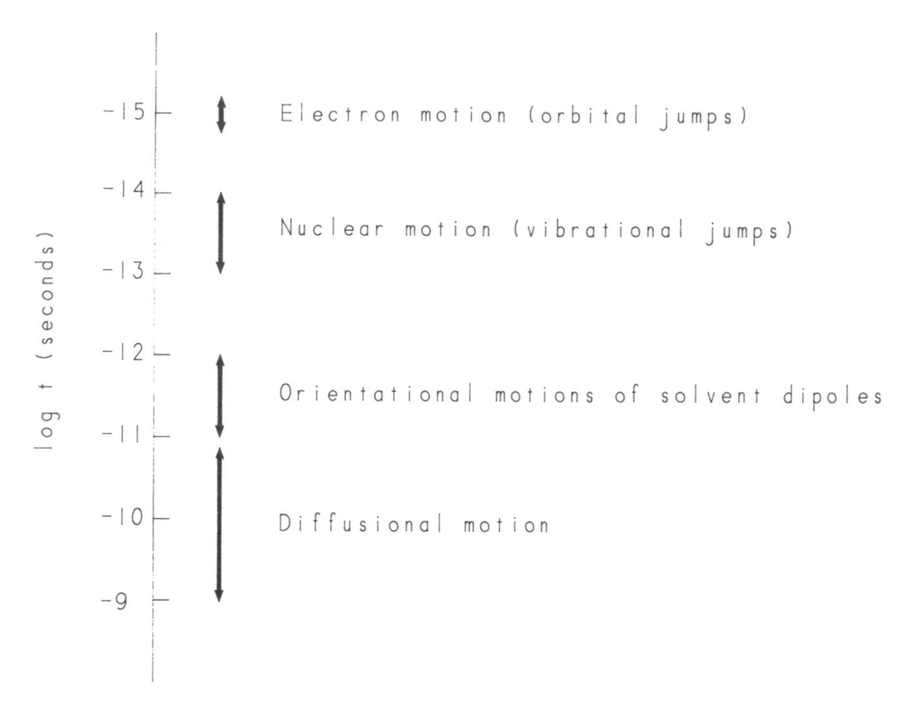

Figure 6.2 Timescales of chemical events.

introduced in Chapter 1, and is invoked in many theoretical treatments to explain chemical reactivity.[5]

This is an appropriate place to consider the time scales of electron and nuclear motion (Fig. 6.2). Timescales of processes involving electronic and nuclear motion can be examined in terms of their frequencies. Thus, an electron moves with a frequency of $\sim 10^{15}$ s^{-1} whereas nuclei vibrate more slowly with a frequency that may range between 10^{12} and 10^{14} s^{-1}. This large difference is at the core of the Born–Oppenheimer approximation and allows one to consider electronic and nuclear motion as separate events occurring on different timescales. In electron transfer, nuclear movements typically take place on a longer timescale than the transfer of an electron. It is this fundamental feature of electron transfer that is the basis of a powerful graphic procedure used to depict the energy changes that take place during nuclear motion. This procedure involves *potential energy surfaces* and is examined in the next section.

6.2.2. Potential Energy Surfaces

To adequately describe an electron-transfer process, it is useful to resort to the use of *potential energy surfaces,* a graphic representation that allows one to dissect the details of a complex mechanism. A potential energy surface is a topological representation of a chemical reaction. The surface is a unique approximation of an electronic wavefunction, Ψ, and the energy of any number of electrons moving in a field of nuclei with fixed positions relative to the electrons. Potential energy surfaces have multidimensional character and are occasionally called *hypersurfaces* with a dimensionality given by $3N - 5$, where N is the number of nuclei (particles) contained in the system. The depiction of potential energy hypersurfaces can be complicated since many nuclei are usually involved. For this reason, it is useful to reduce complex hypersurfaces to simpler two- or three-dimensional representations (for our purposes, total energy is one of the dimensions).

To construct a potential energy curve, we consider a simple Cartesian system where the total energy is coincident with the y-axis and the reaction coordinate is coincident with the x-axis. Any (x,y) point represents a nuclear position with a specific total energy. Figure 6.3 shows an idealized energy curve, shaped like a parabola, representing nuclear motion. Each point on the parabola represents a discrete nuclear geometry and energy associated with this nuclear configuration. From the energy conservation law, the total energy of a nuclear system is

$$\text{Total energy} = \text{P.E.} + \text{K.E.} \tag{6.20}$$

where P.E. is the potential energy of the stationary of nuclei and K.E. is the kinetic energy of the moving nuclei. Physically, the reaction coordinate consists of changes

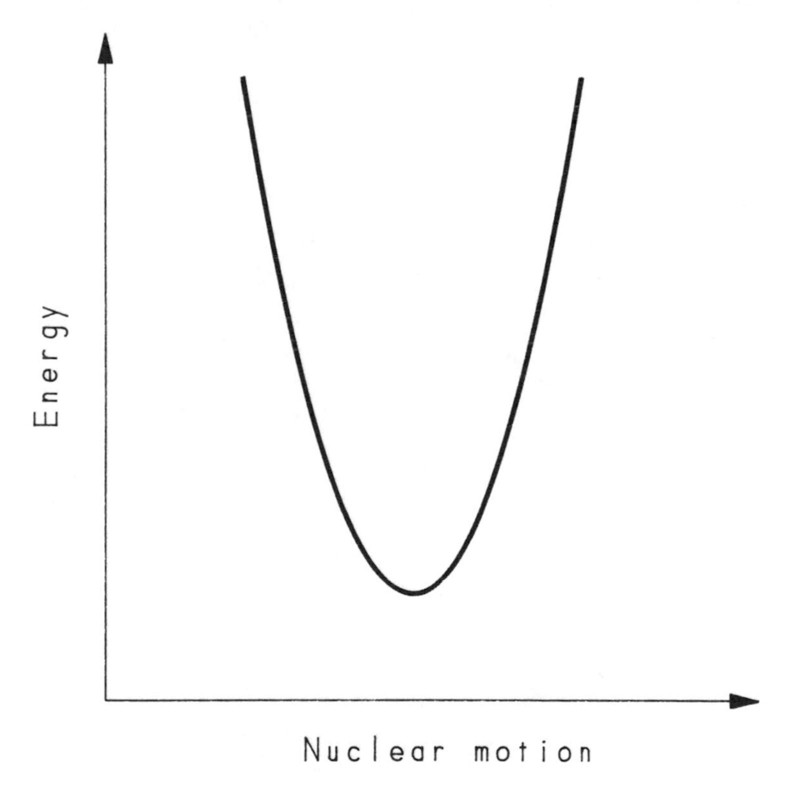

Figure 6.3 A parabolic potential energy curve.

in nuclear position and momenta with time. These nuclear changes are expressed as changes in bond angles and bond lengths. Any nuclear configuration during the reaction is associated with a unique potential and kinetic energy.

For a typical reaction, the curve representing nuclear motion displays features resembling valleys, mountains, peaks, and contours. These geometric characteristics of a reaction may be far more complicated than the features of the idealized curve in Fig. 6.3. However, the illustration in Fig. 6.4 conveys some sense of the potential energy curves for a "real" chemical reaction. The energy of the reaction (given by the y-axis) is plotted vs. the time course of the reaction represented by the changes in nuclear motion (in the context of this discussion, "nuclear coordinate" is the "reaction coordinate"). A smooth contour diagram may be suggestive of a *concerted* pathway. The valleys, representing energy minima, are the energy and nuclear configurations of discrete intermediates with defined lifetimes. Peaks represent activation energies and nuclear geometries of transition states.

During the course of a reaction, electrons may undergo major fluctuations. The lighter electrons move much more rapidly than the heavier nuclei (the Born–Op-

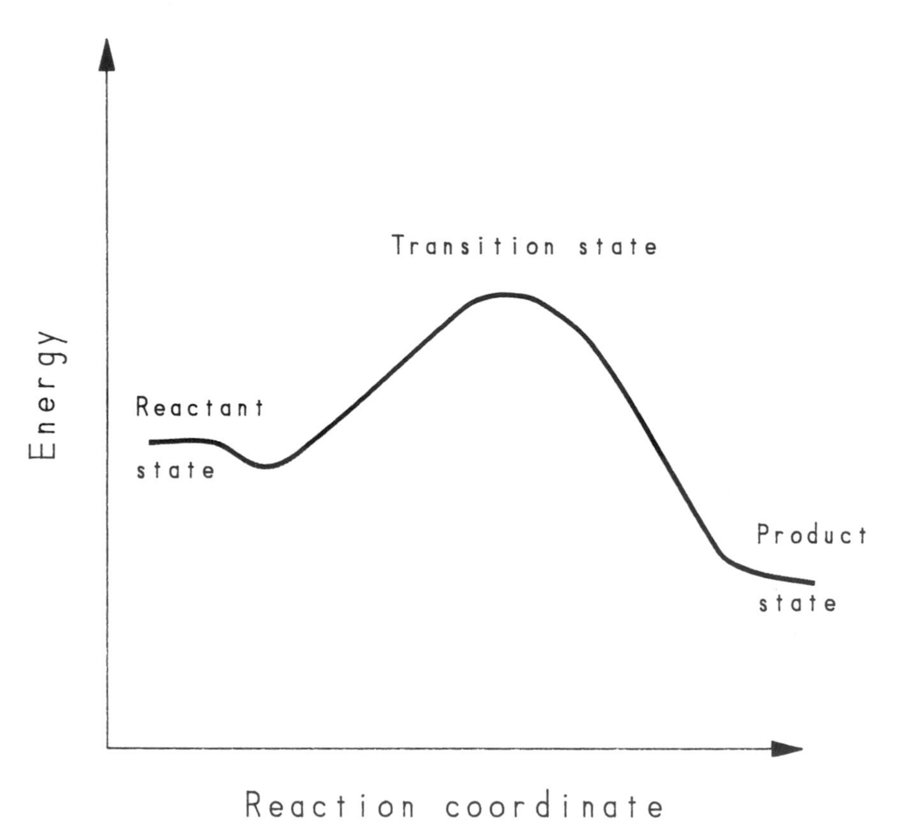

Figure 6.4 A reaction profile with "peaks" and "valleys".

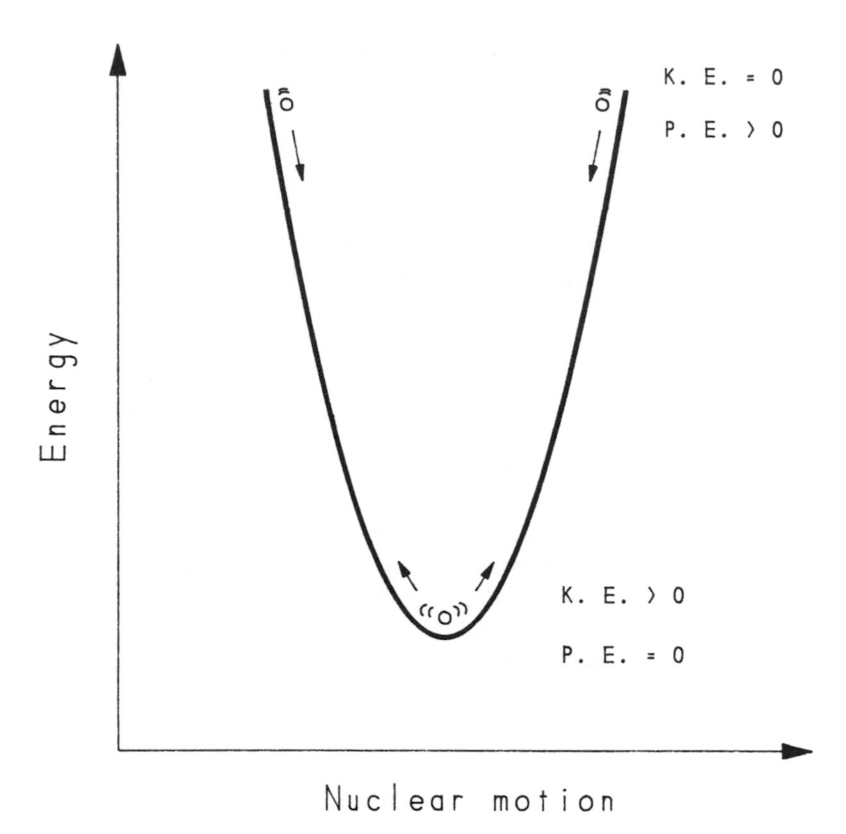

Figure 6.5 The "moving" point on a potential energy curve. For molecular systems consisting of many nuclei, it is practical to specify the center of mass of these nuclei by a "moving" point. This technique permits the visualization of complex systems on a reaction curve. K. E. = kinetic energy; P. E. = potential energy.

penheimer approximation). The heavier nuclei respond belatedly to abrupt changes in electronic configurations.

Nuclear motion can be visualized as a moving point on a potential-energy curve (Fig. 6.5). A "moving" point represents the collective mass of a group of nuclei. The trajectory motion of the point can be represented by a marble rolling on a frictionless surface. Although the total energy of the point remains constant, the proportions of kinetic and potential energy continually change as the point moves up or down on the curve. Let us examine what happens as this imaginary nuclear point moves on the curve. At either end of the curves (i.e., the turning points), the potential energy of the system attains its maximum value. Here, the kinetic energy is zero (i.e., K.E. = 0) (point a in Fig. 6.5) since the velocity is zero ($v = 0$). At the turning point, nuclear motion commences (say, from left to right in Fig. 6.5) and the point begins to move down the slope of the curve. The point looses its potential energy and gains velocity as it travels down the slope of the curve. At the bottom of the curve (the valley), the kinetic energy of the point reaches a maximum (K.E. = $^1/_2mv^2$). Here the point is moving very rapidly. As it travels up the right-hand slope, it gradually slows down until it reaches the top of the right-hand side. The moving point will then return again and repeat its motion. The point eventually slows down since its energy is given off to its surroundings in the sense that a ball rolling on the ground slows down because of opposing frictional forces.

In this manner, potential energy surfaces describing chemical reactions contain information about the changes in total energy as reactants proceed to products. These surfaces can be used to examine the effects of the most important nuclear changes on the total energy. If we are concerned about the effects of two nuclear changes on the reaction, these nuclear changes may be represented by the x- and y-ordinates, and the energy by the z-ordinate. For example, the change in the bond length can be plotted along one axis, whereas nuclear changes resulting from the motion of solvent molecules are represented along the other coordinate.

But what can be plotted when thousands of nuclear changes are taking place? The answer is that one nuclear deformation can often be used to describe the reaction. With this simplification, the potential-energy surface reduces to a simple two-dimensional curve. Thus, if a bond deformation is known to play a prominent role in a chemical transformation, the energy change during the reaction can be plotted vs. the positional changes of the bonded nuclei. Otherwise, if one specific nuclear fluctuation cannot be singled out and identified, one can imagine that the moving point represents a "mean" or composite of all nuclear changes. This approach is useful for heuristic purposes even for quite complicated reactions.

The advantage of using potential-energy curves is that several states and the corresponding energies may be graphed and traced simultaneously. This is particularly useful for photoinduced electron-transfer reactions where ground and excited states, as well as the states of ionic intermediates, may be depicted by their respective potential-energy curves on the same graph.

6.2.3. The Harmonic Oscillator

Potential-energy surfaces have played a vital role in the development of classical theories of photoinduced electron transfer. Intersecting parabolic curves of idealized shape were first used to represent the course of electron transfer from a state just prior to electron transfer, the *precursor state,* to the state just following electron transfer, the *successor state*. The precursor state can be considered a composite of the donor and acceptor molecules together with surrounding solvent molecules. The successor state consists of the products of electron transfer and the solvent molecules that may exist in a different configuration than in the precursor state. The precursor and successor states can each be imagined as a *supermolecule*. Each supermolecule can be described by its own potential-energy curve.

To describe the essential features of potential-energy curves associated with the precursor and successor states, we start with idealized parabolic functions describing the behavior of a *harmonic oscillator*. A harmonic oscillator is defined as a point that, when displaced from equilibrium, experiences a restoring force proportional to its distance from the equilibrium position. An example of a harmonic oscillator is the periodic vibratory motion of two identical atoms in a diatomic molecule. The symmetrical one-dimensional parabolic function shown in Fig. 6.3 describes the behavior of the harmonic oscillator. The restoring force, *f*, is $-kx$, where k is a constant and x is the displacement of the oscillation along the x-axis. The potential energy of the harmonic oscillator is then given by

$$\text{P.E.} = \tfrac{1}{2}kx^2 \tag{6.21}$$

Solution of the Schrödinger equation for a harmonic oscillator leads to

$$E_v = (v + \tfrac{1}{2})\hbar\omega \tag{6.22}$$

where E_v is the energy of the harmonic oscillator, v is the vibrational quantum number (v = 0, 1, 2 . . .), $\hbar = h/2\pi$, and $\omega = (k/m)^{1/2}$. A quantum mechanical treatment can be used to describe the behavior of each vibration of the harmonic oscillator. A nuclear wavefunction, Ψ_v, is assigned to each vibrational level. If we assume the quantization of the vibrational energy of the harmonic oscillator, the parabolic curve assumes the appearance shown in Fig. 6.6. The wavefunction of each vibrational level oscillates and tends to disappear as the displacement of the

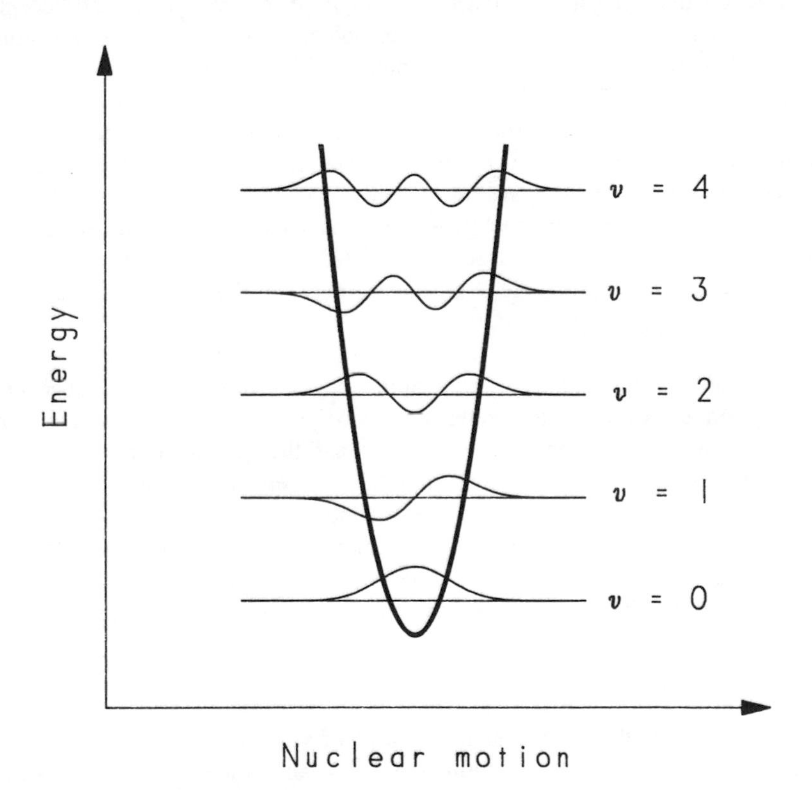

Figure 6.6 The vibrational states of a parabolic curve. The shape of each wavefunction is determined by a function called a Hermite polynomial.

nuclear vibration becomes large. The sign of the wavefunction experiences a change as the horizontal line is crossed. The probability of finding the weighted point of a vibration of the harmonic oscillator is given by the square of the nuclear wavefunction, Ψ_v^2. Note that the energy of the lowest possible vibration ($v = 0$) is a finite quantity (i.e., $E_0 = {}^1/_2\hbar\omega$, where E_0 is the zero-point energy). A nonclassical interpretation of the zero-point energy is that the energy of the system in the lowest state is greater than the energy of the system at complete rest (i.e., even in the lowest vibrational state, the system has a finite quantity of kinetic energy). This tenet is derived from the uncertainty principle. Briefly stated, the uncertainty of the momentum (Δp) and position (Δx) of a point is $\Delta p \Delta x \geqslant \hbar/2$. In Fig. 6.6, there is a finite probability of finding the point somewhere within the space mapped out by the parabola. This leads to the conclusion that $\Delta p \geqslant \hbar/2\Delta x$. That is, the momentum cannot be zero and, therefore, the energy cannot be zero.

6.2.4. Electronic Coupling

We are now in a position to predict what happens to the shape and character of the parabolic curves when an electron donor and acceptor molecule approach distances where electron transfer can take place. We suppose that the donor and acceptor approach from a large distance to form a supermolecule. According to the preceding discussion, a parabola can be used to represent this precursor state. In Marcus theory, the precursor state undergoes a change in nuclear reorganization to the transition state *before* electron transfer. The energy and nuclear configuration of the transition state is represented by the intersection of the two parabolic curves for the precursor and successor supermolecules.

The effects of electronic and nuclear interactions in electron transfer can be better understood if we apply a powerful treatment called *perturbation theory*.[6,7] We will use this approach for a simple electron transfer between the donor and acceptor:

$$\mathrm{D} + \mathrm{A} \rightarrow \mathrm{D}^+ + \mathrm{A}^- \tag{6.23}$$

We imagine that a donor and acceptor approach from an infinite distance until encounter distances where their frontier orbitals begin to interact. In this treatment, we suppose that eigenfunctions (wavefunctions) and the corresponding eigenvalues can be written for the reaction shown in 6.23. The eigenvalues are the energies corresponding to the solution of the eigenfunctions.

Let Ψ_D and Ψ_A be the electronic eigenfunctions of a Hamiltonian operator, and let $\Psi_i = \Psi_D\Psi_A$ be the eigenfunction of the system while the reactants are *initially* separated by an infinite distance. We can then describe the system by an eigenfunction equation:

$$(\mathcal{H}_D + \mathcal{H}_A)\Psi_i = E_{DA}\Psi_i \tag{6.24}$$

Here, $\mathcal{H}_D$ and $\mathcal{H}_A$ are the Hamiltonian operators of separated reactants, and E_{DA} is the initial energy of D and A. Now imagine that D and A are brought together for reaction. For this reactant state, we write a total Hamiltonian operator as a sum of several terms:

$$\mathcal{H} = \mathcal{H}_D + \mathcal{H}_{DA} + \mathcal{H}_A \tag{6.25}$$

where $\mathcal{H}_{DA}$ represents distance-dependent interaction terms between D and A. Equation 6.25 describes the electronic configuration of the reactant state. A parabolic curve can be generated to correspond to the geometry and energy of this state. This energy is designated by H_{ii} and is called the zero-order energy of the reactant state:

$$H_{ii} = \int \Psi_i \mathcal{H} \Psi_i \, d\tau \tag{6.26}$$

where Ψ_i is the electronic eigenfunction integrated over all space. A potential-energy curve is obtained by plotting H_{ii} vs. the nuclear configuration of the reactants.

A similar treatment can be applied to the product or successor state. Thus, $\Psi_f = \Psi_{D+}\Psi_{A-}$ and

$$(\mathcal{H}_{D+} + \mathcal{H}_{A-}) \Psi_f = E_{D+A-} \Psi_f \tag{6.27}$$

The zero-order potential energy of the successor state is then given by

$$H_{ff} = \int \Psi_f \mathcal{H} \Psi_f \, d\tau \tag{6.28}$$

where $\mathcal{H} = \mathcal{H}_{D+} + \mathcal{H}_{D+A-} + \mathcal{H}_{A-} = \mathcal{H}_D + \mathcal{H}_{DA} + \mathcal{H}_A$. A plot of H_{ff} vs. the nuclear reorganizational changes yields a potential-energy curve of the successor state just following electron transfer.

If the reactant and product potential-energy curves are superimposed, they will intersect (Fig. 6.7). The energy represented by intersection is the activation energy and is characterized by a distinct nuclear configuration unique to the transition state for electron transfer. The Born–Oppenheimer approximation is central to our understanding of the transition state (i.e., the nuclear configuration is "frozen" at the point when electron transfer actually takes place). The energy of the intersection, the actual activation free energy of the electron transfer, $\Delta G_{el}^{\neq}$, is reached because of collisions between reacting molecules and neighboring solvent molecules in the surrounding medium. In Marcus theory, energy is supplied by vibrational collisions with solvent molecules *prior* to electron transfer. Once the transition state is reached, the electron is transferred. Energy is thereby conserved. Consider an electron exchange (reaction 6.18) where the nature of the process demands that $\Delta G_{el} = 0$. If the electron were to move from M^{n+1} to M^n before nuclear reorganization, then this would violate energy conservation. That is, the vibrationally excited states of M^n and M^{n+1} would be created before nuclear motions could take place. During decay of these states, heat would necessarily be released into the solution. However, the electron exchange takes place without the release of energy, so that electron transfer *prior* to nuclear reorganization violates energy conservation.

In the transition state, electron transfer is virtually instantaneous as compared to nuclear motion (the Franck–Condon principle). No nuclear motions take place during the transfer. Once the successor state is formed, the latter rapidly undergoes thermal equilibration with the surrounding medium.

What we will attempt now is to show that intersecting potential-energy curves actually interact, and that this interaction leads to the formation of a new set of

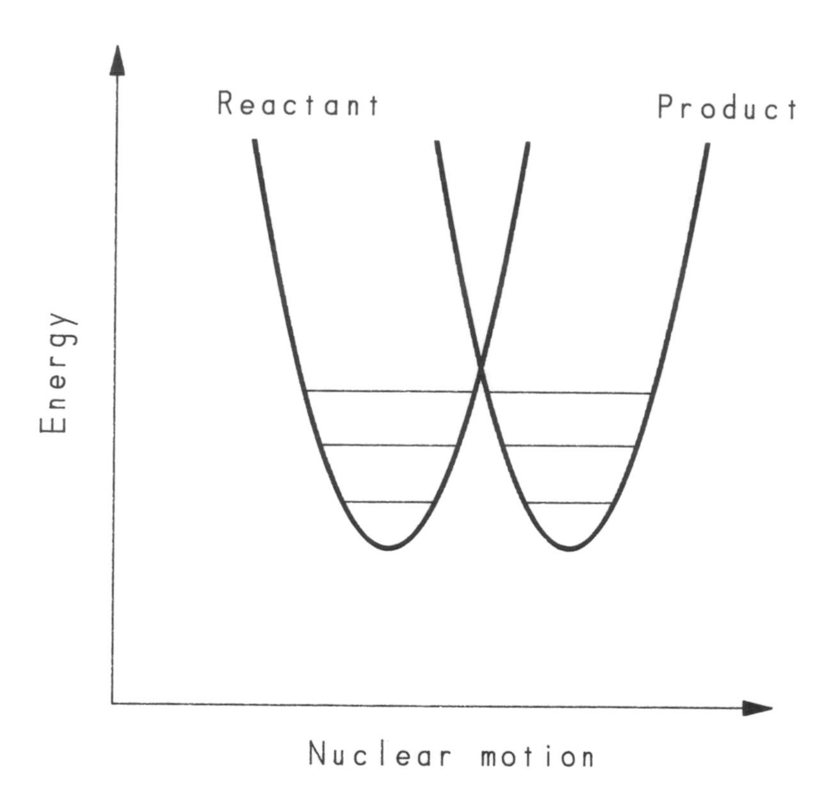

Figure 6.7 Intersecting parabolic curves of a reactant and product state. Each state consists of quantized vibrational states.

curves. When the orbitals of two molecules begin to interact, their eigenfunctions and eigenvalues begin to experience changes. These interactions are designated as *perturbations*. As the zero-order states begin to merge, a new curve is formed (Fig. 6.8). This curve may contain a peak corresponding to the geometry and energy of the transition state. We might say that electron transfer involves a motion of the imaginary nuclear point from the initial state over this peak onto the final state.

To solve the eigenequation of the "new" system, we write new eigenfunctions for the reactant and product states:

$$\mathcal{H}\Psi_+ = E_+\Psi_+ \tag{6.29}$$

$$\mathcal{H}\Psi_- = E_-\Psi_- \tag{6.30}$$

where E_+ and E_- represent the "perturbed" or "shifted" energies. Then the wavefunctions for the eigenequations are

$$\Psi_+ = c_1\Psi_i + c_2\Psi_f \tag{6.31}$$

$$\Psi_- = c_1\Psi_i - c_2\Psi_f \tag{6.32}$$

where c_1 and c_2 are coefficients. Thus, when the donor and acceptor molecules begin to interact, the initial energies of the system, E_i and E_f, are shifted to new "perturbed" energies. This is exemplified schematically in Fig. 6.8. As the curves become closer, the perturbation becomes stronger resulting in a greater repulsion. At strong perturbations, the curves of the initial and final states are finally split into a new set of curves.

It is important to know the shapes of the new curves that are formed from perturbation. We can describe these new curves only if we can calculate the energies corresponding to their shapes. The following derivation shows how in principle these energies can be calculated.[8] From Eqs. 6.31 and 6.32, we have

$$c_1(H_{ii} - E) + c_2 H_{el} = 0 \quad (6.33)$$

$$c_1 H_{el} + c_2(H_{ff} - E) = 0 \quad (6.34)$$

where H_{el} represents the electron-transfer matrix element. The energies, E_+ and E_-, which result from the interaction of the two curves, are the roots of the following secular determinant:

$$\begin{bmatrix} H_{ii} - E & H_{el} \\ H_{el} & H_{ff} - E \end{bmatrix} = 0 \quad (6.35)$$

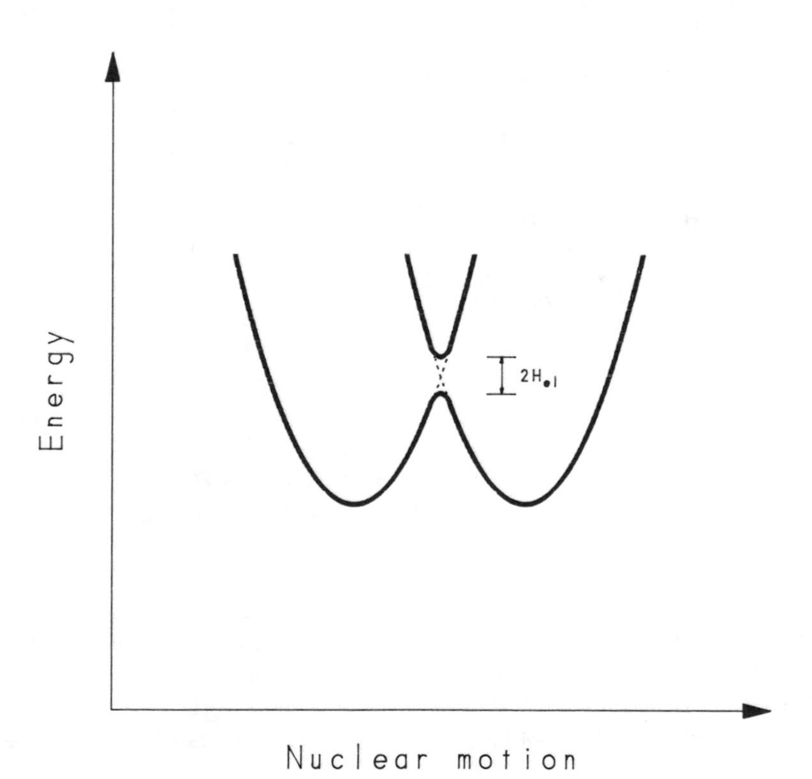

Figure 6.8 A magnified portion of two intersecting curves in electron transfer. The splitting between the curves is related to the interaction energy, H_{el}.

(H_{ii} and H_{ff} can be visualized as diagonal elements that cross at the intersection of the zero-order potential energy curves.) Solving the determinant gives the energies of the perturbed potential curves:

$$E_{+} = \tfrac{1}{2}\{H_{ii} + H_{ff} + [(H_{ii} - H_{ff})^2 + 4H_{el}^2]^{1/2}\} \tag{6.36}$$

$$E_{-} = \tfrac{1}{2}\{H_{ii} + H_{ff} + [(H_{ii} - H_{ff})^2 + 4H_{el}^2]^{1/2}\} \tag{6.37}$$

At the point where the two potential-energy curves intersect, we can assume that $H_{ii} = H_{ff}$, so that $E_{+} - E_{-} = 2H_{el}$. As shown in Fig. 6.8, $2H_{el}$ represents the repulsion between the potential-energy curves. The point of maximum splitting of the two curves represents the point of greatest perturbation.

To calculate the energy on either side of the intersection, we rewrite Eqs. 6.36 and 6.37:

$$E_{\pm} = \frac{1}{2}\left\{H_{ii} + H_{ff} \pm (H_{ii} - H_{ff})\left[1 + \frac{4H_{el}^2}{(H_{ii} - H_{ff})^2}\right]^{1/2}\right\} \tag{6.38}$$

We assume that at small perturbations the interaction is small compared to the energy separation, so that

$$\left|\frac{H_{el}}{H_{ii} - H_{ff}}\right| \ll 1 \tag{6.39}$$

A series expression can be applied to Eq. 6.38 to arrive at

$$E_{\pm} = \frac{1}{2}\left\{H_{ii} + H_{ff} \pm (H_{ii} - H_{ff})\left[1 + \frac{2H_{el}^2}{(H_{ii} - H_{ff})^2}\right]\right\} \tag{6.40}$$

Solving for the perturbed energies, we obtain

$$E_{+} = H_{ii} - \frac{H_{el}^2}{(H_{ff} - H_{ii})} \tag{6.41}$$

and

$$E_{-} = H_{ff} + \frac{H_{el}^2}{H_{ff} - H_{ii}} \tag{6.42}$$

When $H_{el} \ll H_{ff} - H_{ii}$, it follows that $E_{+} \simeq H_{ii}$ and $E_{-} \simeq H_{ff}$.

The term *adiabatic* is used to describe two eigenfunctions that "avoid" one another and do not intersect. "Avoiding" curves do not cross because of strong perturbations between the zero-order states. For our purposes, reactions that take place on a single potential energy surface are classified as adiabatic. Potential-energy surfaces that do not interact but intersect are called *diabatic* or *nonadiabatic*. The interaction between diabatic curves is weak as compared with the strong mixing in the adiabatic case. Figure 6.9 illustrates the conceptual difference between adiabatic and diabatic curves. As a rough rule, $H_{el} \geqq 0.5$ kcal mol^{-1} for an adiabatic process.

In electron-transfer reactions, the interaction between two reactants may range

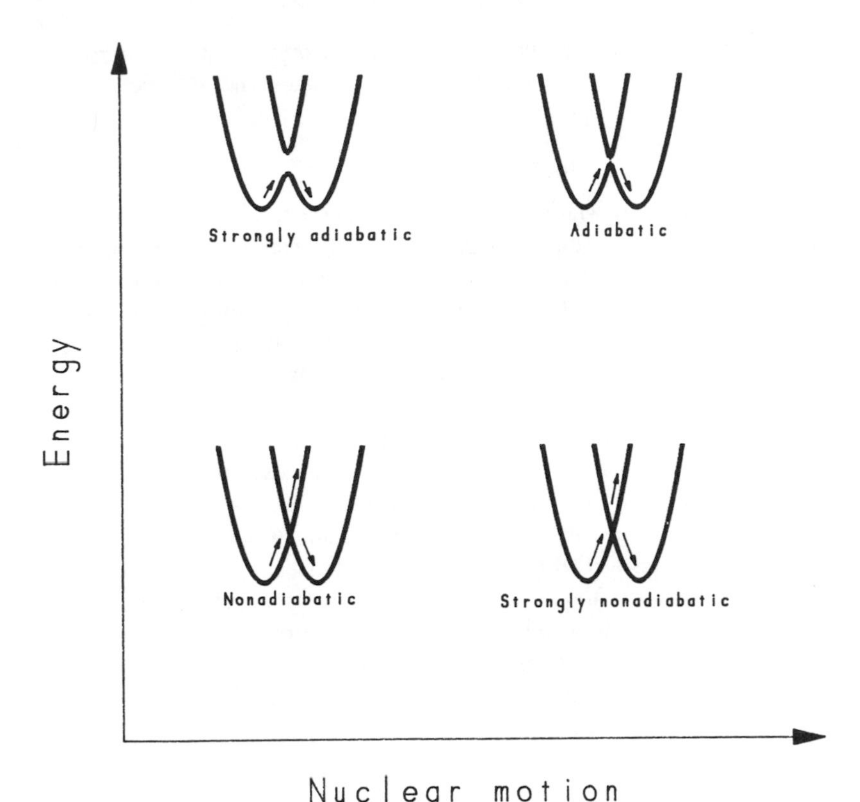

Figure 6.9 Representation of adiabatic and nonadiabatic intersecting curves. In strongly adiabatic electron transfer, H_{el} is sufficiently large so that the reaction "proceeds" on one curve; the crossing is said to be "avoided". In strongly nonadiabatic electron transfer, the "moving" point oscillates many times on the reactant curve before "jumping" to the product curve. The crossing, when it does occur, is "nonadiabatic". (Adapted with permission from Kavarnos, G. J. *Top. Curr. Chem.* **1990,** *156,* 21. Copyright 1990 Springer-Verlag.)

from the strongly nonadiabatic to the strongly adiabatic. The former case may represent reactants separated by large distances. Here two intersecting curves may be plotted for two reactant partners. At large separation distances, there is only a very weak interaction between the curves (i.e., $H_{el} \simeq 0$). If the reactants remain at these distances, the reactant and product curves retain their original shapes and features. However, when the reactant molecules approach to distances within molecular dimensions, the interaction between them increases. The curves then merge and split into adiabatic curves by an amount proportional to the interaction energy. The reactant curve now merges with the product curve at the transition state. The imaginary point describing nuclear motion now can pass smoothly from the reactant state to product state.

For nonadiabatic electron transfers, crossings between the zero-order states are influenced by the shape of the potential-energy curves and by the velocity of the moving nuclear point. The probability of electron transfer at the intersection or crossing of the diabatic curves is given by the Landau–Zener equation[9,10]:

$$\kappa_{el} = 1 - \exp\left(-\frac{4\pi^2 H_{el}^2}{hv\Delta s}\right) \tag{6.43}$$

where v is the velocity by which the system crosses the top of the activation barrier. Δs is the difference of the slopes represented by the intersecting zero-order states. Equation 6.43 links nuclear motion, v, with electronic motion. Consider two reactants approaching one another with a certain velocity from an infinite distance. Rapidly approaching reactants are likely to remain on the reaction curve and not cross at the intersection. This is the nonadiabatic case. Stated another way, the nuclear point moves back and forth many times on the reaction curve, and since it does not spend enough time at the crossing point, electron transfer remains improbable. On the other hand, when the nuclear point moves more slowly, electron transfer becomes more probable and thus more adiabatic. Thus, from Eq. 6.43, we deduce that when the velocity is large, $\kappa_{el} \ll 1$, and when v is small, $\kappa_{el} \simeq 1$. Equation 6.43 also tells us that the slopes of the intersecting curves may influence κ_{el}. With shallow slopes, the electrons constantly respond to the forces imposed by the changing nuclei. A smooth transition from the reactant to product curve is ensured ($\kappa_{el} \simeq 1$). When the slopes are steep, however, the electronic wavefunction responds more abruptly to nuclear changes. There is only a small portion of the nuclear field to which the electrons can respond ($\kappa_{el} \ll 1$). The reaction, when it

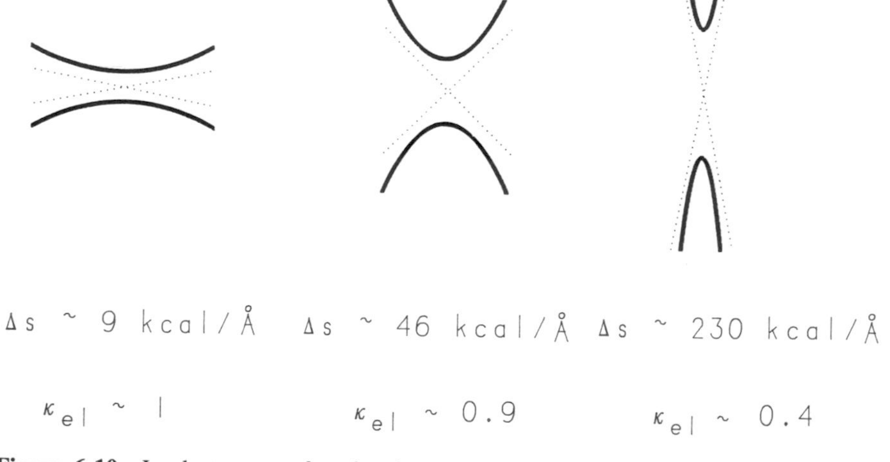

Figure 6.10 In electron transfer, the slopes of the intersecting curves near the crossing point determine the magnitude of κ_{el}. For the case of shallow slopes, the nuclei respond smoothly to changes in the electronic wavefunction. With steeper slopes, the changes along the surface are rapid and abrupt: the electronic interaction is minimal but the probability of a transition to the higher surface is large.

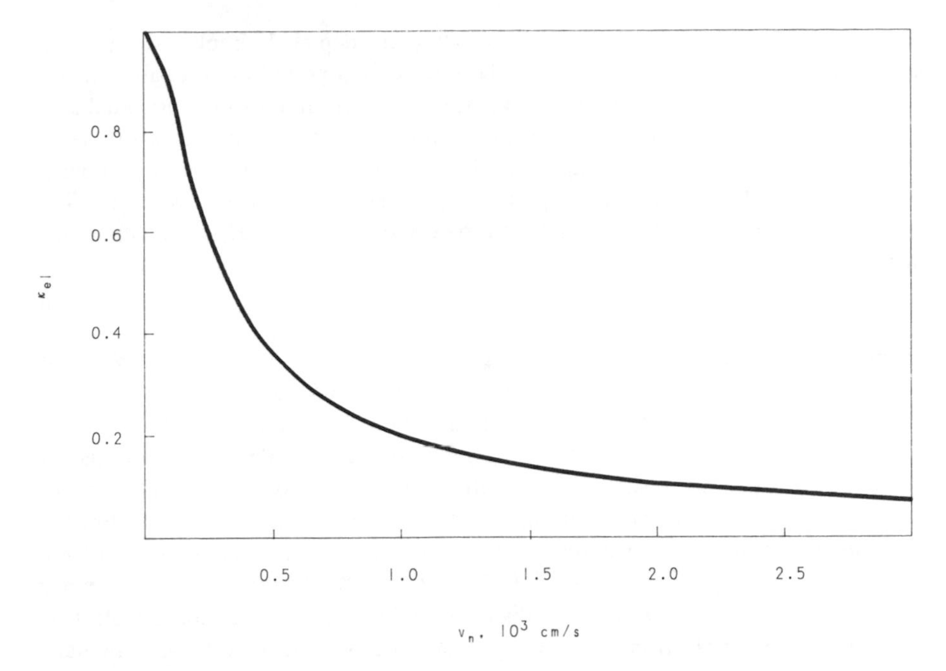

Figure 6.11 Nuclear velocity influences κ_{el}, according to the Zandau-Zener law. With low velocities, the nuclei "feel" the electronic field at all times, enhancing the magnitude of κ_{el}.

occurs, is nonadiabatic. Graphic representations of the effects of nuclear velocities and steepness of the slopes are shown in Figs. 6.10 and 6.11.

An adiabatic electron transfer takes place entirely on one surface. The interaction energy is not so large as to lower the magnitude of the activation energy. However, as H_{el} becomes larger, $\Delta G_{el}{}^{\neq}$ is "squeezed" to a smaller value until the interaction energy between two reactants is very strong (strongly adiabatic), so that resonance stabilization in the transition state becomes the dominant force. This kind of interaction is found in the reaction of planar aromatic molecules to form face-to-face exciplexes. During the formation of exciplexes, electronic interaction is very strong, and charge is transferred at maximum rates. These reactions are thus controlled by diffusion. On the other hand, in classical theories the upper limit of the unimolecular rate of electron transfer in the transition state is $\sim 10^{13}\ s^{-1}$. Nuclear adjustments, which are the focus of classical theories, intervene to reduce this rate to lower than diffusion-controlled values. In the classical theory, the split is large enough so that the curves are adiabatic (in effect allowing electron transfer to proceed smoothly from the precursor to the successor state) yet not so large so that resonance dominates.

What happens if H_{el} approaches zero? The rate of electron transfer falls dramatically. That is, when there is little interaction between the two reactants, the initial and final parabolic curves representing the precursor and successor states, respectively, retain their original shapes. They are depicted simply as two intersecting zero-order curves. Crossing at the intersection point is improbable. In the unlikely event that a crossing does take place, the electron transfer is nonadiabatic. Nonadiabatic electron transfer will occur when there are weak interactions between the donor and acceptor. For example, if the reactants are held rigidly at large distances, then $H_{el} \simeq 0$, and the probability of electron transfer decreases. The following expression can be used to estimate the changes in H_{el} with separation distance:

$$H_{el} = H_{el}^{0} \exp[-\beta(d - d_0)] \tag{6.44}$$

where β is an orbital parameter, d is the actual separation distance, and d_0 is the van der Waals distance or separation distance within the encounter complex. β is inversely proportional to the orbital overlap between donor and acceptor orbitals with values ranging from 0.85 to 2.5 Å^{-1}. The decrease in the interaction energy predicted by Eq. 6.44 is shown graphically in Fig. 6.12 for several values of β. Since β is a measure of the ability of an orbital to extend into space and interact with another orbital, the magnitude of interaction between donor and acceptor orbitals is inversely proportional to β. Equation 6.44 is frequently used when examining the effect of separation distance in biological photoinduced electron transfers where the electron donor and acceptor are fixed relative to one another.

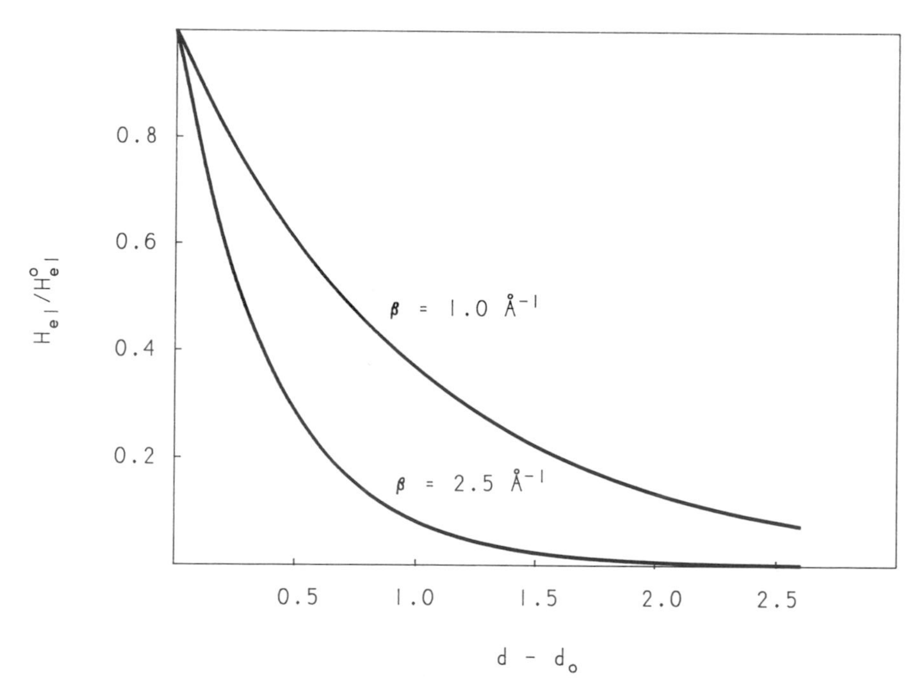

Figure 6.12 H_{el} depends on separation distance, as this figure demonstrates. The orbital parameter, β, has an additional effect on H_{el}.

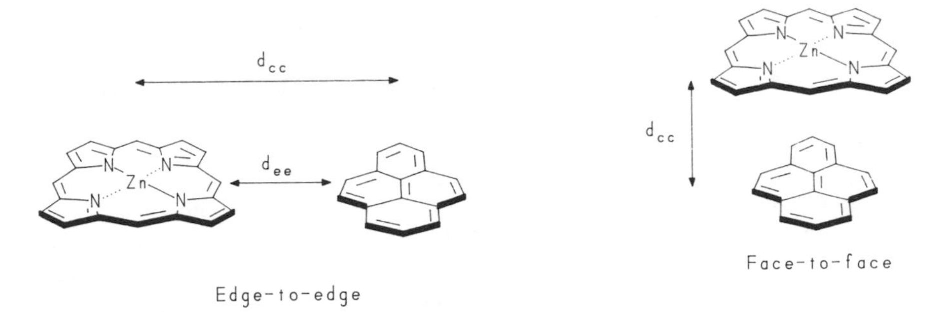

Figure 6.13 In electron transfer, the approach of large aromatic hydrocarbons and porphyrin molecules to one another can be visualized as edge-to-edge or face-to-face.

We can also write Eq. 6.44 as

$$k_{el} = k_{el}^{0} \exp[-\beta(d - d_0)] \tag{6.45}$$

where $k_{el}^{0} \simeq 10^{13}\ s^{-1}$. Provided that other factors beside distance do not influence the rate, Eq. 6.45 can be used to estimate the rate at a known separation distance.

For electron transfer in intramolecular systems, we can use Eq. 6.46[11]:

$$H_{el} \propto \exp[-\beta'(\text{no. bonds} - 1)] \tag{6.46}$$

According to Eq. 6.46, the rate should decrease with increasing number of σ-bonds. As a rule, a rate decrease of a factor of 10 can be expected for a distance of 2.3 Å, which corresponds to about two σ-bonds.

So far we have focussed on the effect of distance. But there are other parameters that affect the interaction between two reactants. Thus, spin changes, symmetry factors, and the relative orientation of two reactants may influence the magnitude of H_{el}. Figure 6.13 demonstrates a possible role due to the mutual orientation of two planar aromatic molecules fixed at a specific orientation. The illustration shows two possible orientations of the reactants: *face-to-face* or *edge-to-edge*. The interaction energy, H_{el}, was shown by detailed calculations to depend on the separation distance, shape, nodal character, and mutual orientation of the orbitals.[12] These calculations are applicable to nonadiabatic electron transfer and suggest that the shape and orbital character (number of nodes) of the reactants influence H_{el} and therefore can control the rate of the electron transfer.

6.2.5. Potential-Energy Curves in Photoinduced Electron Transfer

Although some of the concepts presented in the previous sections were originally intended for electron transfer between ground-state molecules, they are powerful vehicles for evaluating reactivity in photoinduced electron transfer. Figure 6.14 displays potential energy curves for photoinduced electron transfer to form an ion pair from neutral reactants. We assume that the internal energy is temperature independent. This assumption permits us to express the energy coordinate in Fig. 6.14 as a free-energy axis. Each parabola is arranged vertically to reflect the free-energy changes that accompany excitation of the sensitizer and subsequent electron transfer. The free energy corresponding to the intersection of the excited state and ionic curves represents the free-energy barrier of electron transfer.

Photoexcitation of the ground-state molecule proceeds vertically to a vibrational level in the excited state, although in Fig. 6.14, for clarity, the vibrational levels are omitted. After formation of the Franck–Condon state, thermal equilibration to the lowest vibrational level rapidly occurs, resulting in formation of a relaxed state. The next step is formation of the precursor complex. Here the equilibrium constant for the formation of reactant pairs is

$$K_{eq} = \frac{4\pi N d_{cc}^2 \delta d}{1000} \exp\left(-\frac{w_c}{RT}\right) \tag{6.47}$$

where w_c is the work needed to bring the reactants to a separation distance of d_{cc}.[13] In Eq. 6.47, d_{cc} and $d_{cc} + \delta d$ are meant to designate center-to-center distances

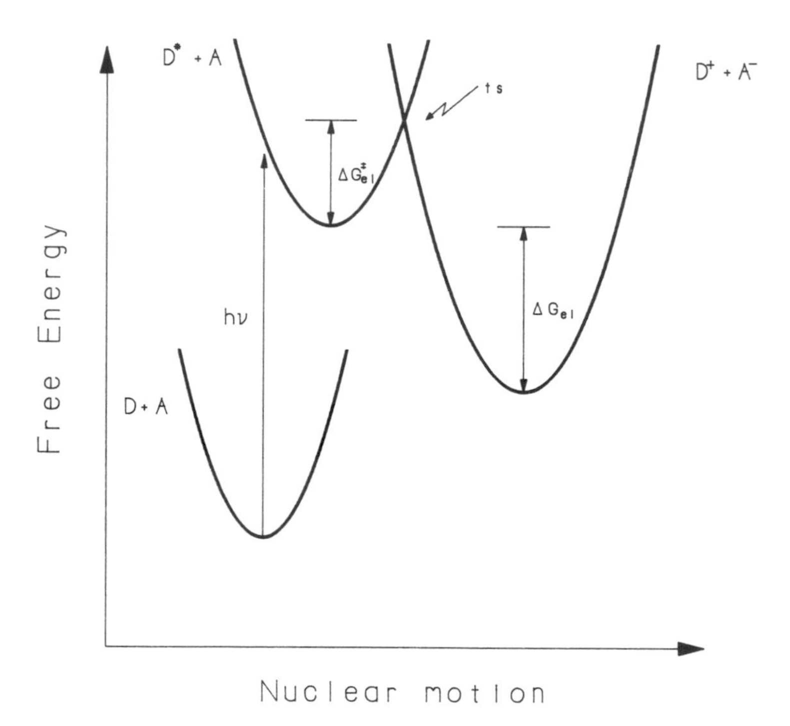

Figure 6.14 In photoinduced electron transfer, the initial step is excitation of the sensitizer-quencher pair to an excited-state curve. This excitation can be regarded as a vertical transition to a Franck-Condon state. This state rapidly undergoes equilibration. Electron transfer takes place at the crossing of the equilibrated excited-state surface and the product surface. (Adapted with permission from Kavarnos, G. J. *Top. Curr. Chem.* **1990,** *156,* 21. Copyright 1990 Springer-Verlag.)

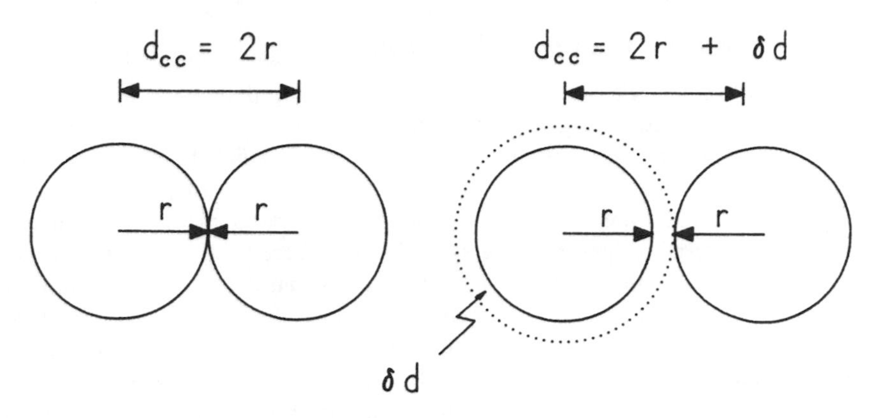

Figure 6.15 The relationship among d_{cc}, $d_{cc} + \delta d$, and r.

over which electron transfer takes place (Fig. 6.15). In general, values of $d_{cc} \sim$ 6–8 Å and $\delta d \simeq$ 2 Å are assumed. The work term, w_c, is represented by Eq. 6.48 with the assumptions that $d_{cc} = r_D + r_A$ and $r = r_D = r_A$ so that $d_{cc} = 2r$:

$$w_c\ (\text{kcal mol}^{-1}) = \frac{331 z_D z_A e^2}{\varepsilon_s d_{cc}(1 + Bd\mu^{1/2})} \tag{6.48}$$

where

$$B = \left(\frac{8\pi N e^2}{1000\varepsilon_s k_B T}\right)^{1/2} \tag{6.49}$$

In polar solvents the work term is generally only a few kilocalories per mole and is negligible ($w_c \sim 0$ kcal mol^{-1}) for neutral molecules. In nonpolar solvents, the work term increases significantly.

Equation 6.47 is formulated assuming that the major contribution to the rate of electron transfer occurs within a small range of separation distances. Actually, the reactants do not have to be in actual contact (i.e., $d_{cc} > r_D + r_A$). For ellipsoidal molecules, d_{cc} can even be smaller than $r_D + r_A$. For these cases, more elaborate work terms may be needed.

Now let us turn to the transition state. Following formation of the precursor complex, nuclear readjustments to form the distorted geometry of the transition state take place. This nuclear configuration is a *nonequilibrium* geometry where the actual electron transfer takes place. Following electron transfer, nuclear relaxation quickly establishes the thermally equilibrated product ions of the successor state. If the products are charged species, then the work involved as the ions separate is $w_c' = -w_c$.

We now want to inquire about the nature of the nuclear changes accompanying the transfer of an electron from the neutral donor–acceptor surface to ionic surface. We turn to a theory developed by Marcus for a physical description of electron transfer between an excited-state and ground-state molecule.

6.2.6. The Free Energy of Activation

As mentioned earlier, classical theories assume that the donor and acceptor are within contact distance during electron transfer and, consequently, experience a weak electronic interaction. This is the so-called weakly adiabatic case. Since κ_{el} is assumed to be approximately unity, the rate-determining factors in weakly adiabatic electron transfer are assumed to involve nuclear reorganization. The major problem is to determine how reacting systems can overcome free-energy barriers and finally proceed to the product state. The factors that determine the nature of the transition state in electron transfer generally include the bond changes that take place during formation of the transition state and the changes in the orientations of the solvent molecules during electron transfer. Each of these processes is characterized by a free energy of activation. Thus, the total free energy of activation is the sum of the individual free energies:

$$\Delta G_{el}^{\neq} = \Delta G_{v}^{\neq} + \Delta G_{s}^{\neq} \tag{6.50}$$

where the subscripts refer to the energies involving bond distortions (v = vibrational) and solvent changes in the ionic sphere surrounding the reactants (s = solvent). It should be noted that electron transfers for which $\Delta G_{el}^{\neq} > 0$ are said to proceed through a nuclear "barrier"—the latter consisting of bond changes and motions of the solvent dipoles. When $\Delta G_{el}^{\neq} \simeq 0$, $\kappa_n \simeq 1$, we speak of a "barrierless" electron transfer (in terms of nuclear changes only). Here $k_{el} \simeq \nu_n \simeq 10^{12}$–$10^{14}$ s^{-1} (assuming that $\kappa_{el} \simeq 1$, see Eq. 6.14). This range of rate constants is roughly equivalent to the vibrational frequencies which "create" and then "destroy" the transition state.

Marcus derived an important equation that takes into account the effect of the driving force, ΔG_{el}, on the free energy of activation[14–16]:

$$\Delta G_{el}^{\neq} = \frac{\lambda}{4}\left(1 + \frac{\Delta G_{el}}{\lambda}\right)^2 \tag{6.51}$$

where λ is defined as the total reorganization energy. This important parameter can be separated into two terms. Thus, $\lambda = \lambda_v + \lambda_s$, where λ_v is the *inner-sphere* reorganization energy and λ_s is the *outer-sphere* reorganization energy. Inner-sphere reorganization refers to the energy changes accompanying changes in bond lengths and bond angles during the electron-transfer step. Outer-sphere reorganization energy is the energy change as the solvent shells surrounding the reactants rearrange. For an electron exchange, since $\Delta G_{el} = 0$, the inner- and outer-sphere reorganization energies are related to the corresponding activation energies as shown below:

$$\Delta G_{v}^{\neq} = \frac{\lambda_v}{4} \tag{6.52}$$

$$\Delta G_{s}^{\neq} = \frac{\lambda_s}{4} \tag{6.53}$$

λ can therefore be estimated from the rate constants of electron exchanges. In turn, the activation barriers for excited-state electron–transfer reactions can be calculated from a knowledge of the reorganization energies. Since the reorganization energies are known for many electron-exchange reactions, the following relationship is useful:

$$\lambda_{DA} = \frac{\lambda_{DD} + \lambda_{AA}}{2} \tag{6.54}$$

where D or A can be an excited state. Let us say that we know λ_{DD} and λ_{AA} for the electron exchanges $[D^m]^* + D^{m+1} \rightleftharpoons D^{m+1} + [D^m]^*$ and $A^{n+1} + A^n \rightleftharpoons A^n + A^{n+1}$, respectively. Then it becomes a straightforward matter to calculate λ_{DA}.

There are also equations for estimating λ. These equations allow one to estimate $\Delta G^{\neq}$ and determine the effects of structure and solvent on the rate of the electron transfer. We shall examine these relationships in the next section.

6.2.7. Nuclear Reorganization

How, one may ask, is the transition state reached in electron transfer? We recall Libby's hypothesis stating that nuclear reorganization occurs *after* electron transfer. In Marcus theory, however, it is acknowledged that energy and momentum must be conserved. The system gains energy from thermal collisions with surrounding molecules until the energy barrier of the transition state is reached. Energy is transferred from the thermal environment to the vibrational modes of the reactants as they undergo structural distortions. Energy is also transferred to the solvent molecules surrounding the reactants. Thus, changes in the bond lengths and solvent molecules are involved in the process called nuclear reorganization.

6.2.7.1. Bond Reorganization

The free-energy barrier associated with changes in bond lengths—the inner-sphere reorganization energy—is given by the following equation[17]:

$$\lambda_v = \sum_i \left[\frac{f(R)_i\, f(P)_i}{f(R)_i + f(P)_i}\right] [\Delta q_i]^2 \tag{6.55}$$

where Δq_i is the difference in equilibrium bond distance between the reactant and product state corresponding to a ith vibration, and $f(R)_i$ and $f(P)_i$ are the force constants for this vibration for a reactant and product molecule. The quantity contained in brackets is referred to as the *reduced force constant*. Equation 6.55 includes the summation of all vibrational modes and applies to the limiting case where all vibrational states (assumed to display harmonic behavior) are populated. This latter condition represents a case where λ_v is assumed to be temperature independent.

Let us demonstrate the calculation of λ_v for the electron-exchange $Fe(H_2O)_6^{2+} + Fe(H_2O)_6^{3+} \rightleftharpoons Fe(H_2O)_6^{3+} + Fe(H_2O)_6^{2+}$.[18,19] First, the Fe-$H_2O$ metal-ligand bond-length change is 0.14 Å, or 0.14×10^{-8} cm. Using the values 1.5×10^5

and 4.2×10^5 dyne cm^{-1} as the force constants for the stretching of the Fe-H_2O metal-ligand bond in $Fe(H_2O)_6^{2+}$ and $Fe(H_2O)_6^{3+}$, respectively, we then use Eq. 6.55 to calculate λ_v:

$$\begin{aligned}\lambda_v &= 12 \times \left(\frac{1.5 \times 10^5 \times 4.2 \times 10^5}{1.5 \times 10^5 + 4.2 \times 10^5}\right)(0.14 \times 10^{-8})^2 \\ &= 0.26 \times 10^{-11} \text{ ergs} = 37 \text{ kcal mol}^{-1}\end{aligned} \tag{6.56}$$

(Conversion from ergs to kilocalories per mole is done by using a factor of 1.439×10^{13}.) Note that since the transition state represents two symmetrical ions, there are *twelve* bonds to consider.

A similar treatment can be applied to the quenching of excited states by an electron-transfer mechanism. If, for example, bond-length changes involving the excited state or the quencher are known to accompany the reaction, then these changes can be used to estimate λ. The only caveat is that it is quite difficult to pinpoint bond-length changes for excited states, which have only fleeting existences. On the other hand, it may be possible to restrict bond-length changes by some structural or environmental feature and compare the rate with the "unrestricted" case.

6.2.7.2. Solvent Reorganization

Solvent reorganization refers to the effects of orientational changes in the solvent molecules surrounding the reactants during electron transfer. In this section we show that solvent plays a special role in establishing the unique nature of the transition state in electron transfer.

The reorganization due to the solvent contribution is closely related to the polarization of the solvent molecules surrounding a reactant pair. The solvent polarization is the sum of an orientational–vibrational and an electronic component[20]:

$$\mathbf{P} = \mathbf{P}_e + \mathbf{P}_u \tag{6.57}$$

$\mathbf{P}$ is the total polarization, $\mathbf{P}_e$ is the electronic polarization, and $\mathbf{P}_u$ is the orientational–vibrational polarization of the solvent molecules.

What is meant by the term polarization? Consider what happens if a substance is placed in a strong electric field between the positive and negative plates of a condenser. Electrons within the substance immediately are pulled toward the positive plate. This polarization is associated with $\mathbf{P}_e$ and the optical dielectric constant, ε_{op}, which is in turn equal to the square of the refractive index, n^2. Should the field within the condenser oscillate rapidly, the electrons respond almost instantaneously to the rapidly changing electric field. $\mathbf{P}_u$ represents the polarization resulting from the alignment of the dipole moments with the negative pole pointing toward the positive plate. $\mathbf{P}_u$ is associated with the static dielectric constant, ε_S, which we introduced in Chapter 1. The orientation polarization, $\mathbf{P}_u$, unlike $\mathbf{P}_e$, responds more slowly at a much lower frequency with an oscillating electric field.

The polarization components, $\mathbf{P}_e$ and $\mathbf{P}_u$, are vector quantities with specified magnitudes and directions. Since $\mathbf{P}_e$ consists of the rapid motions of the solvent electrons, this component always remains in equilibrium with any change in the charge distribution that may take place in reactant molecules in the solvent medium. Order of magnitude estimates of the time required for the electronic, atomic, and orientation adjustments are given in Fig. 6.2.

To understand the role of the orientational–vibrational and electronic polarizations in electron transfer, one should consider the illustration in Fig. 6.16. Initially, a pair of reactant molecules is in a thermally equilibrated state surrounded by a cage of solvent molecules. In the transition state, the electronic polarization of the solvent molecules remains in equilibrium with the charges on the reactants; the orientational–vibrational polarization, however, adjusts to a hypothetical charge in the transition state. The arrangement of the solvent molecules, or the *nonequilibrium polarization,* corresponds to a minimum reorganizational energy. This arrangement of solvent molecules stabilizes the hypothetical charge configuration of the transition state. In the transition state, when the electron is transferred, the electronic polarization responds instantly to electronic motion. But the orientational–vibrational polarization remains fixed. As the successor state is formed from the transition state, the electronic polarization and nuclear polarization adjust to the new charges in the products.

It follows that the transition state can be described by two states, $p^{\neq}$ and $s^{\neq}$, preceding and following the electron transfer. The electronic polarization of these states as well as the equilibrated reactant and product states depends on the charges of each of these states, but the orientational–vibrational polarization of states $p^{\neq}$ and $s^{\neq}$ remains in nonequilibrium with the charge distributions in the transition state. The free energy of the transition state, or in the present context, the nonequilibrium state, is defined as the reversible work performed as the orientational–vibrational component of the solvent molecules adjusts to the hypothetical charge in the transition state. This energy barrier corresponds to the outer-sphere reorganization energy and is given by the following expression:

$$\lambda_s\ (\text{kcal mol}^{-1}) = 332\Delta e^2 \left(\frac{1}{r_D} + \frac{1}{r_A} - \frac{1}{d_{cc}}\right)\left(\frac{1}{\varepsilon_{op}} - \frac{1}{\varepsilon_s}\right) \tag{6.58}$$

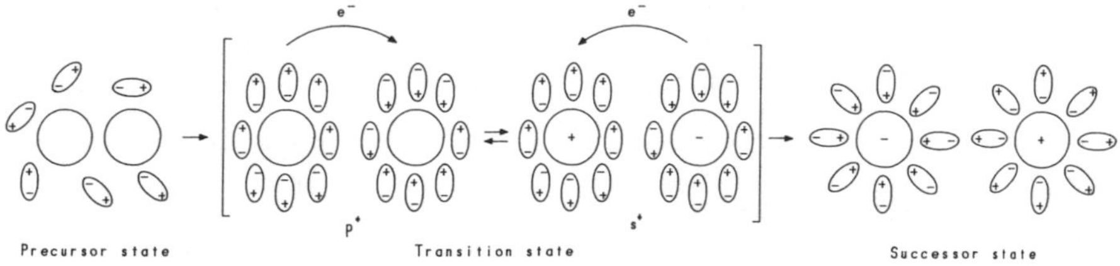

Precursor state

Transition state

Successor state

Figure 6.16 Illustration showing the motion of solvent dipoles during electron transfer (see text).

Table 6.3 PROPERTIES OF SELECTED SOLVENTS[a]

Solvent	ε_s	n
Water	80.2	2.3
Glycerol	42.5	1.5
Acetonitrile	37.5	1.3
Methanol	33.7	2.4
Ethanol	24.45	1.4
n-Propanol	20.6	1.9
Benzene	2.275	1.5

[a] ε_s, static dielectric constant; *n*, refractive index; data from Murov, S. L. *Handbook of Photochemistry*. Marcel Dekker, New York, 1973.

ε_{op}, the optical dielectric constant, is the square of the refractive index of the solvent, and ε_s is the static dielectric constant, sometimes referred to as the relative permittivity. To aid the reader, Table 6.3 lists the static dielectric constants and refractive indices for some solvents.

Equation 6.58 is based on a fairly simple physical picture of two spherical reactants surrounded by a cage of solvent molecules. There are several points to note. First, in polar solvents the reorganization energy is greater than in nonpolar solvents (Table 6.4). In other words, the energy required to reorganize polar solvent molecules from the initial state to the transition state is greater than for less polar molecules such as benzene. Intuitively, this may seem obvious since polar solvent molecules must undergo substantial nuclear changes to adjust to the hypothetical charge in the transition state. It is also apparent from the data in Table 6.4 that the solvent barrier is greater for smaller reactants. This is true for spherical reactants. If the reactants are shaped like ellipsoids, more sophisticated expressions for λ_s are needed. These are given in the references listed at the end of this chapter.

The presence of d_{cc} in Eq. 6.58 implies that the solvent barrier is greater at larger separation distances. This is under the assumption that the reactants are not

Table 6.4 OUTER-SPHERE REORGANIZATIONAL ENERGIES[a]

r_D (Å)	r_A (Å)	Water (kcal mol^{-1})	CH_3CN (kcal mol^{-1})	Ethanol (kcal mol^{-1})	Glycerol (kcal mol^{-1})	Benzene (kcal mol^{-1})
2	2	45.7	43.7	41.4	36.2	0.35
4	2	38.1	36.4	34.5	30.2	0.29
6	2	38.1	36.4	34.5	30.2	0.29
4	4	22.9	21.9	20.7	18.1	0.18
6	4	19.8	19.0	17.9	15.7	0.15
6	6	15.2	14.6	13.8	12.1	0.12

[a] Calculated with Eq. 6.58 with values from Table 6.3.

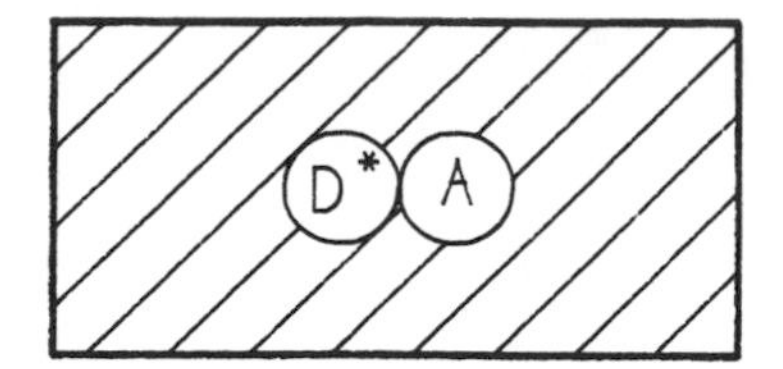

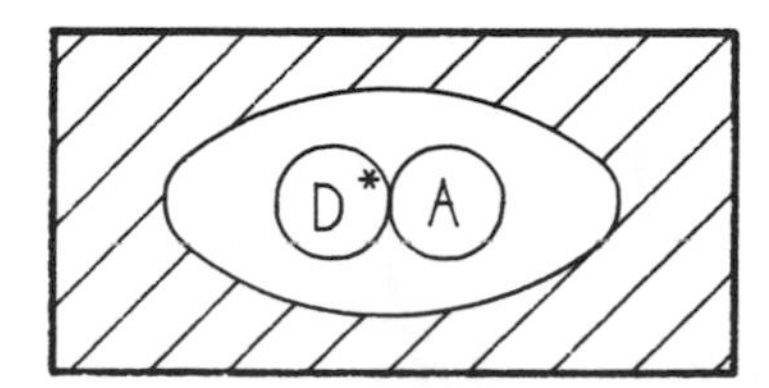

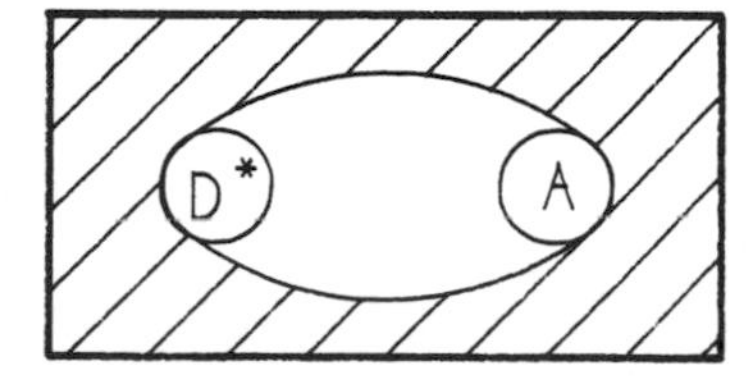

Figure 6.17 λ_s is influenced by a number of factors including the distance separating the reactants, as well as the polarity of the solvent.

in physical contact. A more realistic picture is that the reactants are positioned at contact distance within an ellipsoidal cavity, which itself is immersed in the dielectric continuum. To describe this picture more accurately, we would need rather complicated expressions. We can, however, visualize the effects of the ellipsoidal cavity on λ_s qualitatively (Fig. 6.17). In this conceptualization, the cavity is pictured as a nonpolar volume within a polar medium. At small separation distances, the reactants "feel" the relatively nonpolar environment of the cavity (Fig. 6.17). λ_s then should be small. But when the separation distance increases, λ_s also increases since now the reactants are now closer to the dielectric continuum of the solvent and are now more sensitive to changes in the more polar environment.

6.2.8. Potential-Energy Curves in Optical Electron Transfer

As was mentioned in Section 4.3, potential energy parabolic curves can also be drawn to represent optical electron-transfer transitions taking place in bridged metal–ligand systems. In optical electron transfer, photoexcitation converts the reactants directly into the ion-pair intermediates. Recall that photoinduced electron transfer, unlike optical electron transfer, is actually a thermal electron transfer, which is subsequent to photoexcitation of one of the reactants. Figure 6.18 shows potential-energy curves for an optical electron exchange (reaction 6.19). Photoexcitation is represented as a vertical transition from the reactant to product surface (E_{op} is the energy of the photon that excites the transition). The actual electronic transition occurs at a nuclear geometry midway between the reactant and product states. This crossing point can be regarded as a transient photoexcited complex having

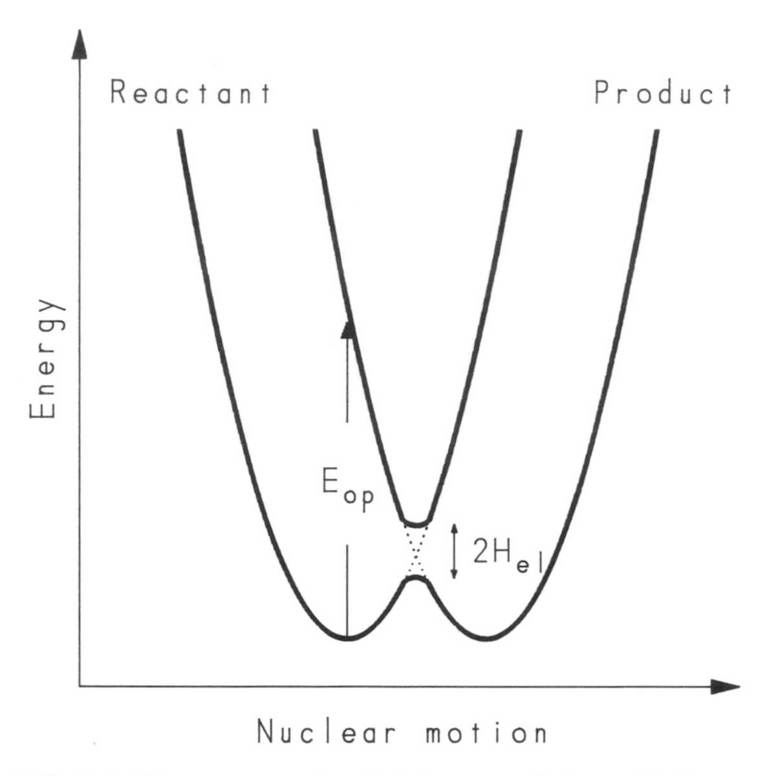

Figure 6.18 Potential energy curves in optical charge-transfer in a multivalence complex. E_{op} represents the photon energy of light-induced electron transfer.

a geometry just between the geometries of the two valence states. The activation energy consists of solvent and bond contributions. The Marcus treatment outlined earlier can be applied to calculate these energy barriers.

6.3. The Inverted Region

In Chapter 4, it was stated that k_{el} increases with increasing $-\Delta G_{el}$ at small exothermicities but eventually at much larger exothermicities begins to decrease with further increases in $-\Delta G_{el}$. In this section, we shall examine the theory behind this trend.

Mathematically, Eq. 6.51 describes a quadratic relationship between the driving force and the free energy of activation. The physical meaning of this relationship will become apparent if we take a closer look at the rate constants of an excited state interacting with a series of homologous quenchers by an electron-transfer mechanism. If the driving force of each one of these electron transfers is plotted vs. the measured rate constants, then, according to Eq. 6.51, the following trends should be observed: initially the rate will increase with an increase in driving force but level off at some value of ΔG_{el} where the rate maximizes. With further increases in the driving force, the rate should progressively decrease again.

Consider the intersecting reactant and product potential energy curves shown in Fig. 6.19. Starting with reactant and product curve a, we note that with increasing exothermicities, the activation energy progressively decreases until at some point (shown by the intersection of the reactant curve and curve b) it reaches zero. Here the rate attains its maximum value. At still greater driving forces, the activation energy begins to increase again (intersection of the reactant curve with curve c).

Because of the parabolic relationship between the driving force and activation energy, two kinetic domains may be defined. The *normal region* is the area in Fig. 6.19 where $-\Delta G_{el} < \lambda$. The second domain where $-\Delta G_{el} > \lambda$ is known as the

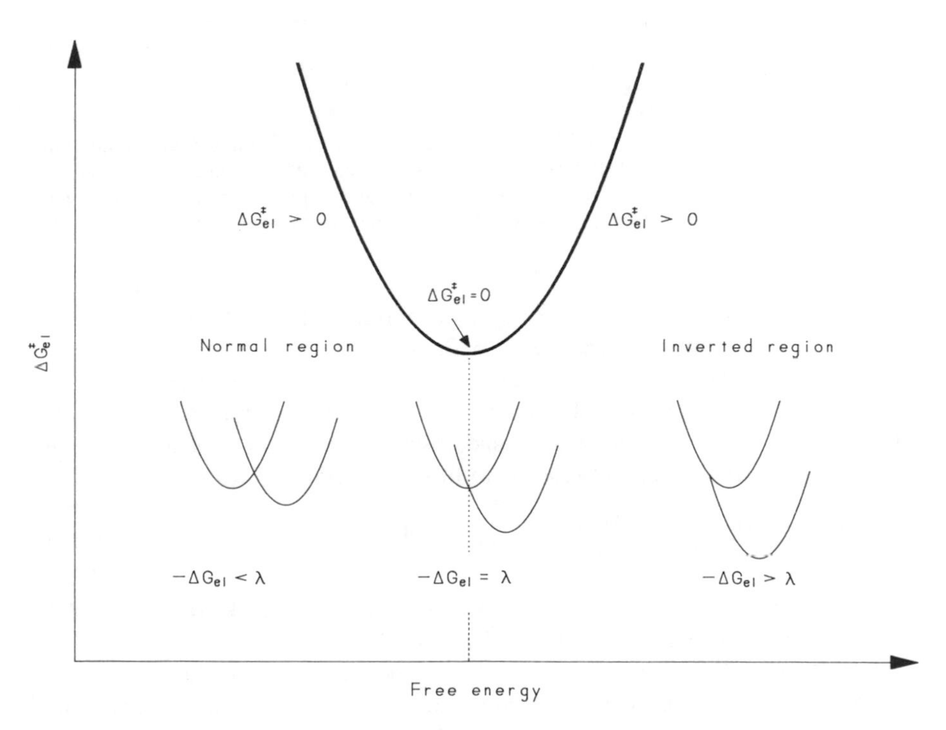

Figure 6.19 The parabolic curve at the top is a plot of $\Delta G_{el}^{\neq}$ *vs*. ΔG_{el}. The normal region is shown on the left; the inverted region is on the right. The rate maximizes where $\Delta G_{el}^{\neq} = 0$. On each side of this point, the activation energy increases (the rate decreases). The intersecting reactant and product curves shown at the bottom of the figure illustrate further the relationships among ΔG_{el}, $\Delta G_{el}^{\neq}$, and λ. In the normal region, $\Delta G_{el}^{\neq}$ decreases with decreasing λ or as ΔG_{el} becomes more negative. The intersecting curves on the lower center portion of the figure show the point where $-\Delta G_{el} = \lambda$, or $\Delta G_{el}^{\neq} = 0$. The intersecting curves on the lower right represent the situation in the inverted region. Here $\Delta G_{el}^{\neq}$ increases with decreasing λ or as ΔG_{el} becomes more negative.

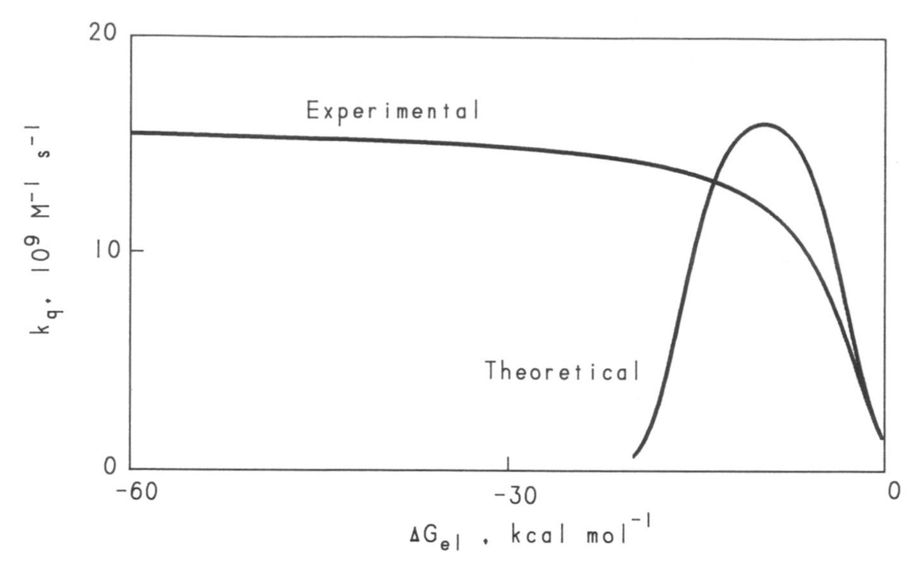

Figure 6.20 The predicted inverted behavior (marked "theoretical") has eluded detection for reactions in solution. Instead, the rate constant usually exhibits a plateau effect (marked "experimental"), leveling off at the diffusional limit. One explanation may be that the "inverted" effect is "hidden" in diffusion-controlled reactions (see text for other explanations).

inverted region. In this middle of these two domains is the point at which $-\Delta G_{el} \simeq \lambda$. Here the rate is maximized since $\Delta G_{el}^{\neq} \simeq 0$.

Experimental evidence for the inverted region has proven to be a formidable and elusive task. Weller, for example, has measured the driving forces for a whole series of homologous electron transfers and observed that the rates plateau at large ΔGs. He proposed an empirical relation to fit his experimental results[21]:

$$\Delta G_{el}^{\neq} = \frac{\Delta G_{el}}{2} + \left[\left(\frac{\Delta G_{el}}{2}\right) + \Delta G^{\neq}(0)^2\right]^{1/2} \tag{6.59}$$

A plot of Eq. 6.59 is shown in Fig. 6.20 with an assumed value of $\Delta G^{\neq}(0) = 2.4$ kcal mol^{-1}. The latter value was obtained empirically.

Why is there a discrepancy between Marcus' prediction and Weller's observations? Several explanations have been proposed. One proposal has been that perhaps more than one mechanism may be involved in the quenching reaction. It was suggested, for example, that for very exothermic reactions the formation of an excited quencher rather than radical ion pairs takes place in the initial quenching step. The formation of an excited state would then presumably proceed faster than the formation of radical ion pairs. However, the existence of an accessible excited state would have to be demonstrated, and so far this has been a difficult task.

Some authors have suggested that the inverted region can be obscured in bimolecular electron transfer by diffusion kinetics.[22] That is, the actual rates of the electron transfer are far more rapid than the rates of diffusional encounters. In

Weller's work, the rates of quenching are measured when what is really wanted are the actual rates of electron transfer. If the latter are more rapid than the usual rates of diffusion, then any variations in k_{el} can be obscured.

To overcome the masking effect of diffusion, we can measure the rates of charge transfer between various acceptor–donor groups attached to the same molecule. This procedure has been successfully employed in a celebrated experiment that provided the first unambiguous evidence of the inverted region. By bombarding a homologous series of steroidal molecules with an electron beam from the Argonne 20-MeV Linac, Gerhard Closs and John Miller were able to demonstrate the parabolic effect predicted by the Marcus equation (Eq. 6.51).[23,24] These workers investigated a series of bifunctional 4-biphenylyl steroids:

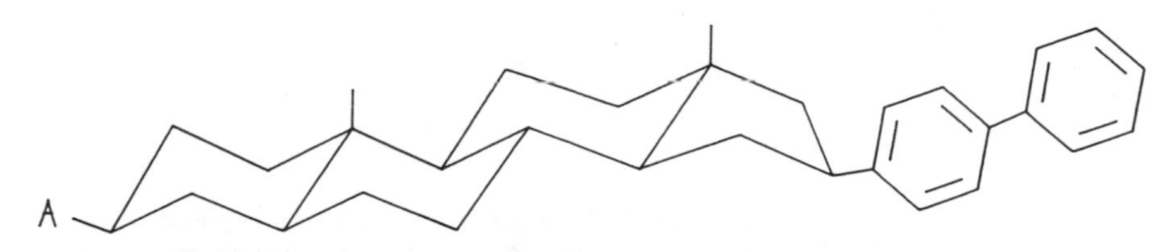

On pulse radiolysis in 2-methyltetrahydrofuran (MTHF), both functional groups can accept the electrons with equal probability to form anions. Electron transfer between the two groups then takes place resulting in a new charge redistribution:

$$\text{A} - \text{Steroid} - \text{B} \xrightarrow{e^-} \text{A}^- - \text{Steroid} - \text{B} + \text{A} - \text{Steroid} - \text{B}^- \tag{6.60}$$

$$\text{A}^- - \text{Steroid} - \text{B} \underset{k_{-el}}{\overset{k_{el}}{\rightleftharpoons}} \text{A} - \text{Steroid} - \text{B}^- \tag{6.61}$$

A and B are the substituted groups, and k_1 and k_{-1} are the rates of electron transfer. The results for several acceptors shown in Fig. 6.21 confirm the presence of the inverted region. On inspection of Eq. 6.51, the maximum rate results when $-\Delta G_{el} = \lambda$. This is the energy where the reaction is *barrierless* since $\Delta G_{el}{}^{\neq} = 0$. The largest contribution to λ was shown to be due to solvent reorganization, so that we can assume that $-\Delta G_{el} \simeq \lambda_s$.[11] The following limiting situations describing the effect of λ_s on the rate of electron transfer were noted. In the normal region, $-\Delta G_{el} < \lambda_s$, and $\Delta G_{el}{}^{\neq}$ increases with increasing λ_s; consequently the rate decreases. However, this trend switches in the inverted region where $-\Delta G_{el} > \lambda_s$. Here the rate actually decreases with increasing λ_s.

In photoinduced electron transfer, the inverted region has important implications in charge separation and electron return. By taking advantage of the relationship, $-\Delta G_{el} = \lambda$, one might be able to control the onset of the inverted region for either forward electron transfer or electron return and thereby influence the reaction. If, for example, the reaction is dominated by the effects of solvent rather than bond reorganization, $\lambda \sim \lambda_s$. By varying the solvent, it is then possible to change λ_s. Let us say that we want to increase the lifetimes of ion-pair intermediates by slowing down the rate of electron return, k_{ret}. We chose a solvent so that $-\Delta G_{el} \gg \lambda_s$. This

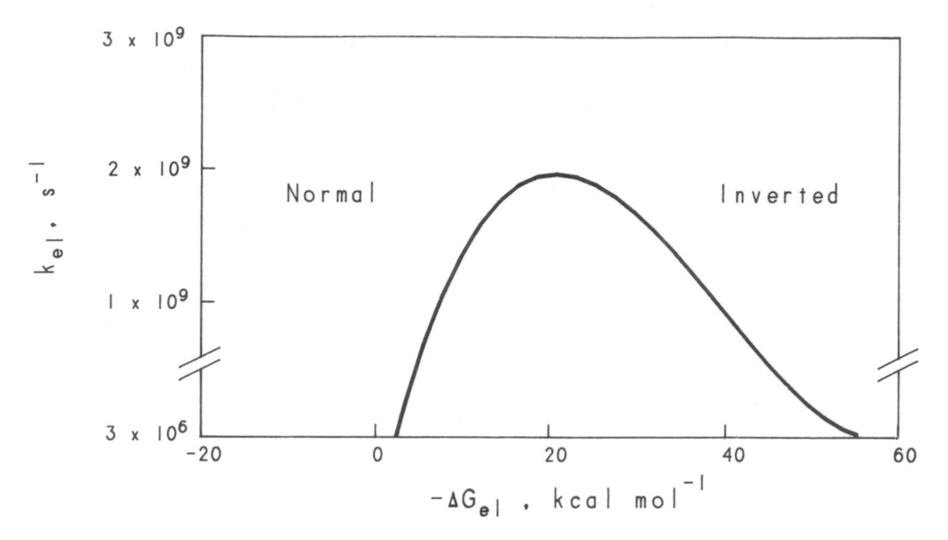

Figure 6.21 The choice of solvent influences the position of the maximum in ΔG_{el} *vs* k_{el} plots. This is due to the effect of solvent polarity on λ_s. (Adapted with permission from Closs, G. L., Calcaterra, L. T., Green, N. J., Penfield, K. W., and Miller, J. R. *J. Phys. Chem.* **1986,** *90,* 3673. Copyright 1986 American Chemical Society.)

places the reaction in the inverted region. In this kinetic region, electron return is slow, and accordingly the lifetime of the ion-pairs is enhanced.

For another example of the relationship between λ and $-\Delta G_{el}$, consider the intramolecular systems **27** to **31** introduced in Chapter 4. As mentioned in Chapter 4, the rates of photoinduced electron transfer in these compounds were studied in a variety of solvents. One of the interesting aspects of these systems is that the rate of forward electron transfer is relatively insensitive to the nature of the solvents employed. For example, for **30** k_{el} is $\sim 0.5 \times 10^{10}$ and 1×10^{10} s^{-1} in benzene and diethyl ether, respectively. In contrast, k_{ret} increases from $\sim 3 \times 10^6$ s^{-1} in benzene to $\sim 2 \times 10^7$ s^{-1} in the more polar *p*-dioxane. Here is a possible explanation. If it is assumed that the forward rate of electron transfer falls in the vicinity where $\Delta G_{el} \sim -\lambda$, then this is the thermodynamic region where the reaction is almost "barrierless." On the other hand, the much smaller rate of electron return is probably due to the much larger magnitude of the energy gap, ~69 kcal mol^{-1}, between the ion pair and ground state. Then electron return falls within the inverted region, and $-\Delta G_{el} \gg \lambda$. Since λ_s increases with increasing ε_s, the effect of an increased λ_s in the inverted region is to accelerate the rate (this argument neglects the effects of λ_v, which are probably negligible anyway).

6.4. Dynamic Solvent Effects

The reader may recall from Eq. 6.16 that the reaction rate of electron transfer is influenced by the nuclear frequency, ν_n. ν_n appears in the preexponential term of

Eq. 6.17. The nuclear frequency consists of molecular vibrations in the reactants and solvent orientations. We consider the case where intramolecular vibrations can be neglected. Thus, in the present discussion, ν_n is largely related to *dynamic* solvent effects.[25]

In the classical electron-transfer theories, the focus of interest is on $\Delta G_{el}^{\neq}$, which is contained in the exponential of Eq. 6.17. In electron transfers where solvent polarity exerts a major effect on the free energy of activation, the solvent influences the outer-sphere reorganization energy. It is also true, however, that solvation exercises a *dynamic solvent effect*. The "dynamic solvent effect" refers to the friction between reactants and polar solvents. Since the solvent environment and the reacting molecules in electron transfer are coupled electrostatically, the rates can be influenced dramatically. The outcome of this coupling turns up in the nuclear frequency, ν_n. ν_n measures the frequency of solvent motion, or how fast the polar solvents can respond to instantaneous charge. In our discussion of classical Marcus theory, we noted that solvent molecules can respond orientationally. Since there are other solvent motions such as vibrational motions, the time required for solvation may span several time scales. However, it is convenient to consider only the orientational motion of solvent molecules. There may be a certain "sluggishness" associated with these motions, which ultimately may affect the rate. For any solvent, this orientational response can be represented by the *longitudinal* relaxation time, τ_L, where

$$\tau_L = (\varepsilon_{op}/\varepsilon_s)\tau_D \tag{6.62}$$

The ratio $\varepsilon_{op}/\varepsilon_s$ measures the degree of coupling between the solvent and the reaction. ε_{op} and ε_s represent dielectric constants measured at different frequencies. τ_L generally falls within the range of 10^{-13} to 10^{-10} s. τ_D, the dielectric relaxation time, represents the rotational diffusion time of a single particle. It is related to the viscosity of the solvent. Consequently, the longitudinal solvation time can be about an order of magnitude smaller than τ_D.

If an adiabatic, outer-sphere electron transfer takes place on a smooth and continuous potential energy surface, then, according to solvent dynamic theory, the rate of electron transfer should be proportional to the inverse of the longitudinal relaxation time (i.e., $k_{el} \propto \tau_L^{-1}$). This rate represents the maximum rate of an electron transfer. This electron transfer should then occur much more rapidly in acetonitrile than the less polar and more viscous *n*-propanol. It should be pointed out, however, that this correlation holds only for adiabatic reactions where dynamic solvent motions are important. However, the possibility of significant nuclear and electronic barriers must be carefully sorted out before concluding that solvent dynamic effects dominate.

Experimentally, the response of a solvent to instantaneous charge can be estimated from dynamic Stokes shift measurements.[26] When a ground-state molecule in equilibrium with its solvent environment is excited with an ultrashort pulse of light, the molecule is converted to its excited state, which still has the ground-state solvent orientation. During its brief lifetime, the nonequilibrated excited state emits fluorescence. Eventually, as the solvent molecules reorient to the charge distribution of the excited state, the fluorescence emission shifts. The change in the fluorescence spectrum is related to the solvent response.

The experimental observation of a correlation between the electron-transfer rates and the longitudinal time constitutes evidence for dynamic solvent effect. An example of a dynamic solvent effect is the intramolecular charge transfer in bianthryl. Photoexcitation of this molecule first leads to a locally excited singlet state followed by an electron transfer to give an equilibrated charge-transfer state[27]:

Locally excited state ⇌ Charge-transfer state (6.63)

Spectroscopic studies over a range of solvent polarities suggest that the activation energy of the electron transfer is small (i.e., the reaction is barrierless). However, subpicosecond fluorescence studies in addition to computational simulations revealed dynamic solvent motion. These investigations demonstrated that $k_{el} \propto \tau_L^{-1}$, implying that solvent motion is rate limiting.

An improved picture of solvent dynamic motion continues to emerge. For example, one model treats the rate of electron transfer in terms of both low-frequency (or classical) modes and classical solvent modes.[28] If solvent motion is frozen so that $\tau_L \sim \infty$, then there are a large number of solvent distributions to consider. Accordingly, electron transfer takes place over a wide distribution of rates. On the other hand, when solvent molecules are allowed to relax rapidly, τ_L approaches zero, and the rate can be treated with classical theory.

It should be noted in many photoinduced electron-transfer reactions, the solvent influences "exponential" $\Delta G_{el}^{\neq}$ rather than "preexponential" ν_n. The reason is that solvent effects depend in a complex way on the molecularity of the solvent and in the way that the reactants couple with the solvent. The precise nature of these interactions remains obscure.

6.5. Quantum Mechanical Theories

The classical theories are intended for *weakly* adiabatic electron transfer. By weakly adiabatic, the reader will recall that we mean that electronic coupling between the reactants in the encounter complex is sufficiently large that $\kappa_e \simeq 1$ in Eq. 6.16, but not so large that resonance effects dominate the electron-transfer process. There is sufficient interaction between the reactants to ensure a smooth passage from the reactant to product surface. The reactants are close enough and there is sufficient orbital overlap.

Nonetheless, there are electron-transfer processes where the electronic coupling is so weak that $\kappa_{el} \ll 1$. These processes are classified as *nonadiabatic*. As we saw in previous sections, the passage from the reactant to the product surface is abrupt for electron transfer between reactants where the electronic coupling is very small. Here the reactants may be well separated or so positioned that there is at best only minimal orbital interaction. Such reactions may take place in rigid environments where the donor and acceptor may be separated by large distances.

Electron transfers in solid environments have an important feature that distinguishes them from weakly adiabatic electron transfers in solution: in the solid state, the available thermal energy may be insufficient to overcome the free-energy barriers due to nuclear reorganization, and therefore to allow passage over the nuclear barrier. The solid environment may be a rigid glass or similar kind of solid matrix where the nuclei are "frozen." The temperature of the system may also be so low as to preclude nuclear motion in the reactants or solvent molecules. Under such conditions, the thermal energy may not be sufficient to induce, say, a bond deformation so that an initial nuclear configuration can be transformed into a transition state. In a solid environment or at cryogenic temperatures, the electron transfer is said to be "classically forbidden."

Photoinduced electron transfers that take place between well-separated reactants, in solid environments, or at cryogenic temperatures may take place at rates faster than one would normally predict on the basis of simple classical treatments. These classically forbidden reactions take place in spite of considerable electronic and nuclear barriers. One may ask why these reactions should take place at all when the barriers—electronic or nuclear—are so large? The answer to this question has to do with a quantum mechanical phenomenon called *tunneling*. We can distinguish two kinds of tunneling—nuclear and electron tunneling. In each of these "tunnelings," electrons or nuclei are capable of penetrating a barrier having a specific width. Nuclear tunneling refers to a pathway where the reaction coordinate passes through the transition-state barrier (Fig. 6.22). A transition can be traced where the imaginary point of a reaction starts from the well of the reactant curve and cuts through the activation barrier. In this way, nuclear tunneling provides a pathway for a reaction otherwise prohibited by a classical barrier. For the most part, nuclear tunneling will occur only at low temperatures or in the solid state.

Electron tunneling is a pathway where an electron moves between weakly interacting sites through a high potential energy barrier (Fig. 6.23). A valence electron can be visualized as moving between two potential fields defined by the nuclei of the donor and acceptor molecules separated by a horizontal distance that we will designate as Δ_t, the *tunneling distance*. The potential energy at each nucleus results from the binding forces between the electron and nucleus. As the electron moves from the energy well of the donor nucleus to the energy well of the acceptor nucleus, the wavefunction of the electron changes as it moves through an energy wall separating the two nuclei. While approximately midway between the donor and acceptor, there is an equal probability that the electron will be on either molecule.

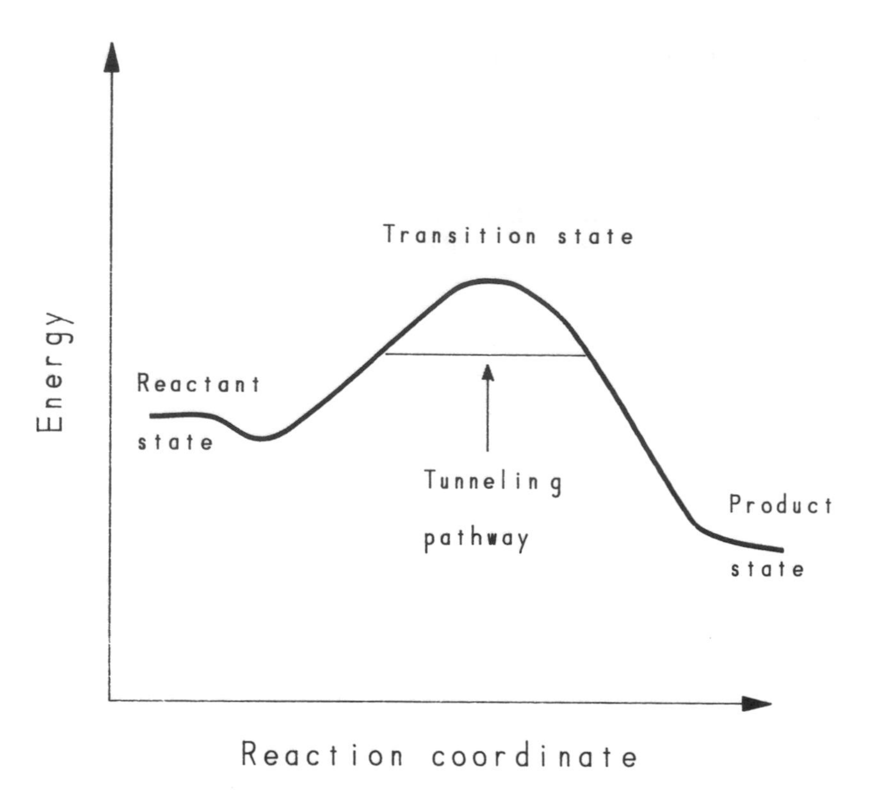

Figure 6.22 Nuclear tunneling can take place when $\hbar\upsilon_{eff} >> k_BT$, i.e., under conditions where the available thermal energy is insufficient to allow for passage over the transition barrier.

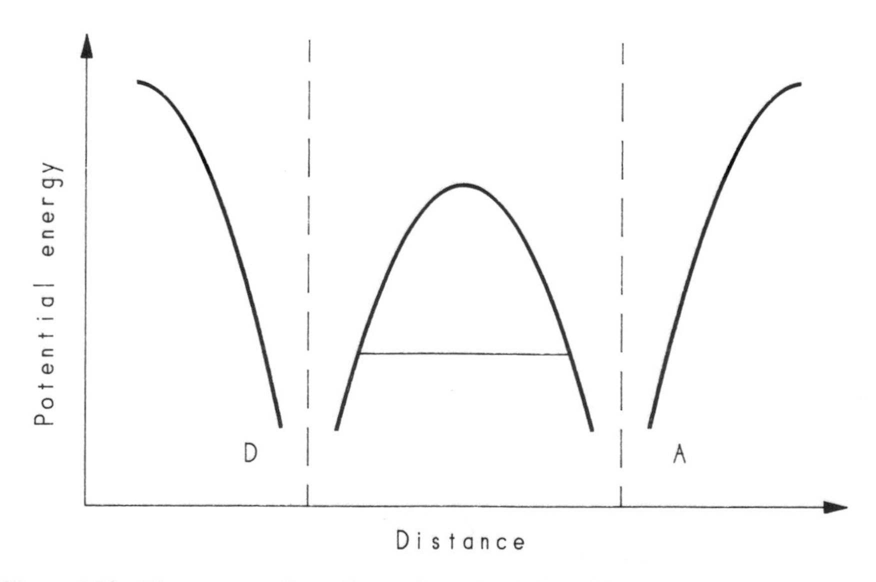

Figure 6.23 Electron tunneling refers to the passage of an electron between two sites, each defined by a potential well.

At various stages of the electron transfer, the energy of the electron never exceeds the maximum energy defined by the barrier at the transition state. Tunneling determines the probability that an electron will oscillate between these states. Unlike nuclear tunneling, electron tunneling is independent of temperature.

There is a close relationship between nuclear and electron tunneling. To understand this connection and its physical consequences, we need a formalism based on quantum mechanics. In quantum mechanical theories, the donor, acceptor, and surrounding solvent or medium molecules comprise a *supermolecule,* a term we have introduced earlier in this chapter in connection with the classical formalism. Within the supermolecule, all atoms of the reactants and medium are constantly undergoing a large number of vibrational oscillations. The oscillations may be in-phase acoustical waves or out-of-phase optical waves (phonons). Nuclear-bond deformations correspond to high-frequency phonons whereas the low-frequency vibrations and rotational motions (librations) of the solvent molecules are associated with low-energy phonons. The optical waves created by nuclear deformations control the rate of electron transfer.

The transition between the reactant and product surfaces can be visualized as an isoenergetic crossing involving the coupling of the nuclear wavefunctions of the reactant and product vibrational modes (the isoenergetic crossing is analogous to the intersecting parabolas shown in Fig. 6.7 for the classical case). Fluctuations of the classical degrees of freedom precede the act of electron transfer. The transition representing the coupling of two vibronic states can be described in terms similar to the classical phenomenon. For a system with many high-frequency modes, the coupling strength at a vibronic curve crossing depends on the electron coupling matrix and the vibrational overlap integral. The isoenergetic crossing is followed by a rapid vibrational relaxation to the lowest vibrational state of the product state. The transition depends on the availability of vibrational modes in the product state, which accept or "soak up" the exothermic energy resulting from the electron transition. These vibrations can be considered the overtones of a fundamental vibrational frequency.

Given this background, we may describe electron transfer as a *nonadiabatic multiphonon radiationless transition*. In the quantum mechanical theory, the rate is treated as a radiationless transition between two eigenstates, each of which is described by its own potential energy surface[29]:

$$k_{el} = \frac{2\pi}{\hbar} |H_{el}|^2 \mathrm{FC} \tag{6.64}$$

Equation 6.64 is known as the *Fermi Golden Rule* expression. The Golden Rule approximates the rates of elementary reactions such as electron transfer. As used in Eq. 6.64, H_{el} is the quantum mechanical counterpart of the classical electron-transfer matrix that couples the reactant and product electronic wavefunctions. Thus, the quantum mechanical matrix, H_{el}, displays an exponential dependence with distance:

$$H_{el}\ (\mathrm{cm}^{-1}) = 10^5 \exp(-\beta d) \tag{6.65}$$

H_{el} values range from $\sim 10^{-5}$ cm^{-1}, which is indicative of weak electronic coupling, to greater than ~ 1 cm^{-1}, the strong coupling limit. Importantly, H_{el} is proportional to Δ_t, the distance separating two reactants (as well as their mutual stereoelectronic orientation), and thus by inference is also proportional to the probability that two states couple at a fixed nuclear configuration. The Golden Rule is valid for small Δ_t. If $\Delta_t \gg 0$, the rule breaks down.

FC, the Franck–Condon factor, represents the weighted density of final states at the initial energy. This term is proportional to the matrix element describing the overlap of nuclear wavefunctions between the initial and final thermally averaged vibronic states:

$$\text{FC} \propto \sum_{v} \sum_{w} \rho_v \, |\langle \chi_i | \chi_f \rangle|^2 \delta(E_{iw} - E_{fw}) \tag{6.66}$$

where χ_i and χ_f represent the nuclear wavefunctions of w and v (the vibrational levels of the initial and final states), ρ_v is the population density of vibrational level **v**, and the last term is the energy difference between these levels.

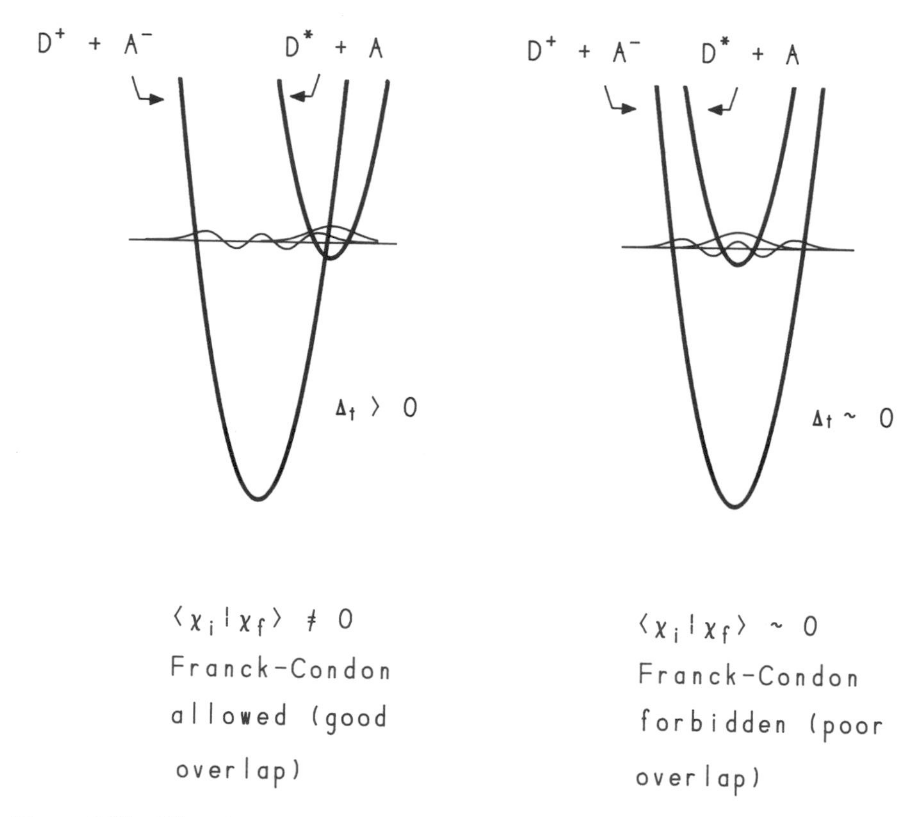

Figure 6.24 When the reactant and product curves are displaced horizontally with respect to one another, the nuclear overlap between these states is affected. Nuclear overlap is more effective for the intersecting curves on the left of the diagram.

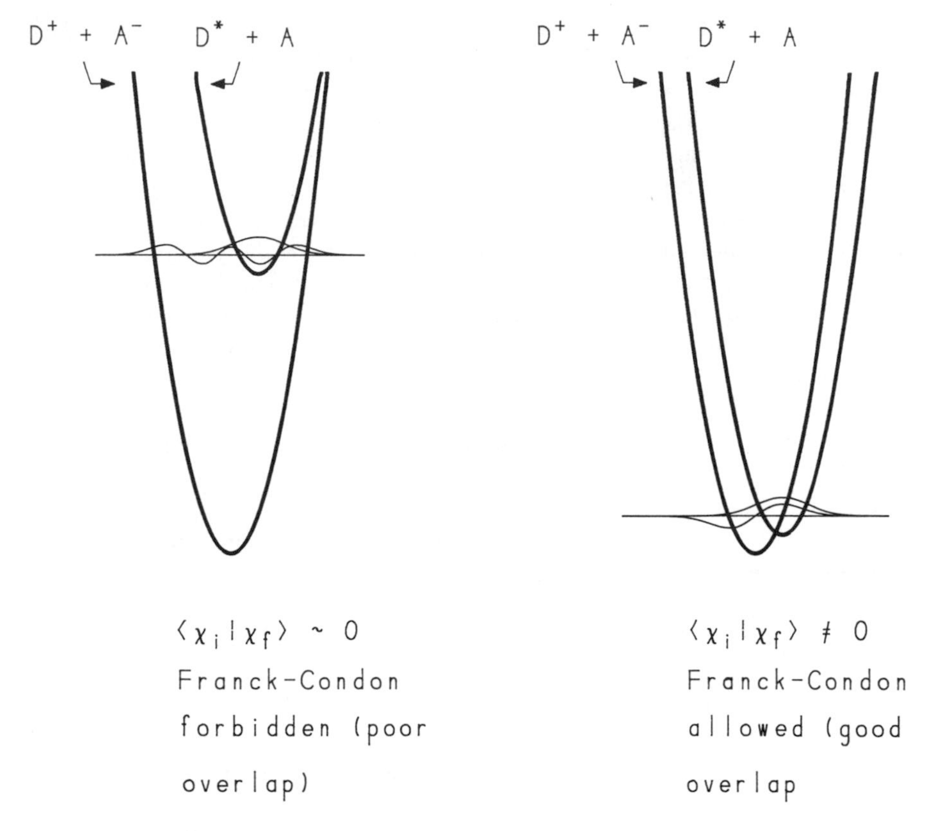

Figure 6.25 The effect of vertical displacement of potential energy curves on nuclear overlap.

As shown in Figs. 6.24 and 6.25, the horizontal and vertical displacement of potential-energy curves influences the overlap of nuclear wavefunctions, which in turn is related to the tunneling distance. These figures depict the oscillating functions of the wavefunctions of each vibrational state. Note that the higher vibrational states correspond to highly oscillating wavefunctions. The horizontal displacement between the minima of these potential energy curves is given by Δ_t. If the curves are so displaced that their minima correspond to the same nuclear geometry, then $\Delta_t \simeq 0$. For a specific nuclear geometry, the wavefunctions for the reactant and product vibrational states are multiplied. The function $|\langle\chi_i|\chi_f\rangle|$ is then evaluated. Thus, Franck–Condon factors are related to the overlap between reactant and product wavefunctions so that the transition corresponding to the greatest overlap between reactant and product vibrational wavefunctions is favored. The overlap between the reactant and product states shown on the left of Fig. 6.24 ($\Delta_t > 0$) is greater than the overlap between the states shown on the right ($\Delta_t \simeq 0$). Accordingly, the case where $\Delta_t > 0$ represents a situation where the nuclear coupling between the vibrational wavefunctions of the reactant and product state is strong.

Similar arguments show that the nuclear overlap between reactant and product curves can be affected by the magnitude of vertical displacement between their energy wells (in this case Δ_t is identical). This is evident from inspection of Fig. 6.25. The orbital overlap between the reactant nuclear wavefunction and the highly oscillating wavefunction of the product state represented in the left portion of the figure is poor. On the right, however, there is positive or constructive interference between the orbitals. This is a reflection of the *energy gap law,* which states that the rate of a radiationless transition rapidly decreases as the energy difference between the initial and final v = 0 states becomes larger. This is another way of saying that if electron transfer is regarded as a radiationless transition, say, within a supermolecule, then the rate will decrease with an increase in energy. A detailed discussion of the energy gap law and its implications in photochemistry can be found in standard textbooks on photochemistry.

Starting with Eq. 6.64, we will now describe two explicit equations for calculating k_{el}. When the assumptions are made that (1) nuclear vibrations can be approximated by an effective nuclear vibration having a frequency ν_{eff}, and (2) this vibration can be approximated by the wavefunction of a harmonic oscillator, it can be shown that

$$k_{el} = \frac{2\pi}{\hbar^2 \nu_{eff}} |H_{el}|^2 \frac{S^P \exp(-S)}{P!} \tag{6.67}$$

where $P = -\Delta G_{el}/\hbar\nu_{eff}$ and $S = \lambda/\hbar\nu_{eff}$. P is the characteristic frequency normalized to the free energy. S is the characteristic frequency normalized to λ and is related to the strength of vibronic coupling and accordingly measures bond-length changes. $\hbar\nu_{eff}$ can be taken as the quantum energy of the harmonic oscillator corresponding to a vibrational state. Equation 6.67 is intended for reactions where the thermal energy is lower than the energy required for the vibration ($k_B T < \hbar\nu_{eff}$).

Table 6.5 REPRESENTATIVE FREQUENCIES OF SOLVENT AND BOND VIBRATIONAL MODES[a]

	J	cm^{-1}	s^{-1}
H_2O	$\sim 2 \times 10^{-23}$	1	$\sim 3 \times 10^{10}$
H_2O (ice)	$\sim 2 \times 10^{-21}$	100	$\sim 3 \times 10^{12}$
Protein	$\sim 3 \times 10^{-21}$	150	$\sim 4.5 \times 10^{12}$
H_2O	$\sim 3 \times 10^{-21}$	170	$\sim 5 \times 10^{12}$
Fe—O in $Fe(H_2O)_6^{2+}$	$\sim 8 \times 10^{-21}$	389	$\sim 1 \times 10^{13}$
Fe—O in $Fe(H_2O)_6^{3+}$	$\sim 9.7 \times 10^{-21}$	490	$\sim 1.5 \times 10^{13}$
C—N	$\sim 2.6 \times 10^{-20}$	1324	$\sim 4 \times 10^{13}$
C—C in aromatic hydrocarbons	$\sim 3 \times 10^{-20}$	1520	$\sim 4.6 \times 10^{13}$
C—H in CH_3	$\sim 5.6 \times 10^{-20}$	2800	$\sim 8.4 \times 10^{13}$
O—H	$\sim 7 \times 10^{-20}$	3600	$\sim 1 \times 10^{14}$

[a] Data obtained from references in Marcus, R. A., and Sutin, N. *Biochim. Biophys. Acta* **1985,** *811,* 265; Brunschwig, B. S., Logan, J., Newton, M. D., and Sutin, N. *J. Am. Chem. Soc.* **1980,** *102,* 5798; and Grampp, G., and Jaenicke, W. *J. Chem. Soc., Faraday Trans. 2* **1985,** *81,* 1035.

Table 6.6 EFFECT OF H_{el} ON THE PREEXPONENTIAL FACTOR[a]

H_{el}, cm^{-1} ($kcal\ mol^{-1}$)	$\frac{2\pi\|H_{el}\|^2}{\hbar(4\pi\lambda k_B T)^{1/2}}$ (s^{-1})
10 (0.03)	5.4×10^{10}
100 (0.3)	5.4×10^{12}
200 (0.6)	2.2×10^{13}
400 (1.2)	8.7×10^{13}
600 (1.8)	2.0×10^{14}
1000 (2.9)	5.4×10^{14}

[a] Preexponential factor in Eq. 6.68 calculated with $\lambda = 5.8$ kcal mol^{-1} and $T = 270$ K.

This corresponds to the situation where a bond-length change may involve a high-frequency nuclear vibration in the reactant or solvent molecules. If the available thermal energy (k_BT) is unable to permit passage over the activation energy, then the high-frequency mode assumes some significance. When high-frequency modes are important, tunneling becomes a factor to consider. This will result in a temperature-independent reaction. It is worthwhile to examine Table 6.5, which contains representative frequencies of solvent and intramolecular vibrations. These values can be compared with the k_BTs at various temperatures. At 5 K, where $k_BT \simeq 3\ cm^{-1}$, the nonclassical expression, Eq. 6.67, can be used, since there is little if any thermal energy available to excite any vibrational mode.

If the temperature is raised so that $k_BT > \hbar\nu_{eff}$, then a semiclassical version of the Marcus expression can be used:

$$k_{el} = \frac{2\pi}{\hbar}|H_{el}|^2 \frac{\exp[-(\Delta G_{el} - \lambda)^2/4\lambda k_B T]}{(4\pi\lambda k_B T)^{1/2}} \tag{6.68}$$

Here there are several key parameters to consider: ΔG_{el}, λ, H_{el}, and T. When $-\Delta G_{el} = \lambda$, the rate is predicted to reach a maximum. This is, in fact, the same conclusion that is reached on the basis of the classical expression of Marcus, Eq. 6.51. Furthermore, the preexponential factor in Eq. 6.16 is usually taken to be $\nu_n\kappa_{el} \sim 10^{12} - 10^{13}\ s^{-1}$, which is typical of an adiabatic reaction. In Eq. 6.68, the preexponential factor is influenced by the magnitude of H_{el}, as shown in Table 6.6. For values of $H_{el} \gtrsim 200\ cm^{-1}$, the preexponential factor approaches $\sim 10^{14}\ s^{-1}$, so that the reaction becomes adiabatic. As the preexponential factor falls below $H_{el} \simeq 200$ cm^{-1}, the reaction becomes increasingly nonadiabatic.

6.6. Applications of Classical and Quantum Mechanical Theories to Biological Photosynthesis

To this stage, this chapter has been concerned with classical and quantum mechanical descriptions of electron transfer. As these theories have continued to experience refinement, it has become possible to apply them to various systems. One of the most notable attempts has been the effort to analyze the events in biological photosynthesis. In fact, theoreticians and experimentalists have devoted much attention to applying classical and nonclassical descriptions to electron transfer in the photosynthetic reaction center. With improved structural knowledge of the bacterial photosynthesis center, our understanding of the complex pathways in photosynthesis has greatly improved over the last 10 years. The remainder of this chapter will show how some of the theoretical models introduced in earlier sections have been used to explain the experimental data obtained from biological photosynthetic systems.

6.6.1. The Photosynthetic Reaction Center

Nature has endowed green plants, algae, and some bacteria with an elaborate and wonderful mechanism to convert sunlight and carbon dioxide into useful chemical energy. Our current understanding of photosynthesis has been acquired through a worldwide scientific effort, applying such methods as optical absorption, electron-spin resonance, X-ray crystallography, electronic structure calculations, and molecular modeling simulations.[30–34]

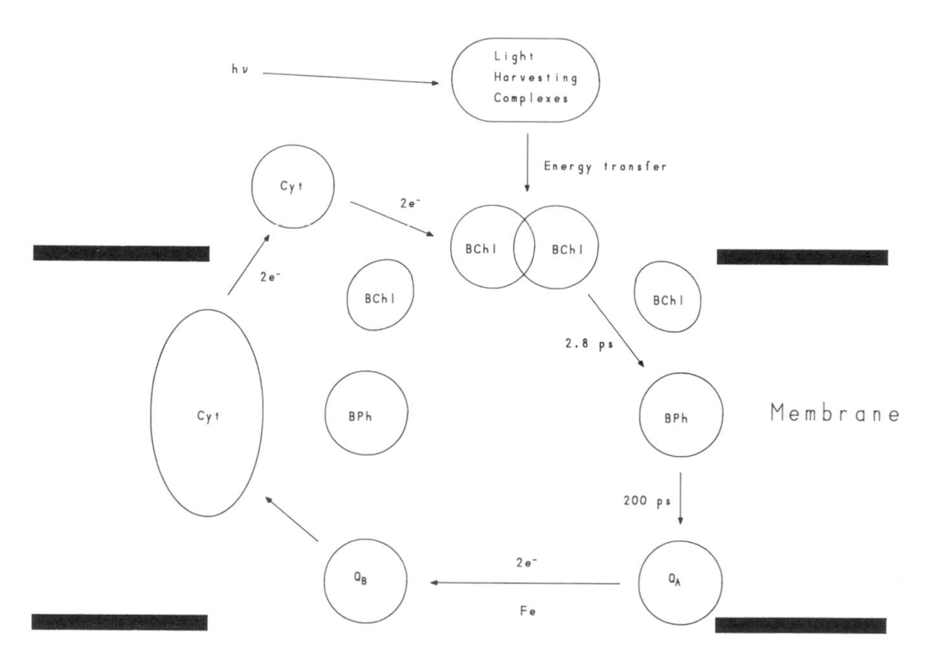

Figure 6.26 The diagram shows the events in the photosynthetic reaction center in the bacterial membrane. Photoexcitation is followed by a series of vectorial, exothermic electron transfers in the forward direction. The result is the creation of long-lived charge separation across the membrane. Energy-rich, charge-separated intermediates are required for energy storage and metabolism by the organism. BCh1 = bacteriochlorophyll; BPh = bacteriopheophytin; Q_A = menaquinone; Q_B = ubiquinone; Cyt = cytochrome.

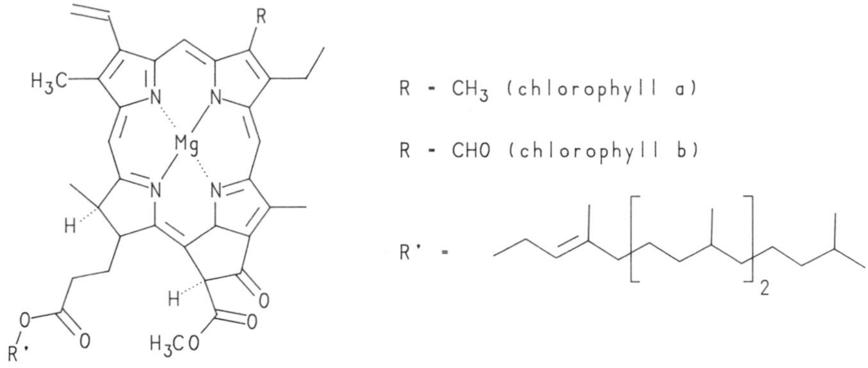

R - CH_3 (chlorophyll a)

R - CHO (chlorophyll b)

Figure 6.27 Structure of chlorophyll.

The primary event of photosynthesis takes place in *photosynthetic reaction centers* distributed in chloroplasts. Within the chloroplasts are folded sacs called thylakoids. These sacs contain a variety of proteins, lipids, glycolipids, and organic pigments. These molecules collectively harness and utilize sunlight for energy requirements. The nature of this unique molecular environment exerts a profound effect on the rates and efficiency in photosynthesis. A schematic representation of the photosynthetic reaction center, shown in Fig. 6.26, is based on the X-ray crystal structure of the reaction center in purple bacterium *Rhodopseudomonas viridis*. This structure was elucidated by Johann Deisenhofer and Robert Huber. For this notable accomplishment, they were awarded the 1988 Nobel Prize for chemistry. Here are the salient features of the reaction center. Its molecular weight is about 145,000. Each reaction center can be described as a tetragonal crystal of cell dimensions approximately 224 Å × 224 Å × 114 Å, stretching from one end of the membrane's surface to the other. Within each unit, there are four molecules of chlorophyll, two molecules of pheophytin, and two quinones (Figs. 6.27 and 6.28). Pheophytin is chlorophyll lacking a magnesium ion.

Long helical polypeptides traversing the membrane align the chlorophylls and quinones into a symmetrical complex with two dangling spiral groups. At the head of this complex, on the periplasmic side of the bacterial membrane, is a *special pair* of two chlorophyll molecules. The separation distance between each partner in this dimer is quite small—only ~3.5 Å. Consequently, these chlorophylls are strongly coupled. Surrounding the special pair on each side of the symmetry axis are two so-called "spectator" chlorophylls. The significance of these chlorphylls will become apparent shortly.

The primary event in photosynthesis is the "capture" of a photon. This is performed by chlorophyll and light-absorbing pigments called carotenoids (Fig. 6.29). These molecules are packed near the inner surface of the membrane. Because they possess a extensive network of conjugated double bonds, these pigments are collectively capable of capturing a significant portion of the visible spectrum between 400 and 800 nm. Efficient singlet energy transfer takes place from these "antenna"

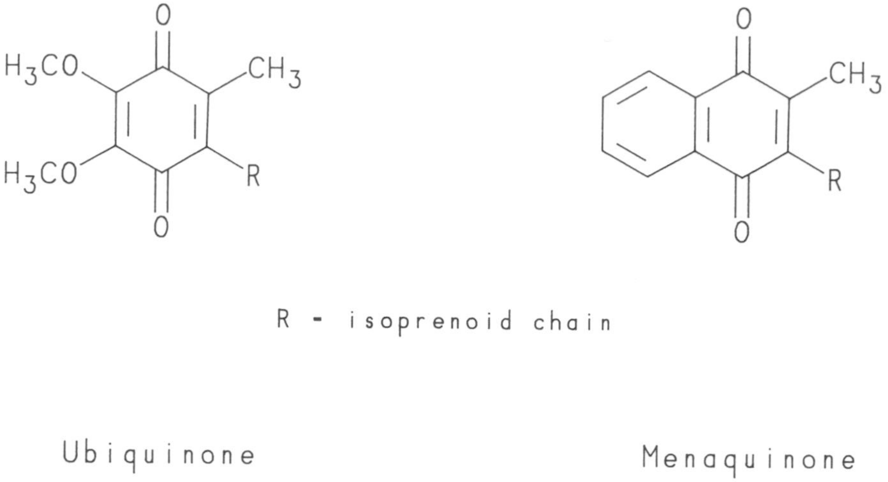

R - isoprenoid chain

Ubiquinone Menaquinone

Figure 6.28 Structure of menaquinone and ubiquinone.

pigments to neighboring chlorophyll molecules by the dipole–dipole mechanism.[35] As a highly organized molecular network, the antenna system channels its excitation energy to the special pair (Fig. 6.30). A series of electron transfers then ensues across the membrane. Initially, singlet chlorophyll donates an electron to neighboring pheophytin within 2.8 ps, leaving a positive charge on the special pair. Within ~200 ps, the electron is subsequently transferred from $BPh^{\cdot -}$ to menaquinone and finally to ubiquinone. These terminal electron acceptors are situated on the cytoplastic side of the membrane.

Now, following the initial electron transfer, the special pair is left in an oxidized state. It must be returned to its original state before it can donate an electron again. For this purpose, a nearby soluble cytochrome neutralizes the oxidized donor. After the special pair is neutralized, it can absorb a second photon and the cycle repeats itself. Following the second cycle, cytochrome is left with a positive charge while ubiquinone has gained two electrons. The net result is a charge separation across the membrane. The reduced quinone then acquires two protons and leaves the reaction center, while another quinone fills the empty site. The electrons and protons in the reduced quinone are shuttled back into soluble cytochrome to participate in the later stages of photosynthesis. The energy stored in the charge-separated state subsequently drives metabolical reactions required by the organism. (In higher plants, the sequence of events just outlined here is more complex, as there are two coupled photosynthetic systems, Photosystems I and II. Each photosystem plays a distinct role in the biosynthesis of molecules to satisfy the energy requirements of the organism. Functionally, the two photosystems are coupled by an elaborate network of electron carriers. In Photosystem I, carbon dioxide is converted into complex carbohydrates and sugars. In Photosystem II, molecular oxygen is generated via the oxidixation of water. Details can be found in standard texts on biochemistry.)

During the cyclic flow of electrons across the photosynthetic membrane, there are several coexisting, dissipative pathways. There is electron return to ground-state reactants and ion-pair recombination leading to triplet chlorophyll (Fig. 6.30).[36] Triplet chlorphyll may participate in triplet recombination. A potential problem, however, is the reaction of triplet chlorophyll with oxygen to form singlet oxygen. Since singlet oxygen can oxidize polyunsaturated fatty acids to generate undesirable lipid peroxides, lipid peroxidation can severely comprise the integrity of the chloroplasts. However, β-carotene can scavenge singlet oxygen before the latter destroys the membrane molecules. Accordingly, the carotenoids play a dual role in the sense that they are not only light absorbers but they also can protect the chloroplast from the deleterious effects of singlet oxygen.

6.6.2. The Efficiency of Photosynthesis

Now we ask, "How is efficient charge-separation in photosynthesis maintained?" To answer this question, we examine the relationship between the forward rates of electron transfer (k_{el} and k_{e1}') and electron return (k_{ret} and k_{ret}'), as shown in Fig. 6.30 (We will neglect reversible electron transfer, since the backward flow of electrons is endothermic.) At the outset, it is useful to remember from earlier chapters that the relative magnitudes of the rate constants determine the efficiency of photosynthesis. The efficiency of charge separation approaches 100%, although the net storage efficiency is only a few percent when loss factors are considered. The rate of the forward reaction, k_{el}', is ~100 times faster than k_{ret} (Fig. 6.30). Here the reader should recall that $\phi_{IP} = k_{el}'/(k_{el}' + k_{ret})$. The disparity between k_{ret} and k_{el}' is essential for the production of energy-rich molecules.

Now let us examine the individual steps more closely. After excitation of the special pair to its singlet state, the subsequent step is electron transfer from the special pair to BPh:

$$^1(\mathrm{BChl})_2^* + \mathrm{BPh} \xrightarrow{k_{el}} (\mathrm{BChl})_2^{+\cdot} + \mathrm{BPh}^{-\cdot} \quad (6.69)$$

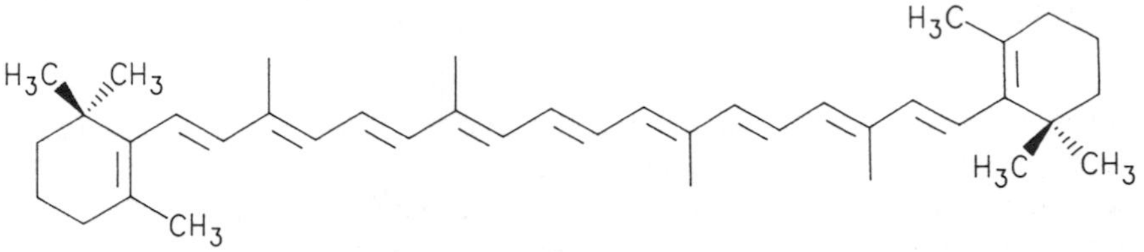

β-carotene

Figure 6.29 Structure of carotenoid.

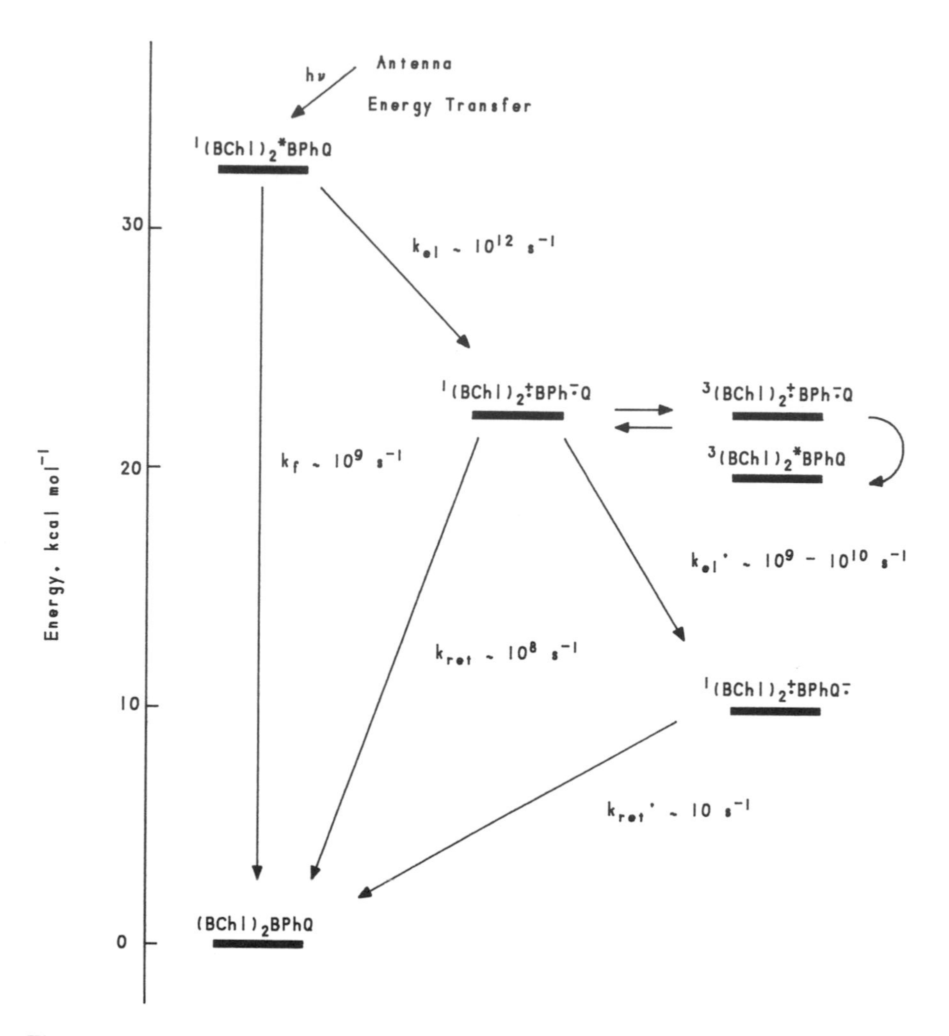

Figure 6.30 Energy diagram for photosynthesis. Light-harvesting molecules capture solar energy and transfer this energy to a "special" pair of chlorophyll molecules. Electron transfer proceeds within the photosynthetic reaction center in competition with fluorescence, electron return, and triplet recombination reactions. Not shown is the possible involvement of nearby BChl in the transfer of an electron from the "special" pair to BPh.

In a sense, this reaction does not convey the complete story, for in fact, electron transfer probably proceeds by sequential hopping between nearest neighbors. This is because the donor–acceptor pairs in the forward chain are spaced fairly far apart to allow direct electron transfer from the special pair to BPh. The special chlorophyll pair is estimated to be about 11 Å (edge-to-edge) from pheophytin. Thus, a direct electron transfer is in conflict with the observed rate of $k_{el} \sim 10^{12}\ s^{-1}$, a value unrealistically fast for a direct electron transfer over a large separation distance of 11 Å. Two possible explanations have been invoked to explain this disparity. One line of reasoning is based on a superexchange mechanism in which the electron passes through the orbitals of the nearby BChl molecule.[37] The idea is that the supermolecule enhances electronic coupling in the forward reaction. The second explanation is that the electron transfer may occur in two steps. The electron first travels from the excited dimer to BPh to give an intermediate $BChl^{\overline{\cdot}}$. The latter in turn rapidly shuttles the electron to BPh. Although this mechanism sounds attractive, attempts to detect $BChl^{\overline{\cdot}}$ have not proven conclusive.

ΔG_{el} and λ have been estimated for 6.69 to be about ~ -4 and ~ 5 kcal mol^{-1}, respectively. This means that the reaction is close to "barrierless," since $-\Delta G_{el} = \lambda$. The rate should then represent a maximum value. The most important contribution to λ probably is due to the nature of environment in the reaction center. There are several reasons why this is so. First, λ_v is probably quite small because of negligible bond-length changes in the porphyrin. That is, no significant bond-length changes involving bonds connected to Mg^{2+} are likely since this ion suffers neither oxidation or reduction. Thus, inner-sphere barriers can be neglected. Second, since the environment of the protein-rich membrane is hydrophobic, the dielectric constant should be small. This results in a small value of λ_s. To determine the significance of the molecular arrangment between chlorophyll and pheophytin in the protein environment, one study has attempted to simulate the rate of electron transfer from pheophytin to a quinone using a *dispersed polaron* approach (a polaron refers to the electron moving with the fluctuations in the surrounding medium).[38] This work suggests that numerous low-frequency protein modes are probably coupled with the reaction.

With the formation of $BPh^{\overline{\cdot}}$, the subsequent step is electron transfer between $BPh^{\overline{\cdot}}$ and Q:

$$BPh^{\overline{\cdot}} + Q \xrightarrow{k_{el}'} BPh + Q^{\overline{\cdot}} \tag{6.70}$$

In contrast to the electron transfer from the dimer to BPh, reaction 6.70 is relatively slow. This is partly due to the large separation distance, estimated to be ~ 10 Å, between $BPh^{\overline{\cdot}}$ and Q. $\Delta G'_{el}$ is ~ 14 kcal mol^{-1}, which is nearly equal to λ for this reaction. Thus, since $\Delta G_{el}^{\neq\,\prime} \simeq 0$ kcal mol^{-1}, we can use Eq. 6.45 to calculate k_{el}'. Assuming that $\beta = 1.2$ Å and $d_0 = 3$ Å, we obtain a value of $k_{el}' = 5 \times 10^9$ s^{-1}. This corresponds very well with the observed rate shown in Fig. 6.30.

The rate of the electron return, k_{ret}, is relatively slow and does not compete effectively k_{el}' (Fig. 6.30). The large driving force of electron return from $BPh^{\overline{\cdot}}$ to the ground state is ~ -28 kcal mol^{-1}. Furthermore, λ is small because of the hydrophobic character of the region. This places this reaction within the inverted thermodynamic region where $-\Delta G_{el} \gg \lambda$. The effect of the large driving force is to "block" electron return.

Strong barriers also block electron return (k_{ret}') between $(BChl)_2^{\overset{+}{\cdot}}$ and $Q^{\overline{\cdot}}$. There is probably poor electronic overlap due to the large separation distance between these species. This large electronic barrier coupled with the inverted effect can account for the dramatically reduced k_{ret}'. The net result is the enhancement of forward charge separation. Thus, the consequence of this elaborate interplay of energetics, environment, and separation distance is to enhance the magnitudes of ϕ_{IP} and τ_{IP} leading ultimately to the efficient storage of chemical energy. By "fine-tuning" the nuclear and electronic barriers in photosynthesis, nature has provided clues for the chemist.

Suggestions for Further Reading

Diffusion-Controlled and Electron-Transfer Rate Constants

Balzani, V., Moggi, L., Manfrin, M. F., Bolletta, F., and Laurence, G. S. *Coord. Chem. Rev.* **1975,** *15,* 321.

Clegg, R. M. *J. Chem. Ed.* **1986,** *63,* 571. This paper develops simple diffusion equations. A good introduction for the student inexperienced in this subject.

Einstein, A. *Investigations on the Theory of the Brownian Movement,* Furth, R., ed. Dover, New York, 1959. A good place to start when studying molecular motion.

Keizer, J. *Chem. Rev.* **1987,** *87,* 167. A review of various models used to explain diffusion.

McCammon, J. A., Northrup, S. H., and Allison, S. A. *J. Phys. Chem.* **1986,** *90,* 3901.

North, A. M. *Quart. Rev.* **1966,** *20,* 421. Discussion on theories of diffusion.

Smoluchowski, M. *Z. Phys. Chem.* **1917,** *92,* 129. A classic paper.

Classical Theories of Electron Transfer

Cannon, R. D. *Electron Transfer Reactions.* Butterworths, London, 1980, p. 208. A thorough look at theoretical aspects of thermal electron transfer of inorganic complexes.

Endicott, J. F., Kumar, K., Ramasami, J., and Rotzinger, F. P. *Prog. Inorg. Chem.* **1983,** *30,* 141.

Marcus, R. A. *Annu. Rev. Phys. Chem.* **1964,** *15,* 155. Written from a historical perspective, this paper deals with fundamental issues of electron-transfer theory.

Marcus, R. A., and Sutin, N. *Biochim. Biophys. Acta* **1985,** *811,* 265. An important review dealing with key aspects of classical electron-transfer theory.

Michl, J., and Bonačić-Koutecký V. *Electronic Aspects of Organic Chemistry.* Wiley, New York, 1990, Chapters 1–3.

Newton, M. D., and Sutin, N. *Annu. Rev. Phys. Chem.* **1984,** *35,* 437.

Pauling, L., and Wilson, E. B., Jr., *Introduction to Quantum Mechanics with Applications to Chemistry.* Dover, New York, 1985. Key aspects of the harmonic oscillator are introduced in this classic text.

Pullman, A. In *The New World of Quantum Chemistry, Proceedings of the Second International Congress of Quantum Chemistry,* Pullman, B., and Parr, R., eds. Reidel, Dordrecht, 1976, p. 149.

Salem, L. *Electrons in Chemical Reactions.* Wiley-Interscience, New York, 1982, Chapter 2. A discussion on potential-energy surfaces.

Suppan, P. *Chimia* **1988,** *42,* 320. A provocative and slightly iconoclastic view on classical electron-transfer theories.

Sutin, N. *Acc. Chem. Res.* **1982,** *9,* 275. An excellent synopsis of electron-transfer theories.

Sutin, N., and Brunschwig, B. In *Mechanistic Aspects of Inorganic Reactions,* Rorabacher, D. B., and Endicott, J. F., eds. ACS Symposium Series 198, American Chemical Society, Washington, DC, 1982, p. 107.

The Inverted Region

Closs, G. L., and Miller, J. R. *Science* **1988,** *240,* 440. Discussions on the inverted region.

Dynamic Solvent Effects

Barbara, P. F., Walter, G. C., and Smith, T. P. *Science* **1992,** *256,* 975. A discussion of recent models of the role of solvent dynamics in photoinduced electron transfer.

Fleming, G. R., and Wolynes, P. G. *Physics Today* **1990,** *43,* 36. An excellent article on the role of solvent in reactions such as electron transfer.

Kosower, E. M., and Huppert, D. *Annu. Rev. Phys. Chem.* **1986,** *37,* 127.

Quantum Mechanical Theories

Engleman, R., and Jortner, *J. Mol. Phys.* **1970,** *18,* 145.

Freed, K. F. *Acc. Chem. Res.* **1975,** *11,* 74. An important paper on radiationless transitions.

Freed, K. F., and Jortner, J. *J. Chem. Phys.* **1970,** *52,* 6272. Studies on the energy gap.

Goldanskii, V. I. *Sci. Am.* **1986,** *254,* 46. This article nicely captures the spirit of "tunneling."

Hopfield, J. J. *Proc. Natl. Acad. Sci. U.S.A.* **1974,** *71,* 3640.

Jortner, J. *Phil. Mag. B* **1979,** *40,* 317.

Kuki, A., and Wolynes, P. G. *Science* **1987,** *236,* 1647.

Liang, N., Closs, G., and Miller, J. *J. Am. Chem. Soc.* **1990,** *112,* 5353. This paper provides experimental evidence that suggests that nuclear tunneling provides the mechanism for the high-frequency modes to dissipate the exothermic energy.

Mikkelson, K. V., and Ratner, M. A. *Chem. Rev.* **1987,** *87,* 113. A thorough look at tunneling in the solid state.

Applications of Classical and Quantum Mechanical Theories to Biological Photosynthesis

Burton, G. W., and Ingold, K. U. *Science* **1984,** *224,* 569. The role of carotenoids is the subject of this paper.

Deisenhofer, J., and Michel, H. *Science* **1989,** *245,* 1463.

Deisenhofer, J., and Michel, H. In *Photochemical Energy Conversion,* Norris, J. R., Jr., and Meisel, D., eds. Elsevier, New York, 1989, p. 232.

Marcus, R. A. In *Supramolecular Photochemistry,* Balzani, V., ed. Reidel, Dordrecht, 1987, p. 45.

Stryer, L. *Biochemistry,* 2nd. ed. Freeman, San Francisco, 1981, Chapter 19. An introduction to photosynthesis.

Warshel, A. *Proc. Natl. Acad. Sci. U.S.A.* **1980,** *77,* 3105. Warshel suggests that there is electrostatic interaction between the oxidized special pair and ionized side groups on surrounding protein molecules.

Youvan, D. C., and Marrs, B. L. *Sci. Am.* **1987,** *256,* 42.

References

1. Lin, C.-T., Böttcher, W., Chou, M., Creutz, C., and Sutin, N. *J. Am. Chem. Soc.* **1976,** *98,* 6536.
2. Sutin, N., and Creutz, C. *J. Chem. Ed.* **1983,** *60,* 809.
3. Libby, W. F. *J. Phys. Chem.* **1952,** *56,* 863.
4. Libby, W. F. *Annu. Rev. Phys. Chem.* **1977,** *28,* 105.
5. Born, M., and Oppenheimer, J. R. *Ann. Phys.* **1927,** *84,* 457.
6. Dewar, M. J. S., and Dougherty, R. C. *The PMO Theory of Organic Chemistry*. Plenum, New York, 1975.
7. Sutin, N. In *Bioorganic Chemistry,* Eichhorn, G. L., ed. American Elsevier, New York, 1973, Vol. 2, Chapter. 19.
8. Atkins, P. W. *Molecular Quantum Mechanics,* 2 ed. Oxford University Press, Oxford, 1983, Chapter 8.
9. Landau, L. *Phys. Z. Zowjet.* **1932,** *2,* 46.
10. Zener, C. *Proc. Roy. Soc.* **1932,** *A137,* 696.
11. Closs, G. L., Calcaterra, L. T., Green, N. J., Penfield, K. W., and Miller, J. R. *J. Phys. Chem.* **1986,** *90,* 3673.
12. Siders, P., Cave, R. J., and Marcus, R. A. *J. Chem. Phys.* **1984,** *81,* 5613.
13. Brunschwig, B., and Sutin, N. *J. Am. Chem. Soc.* **1978,** *100,* 7568.
14. Marcus, R. A. *J. Chem. Phys.* **1956,** *24,* 966.
15. Marcus, R. A. *Disc. Faraday Soc.* **1960,** *29,* 21.
16. Marcus, R. A. *J. Phys. Chem.* **1963,** *67,* 853.
17. Marcus, R. A. *J. Chem. Phys.* **1965,** *43,* 679.
18. Brunschwig, B. S., Logan, J., Newton, M. D., and Sutin, N. *J. Am. Chem.* **1980,** *102,* 5798.
19. Eberson, L. *Adv. Phys. Chem.* **1982,** *18,* 79.
20. Marcus, R. A. *J. Chem. Phys.* **1956,** *24,* 979.
21. Rehm, D., and Weller, A. *Israel J. Chem.* **1970,** *8,* 259.
22. Marcus, R. A., and Siders, P. *J. Phys. Chem.* **1982,** *86,* 622.
23. Calcaterra, L. T., Closs, G. L., and Miller, J. R. *J. Am. Chem. Soc.* **1983,** *105,* 670.
24. Miller, J. R., Calcaterra, L. T., and Closs, G. L. *J. Am. Chem. Soc.* **1984,** *106,* 3047.
25. Maroncelli, M., MacInnis, J., and Fleming, G. R. *Science* **1989,** *243,* 1674.
26. Castner, E. W., Jr., Bagchi, B., Maroncelli, M., Webb, S. P., Ruggiero, A. J., and Fleming, G. R. *Ber. Bunsenges, Phys. Chem.* **1988,** *92,* 363.
27. Kang, T. J., Kahlow, M. A., Giser, D., Swallen, S., Nagarajan, V., Jarzeba, W., and Barbara, P. F. *J. Phys. Chem.* **1988,** *92,* 6800.
28. Sumi, H., and Marcus, R. A. *J. Chem. Phys.* **1986,** *84,* 4894.
29. Levich, V. G. *Adv. Electrochem. Electrochem. Engl.* **1966,** *4,* 249.
30. Friesner, R., and Wertheimer, R. *Proc. Natl. Acad. Sci. U.S.A.* **1982,** *79,* 2138.

31. Friesner, R. A., and Won, Y. *Photochem. Photobiol.* **1989,** *50,* 831.

32. Michel, M. E., Bixon, M., and Jortner, J. *Chem. Phys. Lett.* **1988,** *151,* 188.

33. Bixon, M., and Jortner, J. *J. Phys. Chem.* **1988,** *92,* 7148.

34. Plato, M., Möbius, K., Michel-Beyerle, M. E., Bixon, M., and Jortner, J. *J. Am. Chem. Soc.* **1988,** *110,* 7279.

35. Pearlstein, R. M. *Photochem. Photobiol.* **1982,** *35,* 835.

36. Roelofs, M. G., Chidsey, C. E. D., and Boxer, S. G. *Chem. Phys. Lett.* **1982,** *87,* 582.

37. Marcus, R. A. *Chem. Phys. Lett.* **1986,** *133,* 471.

38. Warshel, A., Chu, Z. T., and Parson, W. W. *Science* **1989,** *246,* 112.

Problems

1. The reaction $[Ru(bpy)_3^{2+}]^* + Fe^{3+} \rightarrow Ru(bpy)_3^{3+} + Fe^{2+}$ has been studied in various concentrations of sodium nitrate. Given that $\Delta G_{el} = -15.0$ kcal mol^{-1}, $\lambda = 18.6$ kcal mol^{-1}, $K_{eq}\nu_n = 1.6 \times 10^{11}$ M^{-1} s^{-1}, and $\kappa_{el} = 1$, calculate k_{dif} at room temperature. (Rodríguez, A., De la Rosa, F., Galán, M. Sánchez, F., and Moyá, M. L. *Photochem. Photobiol.* **1992** *55,* 367.)
2. Calculate k_{el} for the quenching of excited zinc porphyrin by a quinone ring separated by 10 Å. Assume that $\beta = 1.2$ Å^{-1}, $d - d_0 = 7$ Å, and $\Delta G_{el}^{\neq} = 0$.
3. The free-energy change in the electron-transfer reaction of $[Ru(bpy)_3^{2+}]^*$ with ferricytochrome *c* is $\Delta G_{el} = -25$ kcal mol^{-1}. The observed rate constant is 2.5×10^8 M^{-1} s^{-1}. Assuming $K_{eq}\nu_n = 10^{11}$ M^{-1} s^{-1}, calculate $\Delta G_{el}^{\neq}$ and λ. In what kinetic region is this reaction?
4. Calculate κ_{el} for the $Ru(bpy)_3^{2+} + Ru(bpy)_3^{3+} \rightleftharpoons Ru(bpy)_3^{3+} + Ru(bpy)_3^{2+}$ electron exchange using the Landau–Zener formalism. Assume that $H_{el} = 0.14$ kcal mol^{-1}, $v_n = 3 \times 10^4$ cm s^{-1}, and $\Delta s = 2.68 \times 10^{10}$ kcal mol^{-1} cm^{-1}. Now calculate κ_{el}, assuming that $k_a = 10^3$ M^{-1} s^{-1} and $\Delta G_{el}^{\neq} = 8.2$ kcal mol^{-1}. Assume also that $K_{eq}\nu_n = 10^{11}$ M^{-1} s^{-1}, $T = 298$ K, and $R = 1.987$ cal deg^{-1} mol^{-1}. How do the values of κ_{el} compare? Do the same for the $Fe(H_2O)_6^{2+} + Fe(H_2O)_6^{3+} \rightleftharpoons Fe(H_2O)_6^{3+} + Fe(H_2O)_6^{2+}$, assuming the same values as above except that $k_a = 4.2$ M^{-1} s^{-1} and $\Delta G_{el}^{\neq} = 17.7$ kcal mol^{-1}. How do the κ_{el}s compare in this case? (Khan, S. U. M., and Zhou, Z. Y. *J.Chem. Phys.* **1990** *93,* 8808.)
5. For the $[Ru(bpy)_3^{2+}]^* + Ru(bpy)_3^{3+} \rightleftharpoons Ru(bpy)_3^{3+} + Ru(bpy)_3^{2+}$ quenching process, calculate k_a assuming the following values: $\lambda = 14.9$ kcal mol^{-1}, $\Delta G_{el} = -48.3$ kcal mol^{-1}, $\kappa_{el} = 1$, and $K_{eq}\nu_n = 10^{12}$ M^{-1} s^{-1}. How does your value compare with the experimental value of $k_a = 2 \times 10^9$ M^{-1} s^{-1}.
6. Which reaction should have a larger inner-sphere reorganization energy—$[Co(bpy)_3^{2+}]^* + Co(bpy)_3^{3+} \rightleftharpoons Co(bpy)_3^{3+} + Co(bpy)_3^{2+}$, which is accompanied by a large Co-ligand bond-length change, or $[Cr(bpy)_3^{2+}]^* + Cr(bpy)_3^{3+} \rightleftharpoons Cr(bpy)_3^{3+} + Cr(bpy)_3^{2+}$, for which the bond-length changes are small?

7. Given the reduced constants and experimental bond-length changes shown below, calculate λ_v for the one-electron oxidation of perylene to its cation:

Bond	f_i (N m^{-1})	Δq_i (Å)
1–2	335	0.024
2–3	346	0.008
3–4	310	0.01
4–5	303	0.001
5–6	304	0.002
1–6	300	0.018
6–7	280	0.01

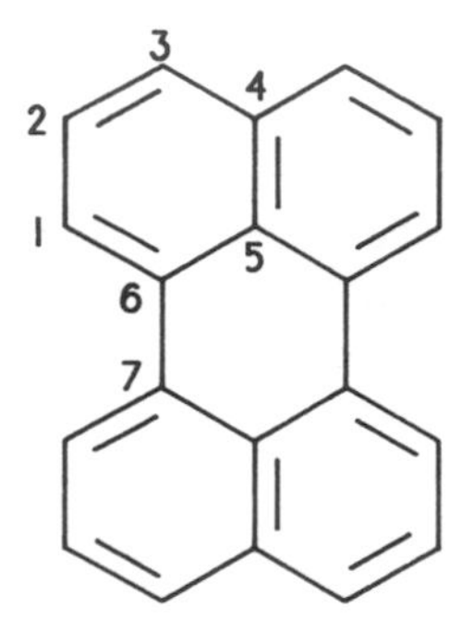

(Grampp, G., Cebe, M., and Cebe, M. *Zeit. Phys. Chem.* **1990,** *166,* 93.)

8. Calculate the maximum possible rates of electron transfer for reactions taking place in each of the following solvents, given their dielectric relaxation times:

	τ_D (ps)
Acetonitrile	4.0
Water	8.3
Methanol	55.6
n-Propanol	435

In what solvent is electron transfer expected to be the slowest? Why?

9. Assuming that $\lambda = 5.8$ kcal mol^{-1} and $T = 270$ K, plot k_{el} vs. ΔG_{el} for values of the latter ranging from 1 to -20 kcal mol^{-1} (at 2 kcal mol^{-1} increments). Assume that $H_{el} = 10$ cm^{-1}. What conclusion can you draw from this plot?

Appendix I

ENERGY CONVERSION FACTORS[a]

	ν (cm^{-1})	$\bar{\nu}$ (s^{-1})	E (kcal mol^{-1})	E (kJ mol^{-1})	eV (volts)
ν (cm^{-1})	1	3×10^{10}	2.86×10^{-3}	1.20×10^{-2}	1.24×10^{-4}
$\bar{\nu}$ (s^{-1})	3.33×10^{-11}	1	9.54×10^{-4}	3.99×10^{-5}	4.14×10^{-5}
E (kcal mol^{-1})	349	1.05×10^{13}	1	4.18	4.34×10^{-2}
E (kJ mol^{-1})	83.6	2.51×10^{6}	0.239	1	1.04×10^{-2}
eV (volts)	8.07×10^{3}	2.42×10^{14}	23.06	96.5	1

[a] The values in the table show the equivalence of the units in the left column with those in the top row. For example, 1 kcal mol^{-1} = 4.18 kJ mol^{-1}.

Appendix II

The entire expression for the exciplex formation energy, ΔH_{ex}, following photoexcitation of one of the reactant partners is given by Eq. II.1:

$$\Delta H_{EX} = IP - EA - \frac{e^2}{d_{cc}} + \Delta E_{orb} - \Delta H_{sol} - \Delta H_{00} \tag{II.1}$$

where e^2/d_{cc} represents the electrostatic interaction between the two partners assumed to be in contact in the exciplex, ΔH_{00} represents the zero–zero energy of the excited-state reactant. ΔE_{orb} is a term representing various electronic stabilizing and destabilizing interactions. For charge-transfer exciplexes, this term is usually taken as $\Delta E_{orb} \simeq 0$. ΔH_{sol}, the exciplex solvation enthalpy, is the energy change associated with transfer of an exciplex or contact ion pair of given radius and dipole moment from the gas phase (g) to solution (s):

$$[DA]_s + [D^+A^-]_g \rightleftharpoons [D^+A^-]_g + [DA]_s \tag{II.2}$$

Since an exciplex (or contact ion) is a dipolar species with a dipole moment, μ, the exciplex solvation enthalpy in a solvent of known polarity can be calculated from Eq. 2.9, which is based on the Kirkwood–Onsager continuum model.

To convert the exciplex formation enthalpy into free energy, we use $G = H - TS$ and obtain

$$\Delta G_{EX} = \Delta H_{EX} - T\Delta S_{EX} \tag{II.3}$$

and

$$\Delta G_{00} = \Delta H_{00} \tag{II.4}$$

For many exciplexes, $T\Delta S_{EX}$ in hexane can be taken as ~ -5 kcal mol^{-1} at room temperature. The argument explaining the meaning of the negative entropy goes as follows. Prior to exciplex formation, solvent molecules surrounding an uncharged donor–acceptor pair are not tightly bound to the reactants. They are free enough to adopt a random number of orientations around both reactants. However, when the exciplex forms, solvent molecules become more tightly bound to the exciplex in order to stabilize its charge. This change from a loosely arranged complex to one much more ordered is associated with a decrease in ΔS.

From Eqs. II.1, II.3, and II.4, we can write the exciplex free energy as

$$\Delta G_{EX} = \mathrm{IP} - \mathrm{EA} - \Delta G_{00} - \frac{e^2}{d_{cc}} - \frac{\mu^2}{\rho^3}\left[\frac{\varepsilon_s - 1}{2\varepsilon_s + 1}\right] - T\Delta S_{EX} \qquad \text{(II.5)}$$

The next step in our goal to formulate a useful expression for the free energy of exciplex formation is to convert the ionization potential and electron affinity into redox potentials, which can be determined experimentally. There are linear relationships (Eqs. II.6 and II.7) that allow us to perform this maneuver:

$$\mathrm{IP} = E^0(\mathrm{D}^+/\mathrm{D}) - \Delta G(\mathrm{D}^+) + \text{constant} \qquad \text{(II.6)}$$

$$\mathrm{EA} = E^0(\mathrm{A}/\mathrm{A}^-) + \Delta G(\mathrm{A}^-) + \text{constant} \qquad \text{(II.7)}$$

$\Delta G(\mathrm{D}^+)$ and $\Delta G(\mathrm{A}^-)$ refer to the solvation energies of ions from the vacuum to solution.

The sum of $\Delta G(\mathrm{D}^+)$ and $\Delta G(\mathrm{A}^-)$, which we can designate as ΔG_{sol}, is given by the Born equation:

$$\Delta G_{sol} = \Delta G(\mathrm{D}^+) + \Delta G(\mathrm{A}^-) = -\frac{e^2}{2}\left(\frac{1}{r_D} + \frac{1}{r_A}\right)\left(1 - \frac{1}{\varepsilon_s}\right) \qquad \text{(II.8)}$$

Equation II.8 includes energy terms for Coulombic and solvent interactions in the exciplex. The term, $1 - 1/\varepsilon_s$, measures the effect of solvent polarity on solvation. In nonpolar solvents, say $\varepsilon_s \sim 1$, $1 - 1/\varepsilon_s \simeq 0$, and $\Delta G_{sol} \simeq 0$. But when ε_s is large, as in polar solvents, $1 - 1/\varepsilon_s \simeq 1$. By combining Eqs. II.1, II.6, II.7, and II.8, we obtain finally Eq. II.9:

$$\begin{aligned}\Delta G_{EX} = {} & E^0(\mathrm{D}^+/\mathrm{D}) - E^0(\mathrm{A}/\mathrm{A}^-) - \Delta G_{00} \\ & + \frac{e^2}{2}\left(\frac{1}{r_D} + \frac{1}{r_A}\right)\left(1 - \frac{1}{\varepsilon_s}\right) \\ & - \frac{e^2}{d_{cc}} - \frac{\mu^2}{\rho^3}\left(\frac{\varepsilon_s - 1}{2\varepsilon_s + 1}\right) - T\Delta S_{EX}\end{aligned} \qquad \text{(II.9)}$$

Equation II.9 expresses the complex effects of energetics, exciplex structure, separation distance, and the solvent dielectric on the exciplex free energy. Since the redox potentials display a dependence on the polarity of the solvent, a semiempirical relationship, similar to Eq. II.9, can be used:

$$\Delta G_{EX}\,(\text{eV}) = [E^0(D^+/D) - E^0(A/A^-)]^{37} - \Delta G_{00} - \frac{\mu^2}{\rho^3}\left(\frac{\varepsilon_s - 1}{2\varepsilon_s + 1} - 0.186\right) + 0.38 \quad (II.10)$$

or

$$\Delta G_{EX}\,(\text{kcal mol}^{-1}) = 23.06[E^0(D^+/D) - E^0(A/A^-)]^{37} - \Delta G_{00} - \frac{\mu^2}{\rho^3}\left(\frac{\varepsilon_s - 1}{2\varepsilon_s + 1} - 4.3\right) + 8.8 \quad (II.11)$$

In Eqs. II.10 and II.11, the redox potentials are those measured in polar acetonitrile, whose dielectric constant is $\varepsilon_s = 37$. This is indicated by the superscript. For details, see Weller, A. *Zeit. Phys. Chem. N. F.* **1982,** *133,* 93 and Chibisov, A. *Russ. Chem. Rev.* **1981,** *50,* 1169.

Appendix III

To determine the solvent dependence of the free energy of formation of a solvent-separated ion pair from neutral reactants, it is convenient to start with the Rehm–Weller equation:

$$\Delta G_{el} = E^0(D^+/D) - E^0(A/A^-) - \Delta G_{00} - \frac{e^2}{\varepsilon_s d_{cc}} \qquad \text{(III.1)}$$

When determining the effects of solvent polarity, it must be recalled that the redox potentials, $E^0(D^+/D)$ and $E^0(A/A^-)$, display their own dependence on ε_s. Actually, $E^0(D^+/D) - E^0(A/A^-)$ increases with increasing ε_s (see Eqs. II.6 and II.7 in Appendix II). However, IP − EA is independent of solvent polarity. With this in mind, we can write

$$[E^0(D^+/D) - E^0(A/A^-)]^{\varepsilon_s} - \Delta G_{sol}{}^{\varepsilon_s} = [E^0(D^+/D) - E^0(A/A^-)]^{37} - \Delta G_{sol}{}^{37} \qquad \text{(III.2)}$$

where the superscripts on the left-hand side of the equation refer to the dielectric constant of the solvent of interest and the superscripts on the right-hand side represent the dielectric constant of acetonitrile. Combining Eqs. II.6, II.7, II.8, and III.2 leads to

$$\Delta G_{SSIP} = [E^0(D^+/D) - E^0(A/A^-)]^{37} - \Delta G_{00} - \frac{e^2}{2}\left(\frac{1}{r_D} + \frac{1}{r_A}\right)\left(\frac{1}{37} - \frac{1}{\varepsilon_s}\right) - \frac{e^2}{\varepsilon_s d_{cc}} \qquad \text{(III.3)}$$

or, in kcal mol^{-1}, to

$$\Delta G_{SSIP}\ (\text{kcal mol}^{-1}) = 23.06[E^0(D^+/D) - E^0(A/A^-)]^{37} - \Delta G_{00} - 166\left(\frac{1}{r_D} + \frac{1}{r_A}\right)\left(\frac{1}{37} - \frac{1}{\varepsilon_s}\right) - \frac{332}{\varepsilon_s d_{cc}}$$

For details, see Weller, A. *Zeit. Phys. Chem. N. F.* **1982,** *133,* 93 and Chibisov, A. *Russ. Chem. Rev.* **1981,** *50,* 1169.

INDEX